Methods and Applications of Autonomous Experimentation

Autonomous Experimentation is poised to revolutionize scientific experiments at advanced experimental facilities. Whereas previously, human experimenters were burdened with the laborious task of overseeing each measurement, recent advances in mathematics, machine learning and algorithms have alleviated this burden by enabling automated and intelligent decision-making, minimizing the need for human interference. Illustrating theoretical foundations and incorporating practitioners' first-hand experiences, this book is a practical guide to successful Autonomous Experimentation.

Despite the field's growing potential, there exists numerous myths and misconceptions surrounding Autonomous Experimentation. Combining insights from theorists, machine-learning engineers and applied scientists, this book aims to lay the foundation for future research and widespread adoption within the scientific community.

This book is particularly useful for members of the scientific community looking to improve their research methods but also contains additional insights for students and industry professionals interested in the future of the field.

Marcus M. Noack received his Ph.D. in applied mathematics from Oslo University, Norway. At Lawrence Berkeley National Laboratory, he is working on stochastic function approximation, optimization and uncertainty quantification, applied to Autonomous Experimentation.

Daniela Ushizima, Ph.D. in physics from the University of Sao Paulo, Brazil after majoring in computer science, has been associated with Lawrence Berkeley National Laboratory since 2007, where she investigates machine learning algorithms applied to image processing. Her primary focus has been on developing computer vision software to automate scientific data analysis.

Chapman & Hall/CRC
Computational Science Series

Series Editor: Horst Simon and Doug Kothe

Methods and Applications of Autonomous Experimentation
Marcus M. Noack, Daniela Ushizima,

Data-Intensive Science
Terence Critchlow, Kerstin Kleese van Dam

Grid Computing
Techniques and Applications
Barry Wilkinson

Scientific Computing with Multicore and Accelerators
Jakub Kurzak, David A. Bader, Jack Dongarra

Introduction to the Simulation of Dynamics Using Simulink
Michael A. Gray

Introduction to Scheduling
Yves Robert, Frederic Vivien

Introduction to Modeling and Simulation with MATLAB® and Python
Steven I. Gordon, Brian Guilfoos

Fundamentals of Multicore Software Development
Victor Pankratius, Ali-Reza Adl-Tabatabai, Walter Tichy

Programming for Hybrid Multi/Manycore MPP Systems
John Levesque, Aaron Vose

Exascale Scientific Applications
Scalability and Performance Portability
Tjerk P. Straatsma, Katerina B. Antypas, Timothy J. Williams

GPU Parallel Program Development Using CUDA
Tolga Soyata

Parallel Programming with Co-arrays
Robert W. Numrich

Contemporary High Performance Computing
From Petascale toward Exascale, Volume 3
Jeffrey S. Vetter

Unmatched
50 Years of Supercomputing
David Barkai

For more information about this series please visit:
https://www.crcpress.com/Chapman--HallCRC-Computational-Science/book-series/
CHCOMPUTSCI

Methods and Applications of Autonomous Experimentation

Edited by Marcus M. Noack and Daniela Ushizima

CRC Press
Taylor & Francis Group
Boca Raton London New York

CRC Press is an imprint of the
Taylor & Francis Group, an **informa** business

A CHAPMAN & HALL BOOK

First edition published 2024
by CRC Press
2385 NW Executive Center Drive, Suite 320, Boca Raton FL 33431

and by CRC Press
4 Park Square, Milton Park, Abingdon, Oxon, OX14 4RN

CRC Press is an imprint of Taylor & Francis Group, LLC

ISBN: 978-1-032-31465-5 (hbk)
ISBN: 978-1-032-41753-0 (pbk)
ISBN: 978-1-003-35959-3 (ebk)

DOI: 10.1201/9781003359593

Typeset in Latin Modern font
by KnowledgeWorks Global Ltd.

Publisher's note: This book has been prepared from camera-ready copy provided by the authors.

Marcus M. Noack dedicates this book to his wife, Kalyn, for enjoying this small volume of space-time together and for her constant support. He also dedicates this book to James Sethian for his mentorship and guidance.

Daniela Ushizima dedicates this book to Antonio for believing, to Iris for watching over, to Jamie for supporting, to Jeff for enduring and to Kerri for illuminating our science work.

"Frequently the messages have meaning; that is they refer to or are correlated according to some system with certain physical or conceptual entities. These semantic aspects of communication are irrelevant to the engineering problem. The significant aspect is that the actual message is one selected from a set of possible messages." (Shannon 1948, p. 379).

"The most dangerous phrase in the language is: We've always done it this way." - Rear Admiral Hopper, United States Navy.

Contents

List of Figures

List of Tables

Foreword

Autonomous Experimentation will change our world. A bold claim to make for the foreword to *Methods and Applications of Autonomous Experimentation*, edited by Marcus M. Noack and Daniela Ushizima, that I will make a case for here. Autonomous Experimentation describes what are essentially research robots, self-driving labs, or otherwise systems that plan, execute, and evaluate their own experiments toward research objectives supplied by human researchers. But why now? What does it mean for (or do to) human researchers? What does it do for science? For society? And finally, who should read this excellent body of work, *Methods and Applications of Autonomous Experimentation*. Autonomous Experimentation multiplies human researcher effort by a staggering 1,000 times or more. By teaming with autonomous research robots, humans are freed from tedious, time-consuming, and de-motivating activities that limit the appeal of scientific research to the future workforce and therein threaten our capability to attack problems facing our world like climate change and resource limitation.

Methods and Applications of Autonomous Experimentation is unique and significant as the first book dedicated to Autonomous Experimentation. It goes deep into artificial intelligence (AI) decision-making methods, including the mathematical underpinnings. Importantly, the chapters on beamline and National Laboratory User Facility experiments, which are high-value, describe time-critical series of experiments that researchers wait months or years to execute in a few short days. Finally, open-source methods speak to the need for science to be accessible to everyone for the good of society. Why should anyone care about Autonomous Experimentation (AE)? Technological solutions are at the vanguard of combatting the many problems that face society today. Global climate change, exhaustion of non-renewable resources and critical materials, food scarcity, and inequitable access to the internet, medicine, and education are active problems pursued by researchers today. From Casadevall and Fang, "Reforming Science: Methodological and Cultural Reforms" [87].

"Regardless of one's stance on the benefits and liabilities of the scientific revolution, science remains humanity's best hope for solving some of its most vexing problems, from feeding a burgeoning population to finding alternative sources of energy to protecting our world against planet-killing meteorites. And yet, notwithstanding the importance of the scientific enterprise, only a tiny fraction of the human population is directly involved in scientific discovery. It is, therefore, crucial for society to nurture and sustain the fragile scientific enterprise and optimize its functioning to ensure the continuing survival and prosperity of mankind".

Regarding the impact of AE on science, it is clear that society needs more, better, and faster scientific progress. By multiplying the effort of scientists by orders of magnitude, AE can make this change. To wrap one's head around the potential of AE, consider the impact of mechanical power in the industrial revolution: Farmers could plow hundreds of acres with the manual labor of a single human. Moreover, the industrial revolution gave rise to factories, assembly lines, and automated manufacturing that fundamentally changed how we lived, worked, and consumed. In the 20th century, the computer gave rise to a cognitive revolution, where human labor to calculate rocket trajectories was replaced by machines, for example. The impact is now with us in the form of GPS, the internet, and the effective leveraging of compute resources to make humans able to do things inconceivable in pre-digital computer society. With AE, hard scientific problems that took decades may take years. Even harder problems that we didn't dare take on become possible. And the speed to solution gives the capability to respond to urgent problems like climate change before it is too late.

What about the scientists? Some worry that AE, as with AI tools in general, will replace humans in the lab, taking their jobs. I argue the opposite: Young scientists will welcome AE, and teams of human and AE research robots will be superior to humans or AE robots working alone (This is well-established in the autonomy community). And since we know from the quote above, there is no over-supply of scientific progress. Perhaps like compute power, the demand for research will persistently outstrip supply. Indeed, the young researchers I talk to are excited and impatient for the arrival of AE robots. Students are driven away from research by the slow, tedious nature of laboratory work that treats them like "hands" or ironically, automations. AE will make research attractive to the workforce by relieving them of work better done by robots.

"One major concern is the imbalance among members of the research workforce and the funding currently available to sustain that workforce. In some respects, the current scientific workforce resembles a pyramid scheme with a small number of principal investigators presiding over an army of research scientists, postdocs, students, and technicians who have little autonomy and increasingly uncertain career prospects". (from "Reforming Science: Methodological and Cultural Reforms" [87])

As the STEM workforce is increasingly choosing computer science and finance careers, and moving away from scientific research, it is imperative that we build laboratories of the future that value human intelligence, creativity, and insight.

Another impact of AE will be increased access to science. When one hundred AE experimental iterations can be done in the time it took to do one experiment, then the labor cost of doing one experiment is reduced 100-fold (even more if the researcher does other work while the robot is operating). Just as Moore's Law increases in compute power reduced costs to the point where we carry personal supercomputers in our pockets as mobile phones, the soaring speed and plummeting cost of Autonomous Experimentation can make scientific research more accessible, especially to under-represented communities.

In closing, *Methods and Applications of Autonomous Experimentation* is for those who want to dive deep into AE methods and decision-making with concrete examples. It is also informative for those wanting to understand more about the value proposition of AE and implementation pathways for this new approach to scientific research.

Dr. Benji Maruyama is the Autonomous Materials Lead
and Principal Materials Research Engineer for the US Air Force Research
Laboratory, Materials & Manufacturing Directorate,
Wright Patterson AFB, Ohio.

Preface

Just like so many other topics that have been adopted into the realm of machine learning and AI—deep learning, digital twins, active learning, and so on—Autonomous Experimentation has become a fuzzy, ill-defined ideal everyone wants, but seemingly no one can deliver. The main reason for that is the missing, inherent meaning of the term "Autonomous Experimentation". In this book, we want to separate the practical methods and applications from the buzz and hype surrounding the term. This effort started a long time before we even thought of writing a book about the subject. In April of 2021, the Center for Advanced Mathematics for Energy Research Applications (CAMERA) organized a workshop titled "Autonomous Discovery for Science and Engineering" to bring scientists and engineers together to present different methods, applications, and perspectives on the topic. One of the key takeaways was that there was no one meaning of the term "Autonomous Discovery" or "Autonomous Experimentation". These are multi-faceted and highly complex topics that mean different things for different people with different backgrounds; and although this makes it an exciting field to work in, that might well shape science for decades to come, it also means that communication and collaborations across vastly different work cultures are necessary to make meaningful progress. The authors like to think of it as a common language that has to be developed to translate challenges in the various applied sciences into mathematical and computation problems without losing scientific meaning and interpretability.

This book is an attempt to lay the groundwork for an organizational paradigm for Autonomous Experimentation and discovery. It is in no way complete, nor can it be, given that we are dealing with a rapidly developing field of research. We tried to cover the most important theoretical concepts that are beneficial for successfully setting up an Autonomous Experiment, and we let applied scientists tell their stories of why and how Autonomous Experimentation was achieved with a variety of instruments. The authors have spent several years working with experimentalists from around the globe on Autonomous Experimentation and invited some of them to share their in-depth knowledge on the subject.

Acknowledgments

We would like to thank all the contributing authors for their insights. We also want to thank CAMERA, Lawrence Berkeley National Laboratory, the Office of Science, and all the contributors and collaborators. The editor-in-chief's times were funded through the Center for Advanced Mathematics for Energy Research Applications (CAMERA), which is jointly funded by the Advanced Scientific Computing Research

(ASCR) and Basic Energy Sciences (BES) within the Department of Energy's Office of Science, under Contract No. DE-AC02-05CH11231.

All color instances mentioned in book are available in ebook version.

Contributors

Marcus M. Noack
Lawrence Berkeley National Laboratory
Berkeley, California, USA

Daniela Ushizima
Lawrence Berkeley National Laboratory
Berkeley, California, USA

Kevin G. Yager
Brookhaven National Laboratory
Upton, New York, USA

Joshua Schrier
Fordham University
The Bronx, New York, USA

Alexander J. Norquist
Haverford College
Haverford, Pennsylvania, USA

Juliane Mueller
National Renewable Energy Laboratory
Denver, Colorado, USA

Mark Risser
Lawrence Berkeley National Laboratory
Berkeley, California, USA

Masafumi Fukuto
Brookhaven National Laboratory
Upton, New York, USA

Yu-chen Karen Chen-Wiegart
Stony Brook University
Stony Brook, New York, USA

A. Gilad Kusne
National Institute of Standards and
  Technology
Gaithersburg, Maryland, USA

Austin McDannald
National Institute of Standards and
  Technology
Gaithersburg, Maryland, USA

Ichiro Takeuchi
University of Maryland
College Park, Maryland, USA

Martin Boehm
Institut Laue-Langevin
Grenoble, France

David E. Perryman
Institut Laue-Langevin
Grenoble, France

Yannick LeGoc
Institut Laue-Langevin
Grenoble, France

Paolo Mutti
Institut Laue-Langevin
Grenoble, France

Tobias Weber
Institut Laue-Langevin
Grenoble, France

Luisa Scaccia
Universit'a di Macerata
Macerata, Italy

Alessio De Francesco
Consiglio Nazionale delle Ricerche
Grenoble, France

Alessandro Cunsolo
University of Wisconsin-Madison
Madison, Wisconsin, USA

Hoi-Ying Holman
Lawrence Berkeley National Laboratory
Berkeley, California, USA

Liang Chen
Lawrence Berkeley National Laboratory
Berkeley, California, USA

Petrus H. Zwart
Lawrence Berkeley National Laboratory
Berkeley, California, USA

Steven Lee
UC Berkeley
Berkeley, California, USA

Aaron N. Michelson
Brookhaven National Laboratory
Upton, New York, USA

Suchismita Sarker
Cornell High Energy Synchrotron Source
Ithaca, New York, USA

Apurva Mehta
SLAC National Accelerator Laboratory
Menlo Park, California, USA

Trent Northen
Lawrence Berkeley National Laboratory
Berkeley, California, USA

Peter Andeer
Lawrence Berkeley National Laboratory
Berkeley, California, USA

Elliot Chang
Lawrence Livermore National Laboratory
Livermore, California, USA

Haruko Wainwright
Massachusetts Institute of Technology
Cambridge, Massachusetts, USA

Linda Beverly
Lawrence Livermore National Laboratory
Livermore, California, USA

John C. Thomas
Lawrence Berkeley National Laboratory
Berkeley, California

Antonio Rossi
Lawrence Berkeley National Laboratory
Berkeley, California, USA

Eli Rotenberg
Lawrence Berkeley National Laboratory
Berkeley, California, USA

Alexander Weber-Bargioni
Lawrence Berkeley National Laboratory
Berkeley, California, USA

Darian Smalley
University of Central Florida
Orlando, Florida, USA

Masahiro Ishigami
University of Central Florida
Orlando, Florida, USA

Phillip M. Maffettone
Brookhaven National Laboratory
Upton, New York, USA

Daniel B. Allan
Brookhaven National Laboratory
Upton, New York, USA

Andi Barbour
Brookhaven National Laboratory
Upton, New York, USA

Thomas A. Caswell
Brookhaven National Laboratory
Upton, New York, USA

Dmitri Gavrilov
Brookhaven National Laboratory
Upton, New York, USA

Marcus D. Hanwell
Brookhaven National Laboratory
Upton, New York, USA

Thomas Morris
Brookhaven National Laboratory
Upton, New York, USA

Daniel Olds
Brookhaven National Laboratory
Upton, New York, USA

Max Rakitin
Brookhaven National Laboratory
Upton, New York, USA

Stuart I. Campbell
Brookhaven National Laboratory
Upton, New York, USA

Bruce Ravel
National Institute of Standards and
    Technology
Gaithersburg, Maryland, USA

Yixuan Sun
Argonne National Laboratory
Lemont, Illinois, USA

Krishnan Raghavan
Argonne National Laboratory
Lemont, Illinois, USA

Prasanna Balaprakash
Oak Ride National Laboratory
Oak Ridge, Tennessee, USA

I

Introduction

Modern instruments at experimental facilities around the globe have the ability to perform increasingly complex experiments with the goal of advancing scientific discovery and understanding across a range of inquiries, including the properties of materials, scientific mechanisms, biological substances, and engineering designs. As these instruments grow in sophistication, complexity, and capability, the measurements and outputs they produce are becoming increasingly multi-faceted. Stemming in part from technological and engineering advances, such as more spatially and temporarily resolved detectors, faster and larger compute capabilities, and robot-controlled experiments, the resulting output is more voluminous and is being generated faster than ever before. One way to characterize both the opportunities and the challenges inherent in these advances is through their effect on the underlying parameter space encapsulated in a suite of experiments.

Increased possibilities for input spanning many options and larger numbers of results measuring a spectrum of outputs reflect this increasing complexity. As scientific questions become more complicated and the underlying parameter space increases in dimensionality, the measurement outcomes depend on an increasing number of user-controllable inputs, such as synthesis, environmental, and processing parameters. Making sense of this requires a systematic way to efficiently ask questions and economically explore this high-dimensional space, probing areas of interest and avoiding time-consuming and costly experiments that may yield relatively little new insight.

Autonomous Experimentation (AE) can be described as an attempt to deal with this challenge. The word "autonomy" is commonly defined as having the right or power of self-government. Clearly, in the world of scientific experiments, self-government can mean many things, from automated systems that know what action to perform because they were set up to do so, to an entire experimental facility that can decide on its own what experiments have to be performed and how to effectively and efficiently reach a scientific conclusion or insight. Throughout this book, we will use the term "autonomy" as the ability to make decisions based on observations without human interaction. However, this does not mean that the experiment was set up entirely without human labor. There is an important distinction between "autonomous" and "automatic" that follows from this definition: a machine that works automatically follows a set of predefined instructions; the important distinction is imposed by intelligent decision-making in every iteration without human input. However, the field is too young, at this point, to fix one definition of "intelligence". Is an algorithm that checks in every iteration of an Autonomous Experiment whether a number satisfies a certain condition intelligent? This is up to the reader to decide. In this book, we will look at methods that do give an experiment loop a deeper level of intelligence and we will look at quite amazing applications of this new technology from around the globe.

The field of Autonomous Experimentation is, without any doubt, in its infancy, with new methods, algorithms, and frameworks being developed and deployed continuously. Due to this novelty and the linked fluidity of a developing field, there is no one correct definition of the term *Autonomous Experimentation*. Different communities have different understandings and perspectives of what it means, and instead of deciding who is right, we want to give leaders of the communities the opportunity to give their perspectives. We will see that many thoughts and visions overlap from different perspectives while other aspects are seen very differently. But they all have one thing in common: the goal to liberate practitioners from micro-managing experiment designs and instead allow for the focus to be on the underlying domain of science.

CHAPTER 1

Autonomous Experimentation in Practice

Kevin G. Yager

Center for Functional Nanomaterials, Brookhaven National Laboratory, Upton, New York, USA

CONTENTS

A UTONOMOUS EXPERIMENTATION (AE) is an emerging paradigm for accelerating scientific discovery, leveraging artificial intelligence and machine-learning methods to automate the entire experimental loop, including the decision-making step. AE combines advancements in hardware automation, data analytics, modeling, and active learning to augment a scientific instrument, enabling it to autonomously explore the search space corresponding to a problem of interest. AE can be deployed quite easily in any context where automation is feasible, by connecting a decision-making algorithm in between data analysis and machine-command modules. AE holds the promise of radically accelerating discovery, by liberating the human researcher to operate at a higher level and focus on scientific understanding rather than experimental management.

DOI: 10.1201/9781003359593-1

1.1 WHAT IS AE?

Autonomous Experimentation (AE) is an emerging paradigm for scientific studies, wherein every step of a traditional research loop is automated, including—crucially—the decision-making step where the next experiment is selected. By its nature, AE is an application of artificial intelligence (AI) and machine-learning (ML) methods to experimental science, since only AI/ML decision-making algorithms can provide the necessary sophistication to direct experimental execution intelligently—that is, in response to the iterative improvement in understanding one accumulates while researching a problem.

Modern scientific problems are increasingly complex. Across a wide range of fields—from biosciences, to materials science, to high-energy physics—one sees common trends toward research teams that are more interdisciplinary, instruments that are more precise and sophisticated, data collection rates that are enormous, and exploration spaces that are vast and high-dimensional. Materials science provides an instructive example. The drive toward multi-component materials, with precise control of ordering across all lengthscales, implies a large and high-dimensional parameter space for exploration, where trends within the space might be arbitrarily complex. Indeed, materials with desired properties might exist only in a small portion of such a space, making brute-force exploration untenable. AE represents an important part of the solution to addressing these research challenges. An autonomous experimental loop can explore a scientific problem tirelessly and without mistakes, greatly increasing instrument utilization (Figure 1.1). Yet this acceleration due to automation is only one component of the promise of AE. The use of a sophisticated decision-making algorithm opens the door to a far more efficient selection of datapoints, allowing particular scientific questions to be answered more quickly. Indeed, the acceleration provided by AE is limited only by the quality of the underlying model-building, which in principle can leverage numerous sources of existing information (past data, relevant theories, researcher-provided constraints, etc.).

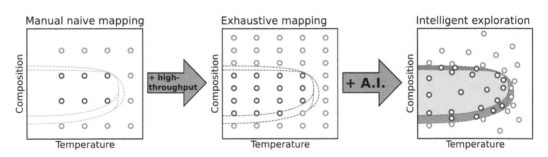

Figure 1.1 Many scientific problems can be thought of as the exploration of a multi-dimensional parameter space. The use of automation and high-throughput methods allows one to map such spaces more thoroughly, but such approaches cannot scale to complex, high-dimensional problems. Intelligent decision-making algorithms can position measurements where they have the most value, vastly improving the probability of discovering hidden features and subtle trends in a scientific problem. By leveraging such methods, Autonomous Experimentation promises to revolutionize research.

AE fits into the proud history of experimental science, which has always developed new generations of experimental tools with improved capabilities (resolution, speed, stability, coverage), and correspondingly more sophisticated and intense data analysis. AE provides another level of experimental sophistication and robustness; the corresponding research acceleration alone would justify its use. But AE also portends a paradigm shift in experimental science. AE promises to radically transform not just the practical way experiments are done but also the way that researchers design experiments, structure their inquiries, and manage their data. If implemented correctly, AE paves the way toward a new science modality where all experiments and resulting data are presumptively integrated into a single scientific model.

1.1.1 Definitions

To better understand AE, it is useful to refine its definition. AE is more than mere automation; it is *autonomous* in the sense that it makes informed decisions. That is, it selects future actions non-trivially based on a constantly expanding dataset and correspondingly continually improves the model. Automation is thus a necessary but not sufficient condition for AE. One can consider progressive levels of automation (Table 1.1). Many experimental tools now incorporate some degree of automation, which streamlines workflows for the user. For instance, robotic sample exchange or controlled sample platforms (for in-situ experiments) represent key experimental automations. Similarly, feedback systems are well-known in experimental

TABLE 1.1 Comparison of different levels of automation. Automated systems can run preprogrammed recipes, a necessary but not sufficient condition for autonomy. Feedback systems respond to changing signals in simple ways. We reserve the word *autonomous* to describe systems that make non-trivial decisions using an underlying model that is continually updated.

	Automated	**Feedback**	**Autonomous**
Context	• Well-defined task • Controlled environment	• Signal changes, but constraints do not	• Fuzzy tasks • Dynamic & uncertain environments
Response	• Scripted, reliable	• Predictable	• Learn, decide, adapt • Builds a model of the problem
Examples	• Scripted acquisition • Auto-analysis • Robotics • Sample environments	• PID • Adaptive scanning	• AI/ML • Reinforcement learning

TABLE 1.2 Different philosophies for autonomous control. One can attempt to re-place human involvement entirely using AI methods, or one can attempt to empower human decision-making by providing advanced assistance tools.

	"Self-driving"	Intelligent assistance
Goal	• Eliminate need for human intervention	• Empower people to do more
Intervention	• When system fails	• By design
Examples	• Self-driving cars • Factory automation	• Art generation • Autonomous discovery

tools, including as part of auto-calibration, auto-alignment, or stability control systems. These typically implement straightforward algorithms, whose behavior is perfectly predictable and depends on the current state (instantaneous instrument signal) but not on past information (world model). Autonomy, by contrast, implies making decisions in a more complex and changing environment, where one's actions must therefore be informed by some kind of learning (model building) [105]. The field of machine-learning has developed a wide variety of techniques to tackle such problems. For instance, reinforcement learning approaches rely on training a statistical model that can select suitable actions, based on the current state of the environment, to maximize a user-specific target metric. Such approaches have been very successful in artificial environments (e.g., game-playing) [382, 483] and natural environments (e.g., robot navigation) [498]. In scientific experiments, the "environment" is the dataset and instrument condition, and the selected "action" is the choice of subsequent experiment or measurement.

AI/ML systems can be deployed in philosophically different ways (Table 1.2). In some cases, the goal of AI is to automate a task so that a human need not be involved. This is especially true when the task is undesirable (tedious and uninteresting), or even unsafe. For instance, self-driving cars and factory automation have as their goal to eliminate the need for humans entirely. Human intervention then correspondingly means the system has failed. On the other hand, AI/ML systems can be used as an intelligent assistant or in an intelligence amplification (IA) mode. Here, the goal is to empower the human to tackle more challenging and sophisticated problems, by automating numerous sub-tasks and providing novel perspectives and suggestions. For instance, recently developed AI art generation systems (such as DALL-E 2 [422], Midjourney [384], or Stable Diffusion [448]) are designed for human interaction. The goal is not to generate images without human intervention but rather to provide a tool that allows the person to generate art more easily and creatively than would otherwise have been possible (Figure 1.2). AE is best thought of in this IA category. The goal is not to replace the human scientist, but rather to liberate them to think about their scientific problem at a higher level. The human is thus directing the AE

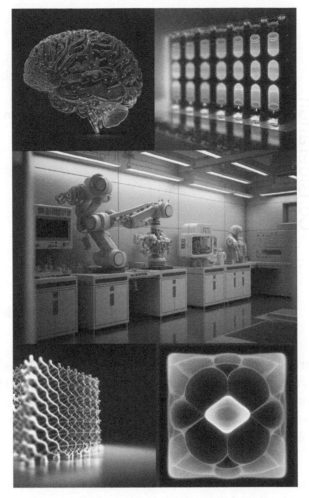

Figure 1.2 A selection of images created using *Midjourney*, an AI art generation system, based on text prompts related to autonomous discovery. Creation using these systems requires collaboration between the AI model and the artist, who progressively refines the prompt and model parameters to achieve a desired outcome.

objective rather than micro-managing experimental execution. This principle applies their unique expertise where it is critically needed—to understand the data, build knowledge, and explain this new science to others. In this sense, some nomenclature—such as "self-driving laboratories"—may, unfortunately, obscure the true potential of AE methods, which is to unleash human scientific creativity.

There are inherent tradeoffs to how one applies AE to a particular experiment. Stein et al. organize experiments based on scientific complexity and workflow automation complexity [493]. Traditional research has attacked problems across a wide range of scientific complexity but has historically used relatively little automation. In recent decades, we have seen increasing use of automation to enable high-throughput experiments, including combinatorial approaches [21, 86, 300, 287]. The most advanced automation systems are quite complex and cover the entire

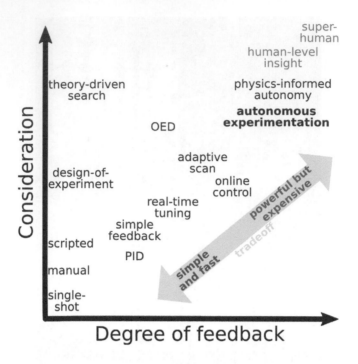

Figure 1.3 Experimental control methods can be organized based on the complexity of the underlying model (consideration) and the richness of the feedback loop. There are inherent tradeoffs between approaches. For instance, more robust modeling provides higher-quality decisions but requires more expensive computation and thus tends to be slower in response time.

workflow of an experiment [91, 266, 587, 179, 510, 5]. Yet such experiments typically target relatively straightforward scientific studies, such as simply generating grids of all possible material variants, and measuring select properties. In parallel, considerable work has gone into applying statistical modeling to guide offline experimental design [38, 285, 284, 286, 84]. Recent progress has seen the development of online autonomous loops that increase the scientific complexity by modeling the underlying problem in some way, and adapting experimental selection accordingly [233, 249, 351, 582, 443, 368, 369, 298, 224, 490, 603, 126, 35, 584]. Decisions may involve the selection of samples, experiments to perform, or instrumental conditions. Looking toward the future, it is clear that next-generation autonomous methods will leverage increasing workflow automation and crucially deploy more sophisticated algorithms to enable more complex scientific reasoning.

An underlying tradeoff in the use of more sophisticated decision-making methods is associated costs—the design cost associated with complexity, and the computational cost to train and invoke a model during experiments (Figure 1.3). As decision algorithms incorporate more feedback data (experimental signals) and use more sophisticated system models, they become more powerful but also more expensive. Thus, in control situations where an extremely fast response is needed, simpler controllers may be preferable. On the other hand, as computational power becomes continually more affordable—owing to increases in hardware capability as well as refinements in machine-learning algorithmic efficiency—we should expect more and more routine deployment of complex control methods. In this sense, it is relevant to research and develop highly complex control methods, especially those that leverage preexisting scientific knowledge, integrate results from simulations, and otherwise leverage auxiliary data channels. A key application area to target is materials discovery, which is well-suited to AE in the sense that the problem (e.g., desired material property) can be cleanly specified, and there exists significant prior knowledge and modeling to draw from [406, 417, 490].

One must also consider the underlying tradeoff of model generality. A more general modeling approach is preferable in the sense that it is generic, and can be easily applied to a wide array of experimental problems. Adapting a system-agnostic decision-making module to a new experimental problem is simple. General models, however, will not perform optimally for any particular scientific problem [575]. Designing highly tailored systems—e.g., that incorporate system physics—should yield increased performance, but at the cost of development time and generality. Both approaches (general and system-specific control methods) are worth researching and developing. There are also opportunities for developing hybrid approaches—flexible models that can be rapidly adapted to specific problems. The future of AE likely involves portfolios of controllers available for the user to select, and machine-learning frameworks, making it easy to retrain models to target specific problems.

1.1.2 Advantages

Many advantages of the autonomous paradigm are obvious. The principal selling point of the method is enhanced efficiency. This can be thought of in terms of increasing utilization of valuable instrument time: an argument especially persuasive for over-subscribed facilities such as synchrotrons, neutron sources, and frontier electron microscopes. A related argument can be made about scientific efficiency. That is, the researcher can answer their scientific question more rapidly with AE methods. Indeed, AE allows one to quantify the desired model quality, and end data collection exactly when that threshold is reached. Overall, this improves the efficiency of resource usage.

The primary output of an autonomous experiment is the collected dataset. However, many AE algorithms generate a system model as part of their operation. This model itself is a scientific product of significant value. In this sense, the AE paradigm can combine both the data collection and the initial model-building steps. The model resulting from AE can provide predictions for the system across the space being probed, allowing the researcher to test hypotheses and more meaningfully propose future studies. Many AE models also perform some form of uncertainty quantification (UQ); thus the model not only yields predictions but quantifies data distribution, allowing one to make clear statements about scientific confidence.

Another way to think about AE is in terms of the effect it has on the human scientist. AE liberates the scientist to focus on high-level concepts, and thereby maximizes the scientific understanding that results from an experiment. AE helps to abstract away much experimental complexity, allowing the human to tackle more challenging scientific problems. The AE paradigm also induces a set of best practices. It forces experimentalists to think more carefully about the experimental plan, to precisely define the objective, and to quantify the uncertainties in the signals they measure. An AE workflow naturally involves digitized data collection, including capturing meta-data, making such data intrinsically suitable for subsequent data sharing or data mining. On the hardware side, AE also encourages more systematic and robotic synthesis and handling of samples. This inherently drives such activities toward being more documented and reproducible, as well as avoiding the dangers of artisanal sample preparation that requires the unique skills of a particular person.

1.1.3 Dangers

As with any emerging paradigm, there are criticisms of AE concepts—some valid, some overblown. An obvious concern with deploying AI/ML is that one is attempting to replace humans. As previously noted (c.f. Table 1.2), AE is better thought of as intelligence augmentation. It frees the human researcher from various compute-heavy tasks and allows them to focus on the meaning of the data. In this sense, AE is merely an augmentation of the ongoing computerization of research workflows. When AE is implemented correctly, the human remains "in the loop" and is always able to intervene. This requires mechanisms of control as well as real-time visualization of the AE execution (surrogate model, uncertainty, objective function, experimental plan, etc.).

Another set of concerns focuses on the biases caused by using an AE methodology. Data-driven AI/ML models inevitably reproduce the biases of their training data, and the unfortunate biases of ML models trained on human data are by now well documented. In scientific research, a valid concern is that an autonomous search will miss an important finding owing to the assumptions built into the search. Similarly, the resulting model will inevitably be limited by the constraints (explicit or implicit) of its design. Of course, these concerns are not unique to AE. Any data-taking strategy (human intuition, formalized design-of-experiment, AE) has a corresponding bias and potential weakness. Every scientific model is limited by the assumptions built into the functional form selected by its designer. Rather than seeking "unbiased" methods, it is arguably better to think in terms of which biases are desirable. For instance, we should seek methods and models that are biased toward known and well-established physics (models inconsistent with known physics have manifestly undesirable biases!). These challenges represent exciting opportunities for future research.

AE practitioners have learned many pragmatic lessons over recent years. There is a cost in converting a system to autonomous operation, in terms of design effort, hardware systems, and software requirements. Increased automation is nearly always desirable (even if AE is not the primary goal); however, the costs may not be justifiable in some cases. A danger in AE systems is the separation it creates between the scientist and the experimental details. Allowing users to treat AE as a "black box" that returns models that are trusted without question would be disastrous. This is not a concern specific to AE; as scientific instruments become increasingly complex, it becomes progressively harder for end users to maintain an understanding of all components and concepts. Similar trends are observed in scientific data analysis, where it is no longer practical for end users to be intimately familiar with the algorithms and implementation details associated with the enormous infrastructure they leverage when using existing software products. In all of these examples, the goal must be to provide efficient and trustworthy infrastructure, while also educating users about the capabilities and limitations of the underlying systems. As AE is developed, we must maintain a set of experts who understand the details of the underlying implementations, and can correspondingly adapt these systems to new problems, and communicate their limitations to end users.

1.2 DEPLOYING AE

1.2.1 Autonomous Loop

There are many ways to design an autonomous experiment. A very useful way to think about AE is as a loop connecting three elements (Figure 1.4): an instrument that makes some kind of measurement, a data analysis pipeline that converts the raw measurement into a meaningful signal, and a decision-making module that suggests the next measurement/experiment to perform, based on the growing dataset. Thus, control passes iteratively between these three modules, with simple messages transmitted from one to the next. This organization is natural and modular. Of course, one can imagine alternate implementations. For instance, arguably the data analysis and decision-making should be merged into a single module. This would allow the

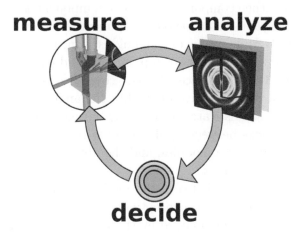

Figure 1.4 An autonomous experiment (AE) can be thought of as a loop combining three automated components. The instrument performs measurements as specified by an external signal; all hardware aspects (sample synthesis, loading, alignment, etc.) must be automated. The resulting raw data is analyzed, and converted into a concise and meaningful signal. A decision-making algorithm models the available data, and selects a new measurement condition, which is sent as a command to the instrument. This loop can thus autonomously explore experimental problems.

decision algorithm to leverage all aspects of raw data, a concerted approach that is more tailored and potentially more optimal. In practice, there are several advantages to cleanly separating the analysis and decision steps. First, one can view the analysis step as a necessary dimensionality reduction. Raw data may contain far too many signals. For example, a high-resolution image contains $> 10^6$ signals; it would require a correspondingly large number of experiments to train a model to understand the dependencies of all signals. Dimensionality reduction also acts as a crucial step in applying domain-specific knowledge, where signals of interest are extracted for consideration. Second, the separation into modules can be viewed as a software engineering best practice. By defining separate steps with a clean interface between them, the associated software development becomes more tractable and robust. One can also replace a module without affecting others. For instance, a given autonomous instrument could offer multiple analysis pipelines, and multiple decision-making modules, which can be selected by the user as needed for a given experiment. Finally, it is worth noting that this separation of capabilities makes the overall system useful beyond the AE context. An independent automated analysis pipeline can be used during conventional (non-machine-guided) experiments; indeed, one could argue that every modern scientific instrument should have an associated analysis pipeline that automatically processes all collected data.

A practical way to implement an autonomous loop is to have three separate software processes, which are passing simple, concise messages to one another in sequence. Such a setup maximizes isolation of the software units, which is desirable both from the perspective of software development, and code execution. Separate software processes allow one to easily stop and modify any of the three components, without affecting the other components. If each component is designed to wait for messages, then stopping and restarting one component will allow the autonomous loop to proceed normally. A crude means of enabling user engagement in the autonomous loop is thus simply allowing the user to stop, modify, and restart any module as suits the ongoing experiment. For instance, the user could decide to modify the decision-making module with a new objective or modify the analysis pipeline to focus on a different signal. A more sophisticated form of user engagement might involve mutable queues managing messages between modules, a set of visualization tools to inspect the inner workings of the three modules, and the ability to send signals to modules for them to update their behavior.

The messages being passed between the modules can be remarkably simple. A broad class of experiments can be conceptualized as exploring a high-dimensional parameter space (whose axes are defined by the variables one can control) and monitoring a finite set of signals (analysis results deemed meaningful for the particular problem). For example, the study of new materials might involve measuring a few properties (crystallinity, elastic modulus, etc.) across a range of preparation conditions (material composition, processing temperature, etc.). Even problems that initially seem infinite-dimensional (e.g., continuous function defining sample processing) can often be parameterized to define a finite-dimensional search space. The corresponding machine-learning problem can then be thought of as modeling the specified signals throughout the specified space. In such cases, the message from the decision agent to the measurement platform consists only of a vector specifying the next location in the space to consider, and the message from the analysis module to the decision agent consists of this parameter space vector and the associated signals at that location. The message from the instrument to the analysis pipeline could be a copy of the raw data (the most recent measurement); but more likely would simply be a unique identifier allowing the analysis pipeline to load and analyze the raw data.

When AE is conceived of as a loop (c.f. Figure 1.4), it becomes clear that it is amenable to any scientific tool whose data collection and data analysis are automated. AE then consists of "closing the loop" by connecting to an appropriate decision-making algorithm. As there are very strong reasons to continue automating experimental hardware and analysis, the opportunities for AE will grow enormously over the coming years. One way to frame AE is as an agent problem, where the decision-making module is an actor operating inside an environment (which changes over time), and the goal is to select actions that maximize an encoded reward. This reinforcement learning approach has been highly successful in game-playing, for instance, and shows promise for rapid decision-making in experimental contexts [298, 13, 92]. A highly generic way to devise the AE decision-making module is to treat the experiment as an optimization problem in a prescribed multi-dimensional parameter space; the optimization goal could be exploration (search the space), understanding (model the entire space as accurately as possible), or more targeted (find the location that maximizes a quantity of interest) [369]. Bayesian mathematical methods provide a rigorous way to quantify uncertainty and thus knowledge and are ideal candidates for AE modeling. A successful approach has been to model the exploration space using Gaussian process (GP) methods [431], treating existing data as known quantities (with corresponding uncertainty) and generating a data-driven model that makes predictions (with uncertainty) throughout the space [369]. The GP model can be thought of as a surrogate model for the true behavior of the system; the GP optimization determines model hyperparameters that best describe the data and then uses these trained parameters to estimate model behavior throughout the space. The GP effectively interpolates and extrapolates the available experimental data, providing predictions one can use in autonomous control. The most straightforward way to define an AE objective is to minimize model uncertainty. In this mode, in each AE loop, the decision-making module would update its model with all available data, identify the location in the model space with the highest uncertainty, and suggest a measurement at that location. Focusing on high uncertainty naturally maximizes knowledge gain per measurement. One can elaborate on this basic procedure to account for any additional experimental constraints. For instance, one can assign a cost to each measurement (instrument time, material use, etc.) and define an objective function that maximizes knowledge per unit cost [358]. Or, one can include additional terms in the objective function related to maximizing a quantity of interest (such as a desired material property). Importantly, the GP approach means that such targeting need not be a naive, greedy maximization; instead, it can naturally provide a blended approach where one is maximizing a quantity while also minimizing uncertainty. That is, the GP naturally balances between exploration and exploitation modes. In addition to the objective function, the GP method can be tailored in other ways to the specific experiment or known scientific constraints. The GP uses a kernel to model the space—effectively a function that describes the presumed correlation between positions in the space. Selecting an appropriate kernel thus affords the chance to encode known physics into the modeling, such as a particular symmetry of the parameter space [365, 34]. Kernel design is an exciting area for future study. Overall, the GP modeling approach is generic since it need not make any assumptions about the space being probed, and naturally provides a surrogate model and associated uncertainty model (both of which are valuable scientific products, at the conclusion of the AE). Yet this approach is also powerful in that it can be tailored to specific problems.

1.2.2 Implementation Example

As a concrete and successful use case, we consider the application of GP methods to autonomous X-ray scattering [368, 359, 358, 369, 584]. A collaboration between the Center for Functional Nanomaterials (CFN) and the National Synchrotron Light Source II (NSLS-II) at Brookhaven National Laboratory (BNL), and the Center for Advanced Mathematics for Energy Research Applications (CAMERA) at Lawrence Berkeley National Laboratory (LBNL), has demonstrated autonomous data collection at multiple synchrotron x-ray scattering beamlines. X-ray scattering enables the quantification of order at the atomic, molecular, and nano-scale by measuring the far-field diffraction and scattering of incident X-rays [585]. Synchrotron beamlines enable rapid data collection, and the possibility for in-situ monitoring, making them ideally suited to AE [565, 95, 190]. This work has demonstrated enhanced efficiency, through increased beamtime utilization, more rapid convergence to high-quality data, and new material discoveries. Three classes of experiments well-suited to autonomous x-ray scattering were identified [584]: (1) 2D imaging of heterogeneous materials by x-ray beam scanning; (2) exploration of physical parameter spaces through sample libraries or gradient combinatorial samples; (3) control of real-time material synthesis or processing environments.

As a specific scientific example, consider the study of block copolymer (BCP) thin films—these materials spontaneously self-assemble into well-defined nanoscale morphologies. The exact morphology that forms depends on the chain architecture [128]. It has recently been shown that blending together different BCPs can give rise to new phases [583, 35], and that substrate chemical gratings can direct this process, effectively selecting from the competing structures that might form.[492] A deeper study of this problem implicates a large and complex parameter space since one can vary the materials being blended, the blend ratio, the structure of the underlying chemical grating, film thickness, subsequent processing conditions, etc. All these parameters play a role in structure formation, and there could be small subsets of this large parameter space exhibiting interesting behavior, making this an ideal challenge for autonomous study. Researchers deployed autonomous x-ray scattering to this research problem (Figure 1.5). Combinatorial arrays of substrate chemical gratings were generated, varying both the grating pitch and duty cycle (i.e., width of chemical stripes). A BCP blend thin film was then cast on top, and the ordered material was studied as a function of position using synchrotron x-ray scattering controlled by a GP algorithm (using the *gpCAM* software). In an initial study, the AE was tasked

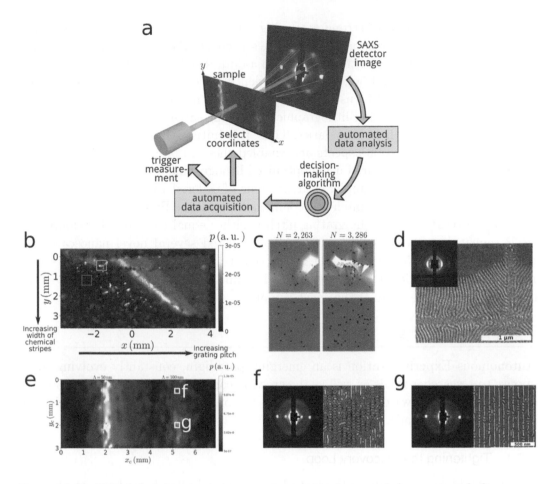

Figure 1.5 Example of autonomous x-ray scattering used for material discovery, based on work described in [369] and [126]. (a) The AE loop directed an x-ray scattering beamline to probe a 2D combinatorial array. The samples consisted of block copolymer (BCP) blend thin films ordering on a chemical grating, where the grating pitch was varied along one direction, and the grating stripe width varied in the other direction. A GP algorithm was used to search the space. (b) An initial experiment mapped the scattering intensity (p) of the primary BCP peak. (c) The AE was tasked with focusing on well-ordered material and correspondingly concentrated more data collection in regions with high scattering intensity (e.g., green box). (d) This particular experiment discovered unexpected four-fold ordering in scattering patterns, which was traced to the boundaries between lithographic patterning regions. Thus, this AE rapidly identified errors in the sample preparation, allowing this to be fixed. (e) A subsequent sample with optimized patterning was then mapped using AE, where researchers identified both expected and novel morphologies. (f, g) The new nanoscale morphologies that were discovered exhibit unexpected ordering arising from the interplay between the directing field of the template and the local rearrangement of polymer chains. Overall, this demonstrates the accelerated material discovery enabled by AE, both by rapidly iterating to optimize experimental conditions, and by identifying novel material ordering that would easily have been missed in a conventional study.

with identifying regions of strong scattering intensity (of the peak associated with BCP ordering); one correspondingly observes a concentration of data collection in sample areas with strong scattering but also a data distribution that improves over-all model quality (Figure 1.5c). This search identified regions of unexpected four-fold symmetry, which enabled focused investigation using microscopy [369]. This uncov-ered errors in the underlying lithographic patterning, giving rise to undesired BCP ordering at patterning field boundaries. Thus, this initial AE rapidly identified prob-lematic sample preparation aspects and enabled these to be immediately addressed. A subsequent sample prepared using optimized lithographic processes was then stud-ied (Figure 1.5e). In this sample, clear domains of ordered BCP phases were found, especially when the grating pitch was commensurate to the BCP's intrinsic repeating lengthscale (that is, when the grating pitch is either equal to or roughly double the BCP repeat-distance). Moreover, this AE discovered several novel nanoscale mor-phologies (Figure 1.5f, g) [126]. These new structural motifs could easily have been missed in a conventional experiment, which would likely have been more selective and arbitrary in the parts of the space explored.

1.3 FUTURE OF AE

Autonomous Experimentation is an emerging paradigm, constantly evolving in re-sponse to researcher needs as well as the growing capabilities of AI/ML methods and computational hardware. It is thus exciting to consider how this field will develop over the coming decade.

1.3.1 Tightening the Discovery Loop

Experimental research can be thought of as a loop, wherein hypotheses are formed, experiments are conducted, data are analyzed, models are improved, and new hy-potheses are formed. The AE loop (c.f. Figure 1.4) can be thought of as an au-tomation of this process. Correspondingly, one can view the over-arching objective of AE to be part of an effort to "tighten" the discovery loop—that is, shorten the iteration time between the successive steps of research. We can view AE as part of a sequence of increasingly sophisticated discovery loops (Figure 1.6), which cor-respondingly makes the needed capabilities more obvious. In moving from entirely manual and ad-hoc research iteration toward data-driven discovery, one could de-velop combinatorial sample synthesis or robotic sample handling methods, deploy more sophisticated (multi-modal, in-situ, etc.) characterization tools, and integrate automated data analytics. More advanced AE methods require the additional in-clusion of machine decision-making algorithms. An even more integrated research loop is required for real-time guiding of material synthesis. In this case, automated material synthesis must be coupled to in-situ/operando measurements, and should exploit decision-making algorithms that incorporate known material physics (e.g., via explicit material modeling). Overall, tightening the discovery loop not only accel-erates iterative experimentation but also yields more sophisticated exploration and modeling, by more tightly coupling synthesis, measurements, theory/simulation, and decision-making.

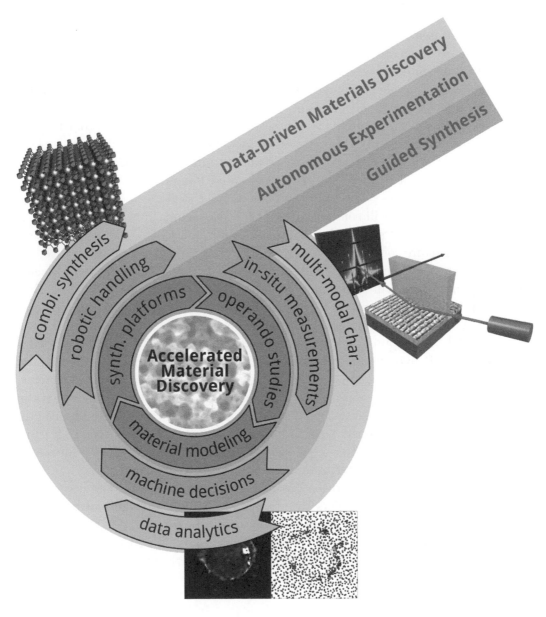

Figure 1.6   Autonomous experimentation can be viewed as part of an ongoing process in experimental science that pushes toward more complex, automated, and machine-controlled experimental tools. The underlying goal is to tighten the discovery loop, where researchers iterate between experiments, data analysis, and improved modeling/understanding.

1.3.2 Diversity of Approaches

As previously described, AE control methods involve a number of tradeoffs. There is a tradeoff between generality and specificity; the former makes it easy to start new experiments, while the latter provides the opportunity for greater efficiency and insight. There is a tradeoff between quality and speed; computationally expensive models can provide more insightful control, but cannot keep pace with fast data collection. In this context, it seems likely that the future of AE will involve a diversity of data collection methods, tailored to different use cases. Instrument users will then be able to select the method best suited to their problem; one can even imagine a higher-level AI/ML agent automatically selecting methods based on problem statements. Some methods will exploit Bayesian modeling of uncertainty (such as GP). Other methods will exploit deep neural networks to model the space (reinforcement learning). Some available models might be pre-trained on prior experimental data, or on a suite of simulations relevant to the system under study.

It is likely that AE will continue to exploit black box machine-learning methods; that is, to use controllers whose internal state is indecipherable. Such controllers can be valuable in many contexts; especially where there is plenty of training data, and where the primary objective is a simple and verifiable optimization. Yet it is also clear that the future of AE will increasingly move beyond black box methods [545]. There is thus much exciting research to be done to develop AE controllers that are physics-informed. This can be accomplished by applying physics constraints to existing methods. In model-based approaches, one can design the models with physics knowledge. For instance, GP hyperparameter optimization can be constrained through known physical relations, or the kernel can be designed to capture underlying symmetries [365, 35]. In deep-learning approaches, training using physically meaningful data inherently captures some aspects of the underlying physics. However, the internal states of these models are notoriously hard to interpret, and they can be prone to learning spurious (unphysical) correlations. An exciting area of ongoing research is focused on making such systems more interpretable, or inherently explainable.

Existing work already provides a possible model for the abstraction necessary to allow a diversity of AE controllers to be available. Each controller can be thought of as a function that takes as input a specification of existing data, in the form of signal values located in an abstract multi-dimensional search space. The output of the controller is then a suggestion (or list of suggestions) for subsequent points to measure in the space.

1.3.3 Integration

The future of AE will likely be marked by increasing levels of integration. As previously described, we expect to see a deeper embedding of AE methods into scientific instruments, thereby tightening the discovery loop and providing the user with a more powerful and streamlined means of taking data. We can also anticipate coupling of several layers of AI/ML control at a single instrument. The AE loop involves an autonomous controller, making decisions at a speed matched to experimental data-taking. One can also imagine a larger loop overseeing the instrument and data collection, making adjustments as necessary. This could involve modifying data collection (instrument parameters), the data analysis pipeline (selecting new signals, or retraining a data-analysis ML model), or even the decision-making algorithm (e.g., modifying model hyperparameters). This approach could be used to mitigate the tradeoffs mentioned earlier. The inner loop can be optimized for high-speed control on the instrument, while the outer loop can more slowly reoptimize strategies based on more complete modeling.

Another integration to consider is between instruments. Once individual instruments are autonomous, it becomes possible to integrate them, so that they are working together to solve a particular scientific problem. Such a multi-modal AE system could take many forms. High-speed measurements on one instrument could be used to select high-value/high-cost measurements on another. Or multiple instruments could be measuring simultaneously, with an overall model being developed to learn the relationship between the modalities. Or the scientific problem could be partitioned between instruments/modalities; for instance one AE could be attempting to explore a phase space, while another focuses on bounding certain model parameters, and another exploits the improving knowledge to optimize material synthesis.

The most ambitious integration would see a shift away from scientific facilities offering a set of disconnected tools, toward a discovery ecosystem where the user codifies their scientific objective, and the platform handles all details thereafter (Figure 1.7). In such a design, a set of AI/ML agents would operate on behalf of

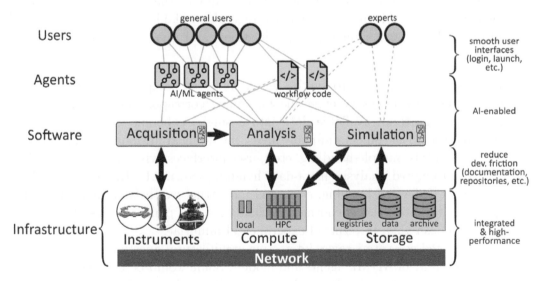

Figure 1.7 Future scientific facilities could be designed as an integrated ecosystem, rather than as isolated tools. Users would correspondingly not micro-manage execution on any particular instrument, or even worry about coordination between instruments. Rather, the user specifies and tracks high-level scientific goals; a distributed set of AI/ML agents then launch experiments and simulations on the user's behalf, and integrate results into a coherent whole. Realizing this vision requires advancements in many areas, including developing user interfaces, AI/ML methods, and integrated research infrastructure.

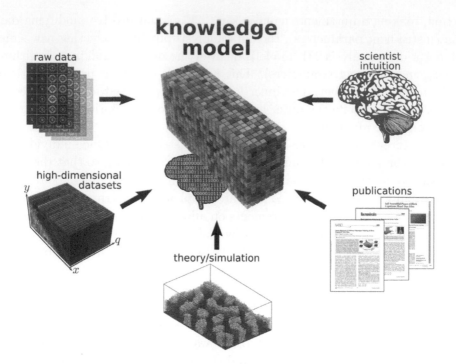

Figure 1.8 A grand challenge in science is to develop methods that integrate disparate sources of information into a concerted model that embodies meaningful knowledge. Ideally, such models would leverage all available sources: raw and processed data, theory and simulations, publications, and myriad inputs from human scientists.

the user, coordinating so as to synthesize the required samples, route them to instruments, perform necessary experiments, launch corresponding simulations, and continually integrate all available data into a single, coherent model. This grand goal will require advancements in many areas, including traditional automation and robotics (standards, sample handling, etc.), user interfaces (visualization, etc.), data workflows (automated analysis, meta-data handling, etc.), AI/ML methods (collaborative agents, federated learning, etc.), and infrastructure (networking, computing, storage). Yet the potential is enormous: scientists could plan in terms of goals and hypotheses, rather than in terms of experimental minutiae.

An integrated ecosystem raises interesting architectural questions. It would likely involve a hierarchy of AI/ML agents and models. Some would be user-facing, and responsible for translating scientific goals into experimental plans, launching processes, and visualizing results. Others would be embedded in the instruments, automating low-level functions. Others would act as intermediaries, integrating data and building models. One school of thought might argue that these myriad agents should be able to probe deeply into each other's state. Much as a human expert might, the agents should be able to analyze the internal workings of other ML models, and make arbitrary modifications. This would provide enormous power, at the cost of significant complexity. An opposite argument might be that one should intentionally bind agents, and force them to interact through narrow and well-defined interfaces. Such abstractions have the usual software engineering advantages of making the system more tractable, maintainable, and modular. The separation of modules may provide an additional advantage in the case of AI systems: it may address alignment [205, 155, 252]. A pernicious problem in AI is that it may game the system, optimizing the target metric in an unintended way that subverts the true goals of the user (i.e., its behavior is not aligned with the user intent). Refining the objective function cannot necessarily address this challenge, since the full set of user goals/values is difficult to specify, or possibly even inconsistent. A speculative but intriguing possibility is that designing AI systems as separate competing modules might address this. The various modules act to oversee, rate, and influence each other; but none has complete knowledge or control. Any given module has limited opportunities for subversion, and the designer can fine-tune interactions so that the emergent system exhibits the right behavior. An analogy can be drawn to human intelligence. Human decision-making arises through a set of competing impulses, desires, and thoughts. Different aspects of thinking vie for attention and control, but no single aspect is afforded primacy. Even the most considered of human deliberation is not ultimately in charge, as other aspects (reflexes, strong emotions, etc.) are able to override under the right conditions. It is the coordinated interaction of these systems that gives rise to the intelligent behavior of humans.

A final aspect of integration to consider is at the level of data and knowledge. The existing applications of ML to science have focused on solving isolated tasks. Yet the emergence of foundation models in AI/ML demonstrates the power of extremely large models trained on all available data in a task-agnostic manner [65]. Such systems can then be focused on a specific task, with enormous success having now been demonstrated across many tasks, including coherent text generation, interactive discussion, and generation of artistic assets (images, 3D models, short video clips). One can correspondingly consider how these approaches would translate to the realm of science. We envision an enormous opportunity for the community to begin building AI/ML models that integrate any and all available sources of data, including raw and processed experimental data, simulation results, system theories embedded into mathematical formalisms, the enormous corpus of scientific papers, and even human scientist intuitions and idea (Figure 1.8). Such models would, by design, capture correlations and connections within and between the training datasets, and compress the complexity of the available data into a more concise (though highly complex) inner representation. Such models could capture the underlying physics of particular fields of study, and act as knowledge oracles for researchers, allowing them to predict new physics or materials, speculate on relationships, and design new experimental campaigns. A provocative suggestion is to consider that intelligence is often an act of compression—humans understand the world by ignoring unnecessary signals and reasoning with an abstracted world model—and science is successful by uncovering the minimal mathematical relation required to explain phenomena. Integrated ML models for science, by compressing all available data into a concise and self-consistent whole, could open the door to more advanced and intelligent decision-making in AE systems.

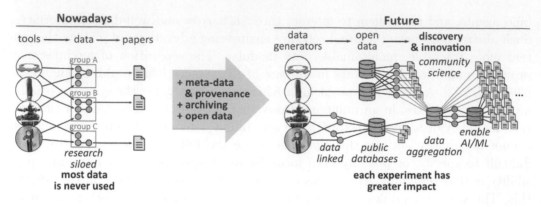

Figure 1.9 Community data practices must evolve in order to extract maximum value from experiments. Currently, most research groups operate in isolation. The data they collect is never released, and only a small subset of data is analyzed or converted into scientific insights. Three elements are required for improved data practices: (1) meta-data (including provenance) must be captured during experiments and linked to raw data; (2) data must be archived indefinitely; (3) data must be openly available to all. Improved practices lead to enormous enhancements for science, including greater scientific output, new kinds of scientific study (data mining, meta-analyses, etc.), and improved efficiency.

1.3.4 Data Practices

Data is fundamental to science. While the goal of science is knowledge and understanding, trustworthy data acts as the substrate upon which these are built. Data management should not, therefore, be considered an afterthought in scientific endeavors. AE methods tend to encourage proper data practices, including digitizing all aspects of data handling, and capturing meta-data. Correspondingly, the true potential of the AE paradigm will not be realized if community data practices are insufficient.

Current data practices in many scientific communities are quite isolated and inefficient (Figure 1.9). Scientists in different groups collect data on multiple instruments. Only a subset of this data is ever analyzed, and a smaller subset is disclosed through publications. Research is siloed, and the majority of data is never used—despite the significant cost associated with each measurement. Three elements are required to improve data management: (1) all meta-data must be collected during experiments or analysis, including provenance information allowing one to trace data to its origin; (2) data should be retained over long time periods; and (3) data should be made openly available to the whole scientific community. Adopting these practices opens the door to an enormous array of increased impacts. Efficiency is increased by avoiding wasted repeated experiments. Reproducibility is improved by facilitating comparison and replication. Data-hungry AI/ML efforts are empowered. A broader and more diverse population of scientists can participate in research. Each individual experiment has a greater chance of contributing to meaningful progress. The total aggregation of open data provides an unprecedented, detailed understanding.

There is growing understanding of the importance of data management and openness. Funders and publishers of science are increasingly encouraging adherence with FAIR principles: guidelines to improve the Findability, Accessibility, Interoperability, and Reuse of digital assets. AE can both help with these efforts, and will benefit enormously as scientific data becomes more systematized and accessible.

1.3.5 Perspective

Autonomous Experimentation is the future of physical science. Indeed, a broader and broader range of research communities is discovering the power of AE, and correspondingly adapting their workflows toward AE methods. Of course, one can make a plausible argument that "not all experiments should be autonomous". It is seemingly obvious that some aspects of current experimental research require the finesse of human hands, or the engaged consideration of a human experimenter at every step of the process. *We posit that these arguments are pragmatic but ultimately flawed; and that every experimental process should indeed be transitioning toward automated handling and autonomous execution.*

The reason is that AE can enable experimental science to be more reproducible and documented; reliance on humans instead means accepting that processes will remain artisanal, undefined, and error-prone. AE enforces the practices we all know to be fundamental to good science: systematization of data, rigorous consideration of errors, strict definition of hypotheses, and the analysis and integration of all data. Most provocatively of all, we might imagine a future in which non-AE approaches are not considered good science, as they cannot adhere to best practices. None of this means that humans are eliminated from the scientific process. Science is fundamentally a human process, since our goal is to derive knowledge that can be communicated to people, increasing their understanding. The goal of AE is to unleash the full potential of the scientific method, and give humans the broadest and deepest possible view of the universe.

CHAPTER 2

A Friendly Mathematical Perspective on Autonomous Experimentation

Marcus M. Noack

Applied Mathematics and Computational Research Division, Lawrence Berkeley National Laboratory, Berkeley, California, USA

CONTENTS

FROM a (this) Mathematician's perspective, autonomous experimentation—at least the intelligent decision-making part—involves solving two math problems in each sequential iteration of an experiment. First, a function approximation from noisy data $\mathcal{D} = \{\mathbf{x}_i, \mathbf{y}_i\}$, $\mathbf{y}(\mathbf{x}) = \mathbf{f}(\mathbf{x}) + \boldsymbol{\epsilon}(\mathbf{x})$, which should provide a model and a measure for "not-knowing", and second, a constrained function optimization to maximize knowledge gain or some predicted reward [389, 390]. Here, $\mathbf{f} : \mathcal{X} \to \mathbb{R}^n$ is the underlying inaccessible data-generating latent (model) function, where $\mathcal{X}$ is the input domain; most often $\mathcal{X} \subset \mathbb{R}^m$, but this is not a requirement. $\boldsymbol{\epsilon}(\mathbf{x})$ is the noise term. An increasing number of experiments require input domains that are non-Euclidean. In fact, $\mathcal{X}$ does not even have to be a vector space [26], but can be some non-linear manifold. But more about that later. In this introduction, we will assume that we are interested in approximating a scalar function f, i.e., $n = 1$. Of course, the learning process can be applied to all components of a vector-valued function. Other, more powerful learning techniques for vector-valued functions are called multi-task learning and will play a role in some chapters of this book, but are out of scope in

DOI: 10.1201/9781003359593-2

this introduction. So let's focus on scalar latent functions $f(\mathbf{x})$. Formally, we want to solve

$$f^* = \arg\min_{f \in \mathcal{F}} \ Loss(f, \mathbf{y}), \tag{2.1}$$

[26] where $\mathcal{F}$ is some set of functions—sometimes called the hypothesis space—and $\mathbf{y}$ is the set of observations. Intuitively speaking, this means finding a function that optimally explains the observed data. Therefore, the *Loss* measures the difference between the observations y_i and the corresponding predictions of the model function f. The loss function is often called the empirical risk, and solving (2.1) is then called empirical risk minimization, and solving it directly leads to poor generalization and over-fitting [550]; of course, the function that explains the data optimally is not necessarily the function that leads to optimal learning. Having determined an approximation for f^*, the function is used for decision-making to guarantee optimal data acquisition. This is most often done by estimating the uncertainty of the predictions and formulating some measure of predicted value—in the sense of worth or utility—across $\mathcal{X}$. In other instances, the function f^* is directly exploited for decision-making. Maximizing this value in the form of an acquisition function $f_a(\mathbf{x})$—sometimes called an objective function—

$$\mathbf{x}^* = \arg\max_{\mathbf{x} \in \mathcal{X}} f_a(\mathbf{x}) \tag{2.2}$$

leads to an optimal "value" of the next measurement $\mathbf{x}^*$. The acquisition function—most often a functional, i.e., a function of a function—can depend on domain-informed expectations of the experimenter. One might, for instance, be interested in regions where f shows particularly high first- or second-order derivatives. A note about terminology here. The notation $f_a(\mathbf{x})$ hides the fact that we are actually interested in a function of our approximation of the latent function. Therefore, the correct notation would be $f_a(\phi(\mathbf{x}))$; in other words, f_a is a functional that eats some approximation of the latent function at $\mathbf{x}$ ($\phi(\mathbf{x})$) and spits out a scalar. ϕ could, for instance, represent a posterior distribution of a stochastic process (Figure 2.1). A more general (but less intuitive) perspective of the acquisition functional is formulating a quantity of interest [390]

$$Q(f) = \int_{\mathcal{X}} f(\mathbf{x}) p(\mathbf{x}) d\mathbf{x}, \tag{2.3}$$

where $p(\mathbf{x})$ is some probability density function, often related to the posterior of an approximation, and then maximizing its information gain.

Arguably, problem (2.1), the function approximation from data, is mathematically more interesting than (2.2) due to the flexibility in choosing $\mathcal{F}$ and the loss-function. Once a candidate function, or hypothesis, f^* is found, the optimization problem (2.2) is mathematically comparably mundane but we will cover some approaches later. In this introduction, we want to start with the intuitive objective (2.1) and follow a chain of causality to recover current methods for function approximation and autonomous experimentation, while omitting unnecessarily technical and distracting details. We will discover that there is an equivalence of the regularization and the Bayesian perspective to function approximation [26, 226].

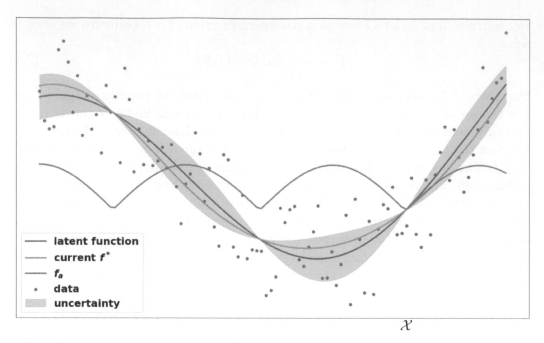

Figure 2.1 The inaccessible, data-generating latent function f, defined on $\mathcal{X}$, can be approximated via some optimality condition (2.1) by f^*. Based on the characteristics of the approximation—here the estimated uncertainty—an acquisition functional f_a is defined whose maxima are considered optimal points for data acquisition.

2.1 REGULARIZATION

On a fundamental level, and ignoring the complexities of instrument control and data analysis, autonomous experimentation is indeed primarily a function-approximation problem, in which the algorithm can choose data-point locations in each iteration to optimize approximation performance and learning. An algorithm following this principle—making sequential optimal decisions in each iteration given the current state—is also called a *greedy algorithm*. Every measurement of an instrument is analyzed and subsequently merged with previous results. In every iteration, the function approximation is repeated in order to find the model that generated the observations. For this, we have to search through an uncountable set of functions $\mathcal{F}$. Without imposing any smoothness on the elements in $\mathcal{F}$, solving (2.1) will most definitely lead to over-fitting and little generalization potential. One could imagine a set of delta functions that explain the observations exactly but no knowledge would be gained from this. The approach to solving this problem is to use regularization, which means adding a bias to the loss that gives priority to smooth functions

$$Loss = \frac{1}{N} \sum_i^N (|f(\mathbf{x}_i) - y_i|) + \mu S(f), \tag{2.4}$$

where S is some measure of smoothness of the candidate function (the regularizer) and $|\cdot|$ is some norm. μ is called the regularization parameter and controls the desired smoothness of the solution. Some authors use λ as the regularization parameter, which bears the danger of being confused with a Lagrange multiplier.

The regularization perspective leads us directly to ridge regression, which is performed by solving

$$\arg\min_{\mathbf{w}} \ Loss(\mathbf{w}) = \arg\min_{\mathbf{w}} \ \sum_i^N (y_i - f(\mathbf{x}_i))^2 + \mu ||\mathbf{w}||^2, \tag{2.5}$$

where we have chosen a quadratic loss function and $||w||$ as smoothness measure, where $||\cdot||$ is the Euclidean norm. We assume that our scalar latent function is of the form $f = \sum_j^N w_j \mathbf{x}_i$, for some weights w_i. Setting the first derivative to zero

$$\frac{\partial Loss(\mathbf{w})}{\partial w_k} = 2\mu w_k - 2\sum_i^N (y_i - w_i \mathbf{x}_i)\mathbf{x}_k = 0 \tag{2.6}$$

reveals the unique solution

$$\mathbf{w} = \mathbf{X}^T(\mathbf{G} + \mu\mathbf{I})^{-1}\mathbf{y}, \tag{2.7}$$

where $\mathbf{G}$ is the Gram matrix of inner products $G_{ij} = \langle \mathbf{x}_i, \mathbf{x}_j \rangle$. Predictions at a new location can then be calculated as

$$f(\mathbf{x}) = \sum_i^N w_i \mathbf{x} = \mathbf{y}^T(\mathbf{G} + \mu I)^{-1}\langle \mathbf{x}_i, \mathbf{x} \rangle. \tag{2.8}$$

The astonishing takeaway here is that for ridge regression, all we have to compute are the inner products of the data-point locations. The regularization parameter μ naturally appears here as noise.

Ridge regression is one of the most popular tools for rigorous deterministic linear regression; however, we might want to consider more complicated relationships than linear models. We can do so by defining a non-linear mapping $\mathbf{\Phi}(\mathbf{x})$, which maps positions in the input space $\mathbf{x} \in \mathcal{X}$ into elements of a so-called feature space. Following the derivation for ridge regression, we can see the beauty of this approach: since for predictions, we access point locations in input space or feature space only via the Gram matrix, we actually don't need the mapping $\mathbf{\Phi}(\mathbf{x})$ explicitly, only the inner product $\langle \mathbf{\Phi}(\mathbf{x}_i), \mathbf{\Phi}(\mathbf{x}_j) \rangle$. This inner product is called the *kernel* and the associated non-linear regression turns into kernel ridge regression [193, 506]. When a kernel is used to calculate the entries of the Gram matrix, it is often called the kernel matrix $\mathbf{K}$. The Gram (or kernel) matrix is, by definition, positive semi-definite; for any vector $\mathbf{u}$ we find

$$\mathbf{u}^T\mathbf{G}\mathbf{u} = \sum_{i,j}^N u_i u_j \langle \mathbf{\Phi}(\mathbf{x}_i), \mathbf{\Phi}(\mathbf{x}_j) \rangle = \langle \sum_i^N u_i \mathbf{\Phi}(\mathbf{x}_i), \sum_j^N u_j \mathbf{\Phi}(\mathbf{x}_j) \rangle = || \sum_i^N u_i \mathbf{\Phi}(\mathbf{x}_i)||^2 \geq 0$$
$$\tag{2.9}$$

2.2 THE REPRODUCING KERNEL HILBERT SPACE

In the last section, we have learned how ridge regression can be written in terms of inner products of data-point locations, and that we can extend this principle to non-linear kernel ridge regression by introducing non-linear maps $\boldsymbol{\Phi}(\mathbf{x})$ to assign every $\mathbf{x}$ in the input space a position in feature space where the regression problem then (hopefully) becomes linear. We therefore need the feature space to be endowed with an inner product $\langle \boldsymbol{\Phi}(\mathbf{x}_i), \boldsymbol{\Phi}(\mathbf{x}_j) \rangle$, which should yield symmetric positive semi-definite Gram matrices.

Let's create such a feature space. An inner product of elements u, v of a vector space V must satisfy

1. Symmetry: $\langle u, v \rangle = \langle v, u \rangle \ \forall \ u, v \in V$

2. Bilinearity: $\langle \alpha u + \beta v, w \rangle = \alpha \langle u, w \rangle + \beta \langle v, w \rangle \ \forall u, v, w \in V, \forall \alpha, \beta \in \mathbb{R}$

3. Positive Definiteness: $\langle u, u \rangle \geq 0 \ \forall u \in V$ (if strict: $\langle u, u \rangle = 0 \rightarrow u = 0$).

A vector space endowed with an inner product is called an inner product or pre-Hilbert space. A Hilbert space is a complete inner product space. Completeness refers to the property that all Cauchy sequences converge to an element of the space. Let's define a pre-Hilbert space

$$\mathcal{H} = \{f : f(\mathbf{x}) = \sum_{i}^{N} \alpha_i k(\mathbf{x}_i, \mathbf{x}) \ \forall \ \mathbf{x} \in \mathcal{X}, \alpha_i \in \mathbb{R}\}, \tag{2.10}$$

endow it with the inner product

$$\langle f, g \rangle_{\mathcal{H}} = \sum_i \sum_j \alpha_i \beta_i k(\mathbf{x}_i, \mathbf{x}_j), \tag{2.11}$$

where $g = \sum_j \beta_j k(\mathbf{x}_j, \mathbf{x})$, and add all sequences that converge under the norm $||f||_{\mathcal{H}} = \sqrt{\langle f, f \rangle_{\mathcal{H}}}$. It can be shown that (2.11) is a valid inner product according to the definition above. From this definition follows directly that

$$\langle f(\mathbf{x}), k(\mathbf{x}_0, \mathbf{x}) \rangle_{\mathcal{H}} = \sum_i \alpha_i k(\mathbf{x}_i, \mathbf{x}) = f(\mathbf{x}), \tag{2.12}$$

which allows convenient evaluation of the function at a new point $\mathbf{x}_0$. This is called the reproducing property of kernels and the reason that $\mathcal{H}$ is called a reproducing kernel Hilbert space (RKHS) [301] (see Figure 2.2). It also alludes to the fact that f can really be represented as a linear function in the RKHS via the inner product. With this introduction to RKHS, we have now the tools to solve equations (2.7) and (2.8) for non-linear functions efficiently.

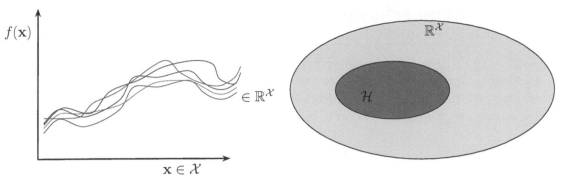

Figure 2.2 Definition of the reproducing kernel Hilbert space (RKHS) [301]. Let $\mathcal{X}$ be an arbitrary set. Denote by $\mathbb{R}^{\mathcal{X}}$ all functions defined on $\mathcal{X}$. If $\mathcal{H} \subset \mathbb{R}^{\mathcal{X}}$ is endowed with an inner product, is complete, and all point-wise evaluations of f are bounded, the space is a reproducing kernel Hilbert space.

2.3 THE BAYESIAN INFERENCE PERSPECTIVE

While the derivation in the last two sections gives us all tools to approximate a large set of functions, it is unable to assess any measure of confidence given the geometry and noise of the data. For this, the Bayesian perspective is more appropriate. We are again assuming data of the form $\mathcal{D} = \{\mathbf{x}_i, y_i\}$ and we are interested in the probability of some function parameters given the data. We are assuming some parameterization $f = \sum_i w_i \beta(\mathbf{x}_i^0, \mathbf{x})$, where w_i are the weight parameters and $\beta(\mathbf{x}_i^0, \mathbf{x})$ are basis functions centered around $\mathbf{x}_i^0$. Following Bayes' theorem, this means calculating

$$p(\mathbf{w}|\mathbf{y}) = \frac{p(\mathbf{y}|\mathbf{w})p(\mathbf{w})}{p(\mathbf{y})}, \tag{2.13}$$

and assuming that all observations $\mathbf{y}$ are drawn from a probability density function (PDF), which is often assumed to be Gaussian,

$$\mathbf{y}|f(\mathbf{x}) \sim N(f(\mathbf{x}), \sigma^2 I), \tag{2.14}$$

where σ^2 is the homoscedastic (constant or identically and independently distributed (i.i.d.)) noise. $p(\mathbf{w}|\mathbf{y})$ is called the posterior probability distribution (the result), which depends on the likelihood $p(\mathbf{y}|\mathbf{w})$ of the observations and a prior probability distribution $p(\mathbf{w})$. The evidence $p(\mathbf{y}) = \int p(\mathbf{y}|\mathbf{w})p(\mathbf{w})d\mathbf{w}$ might be costly to compute for some distributions and sometimes can only be approximated. There are three obvious issues with this approach. First, we are calculating a probability distribution over weights, which is only indirectly linked to the uncertainties of function values—our primary quantity of interest. Second, the parametric definition of the underlying function f is fixed beforehand, carrying the risk of excessive or missing expressibility. Third, a prior belief, in the form of a PDF has to be provided, which can be challenging if little is known about the function beforehand. While workarounds exist, the following shows an ingenious and straightforward way to address those issues.

2.4 MERGING THE RKHS WITH BAYESIAN INFERENCE FOR A GAUSSIAN-PROCESS PERSPECTIVE

Instead of working with probability distributions over weights w_i, it is often more convenient to work with probability distributions over function spaces, or more simply speaking, over function values. See Figure 2.3 for a visualization of the concept. In this case, we are integrating the probability density functions over $\mathbb{R}^N$

$$p(f^*|\mathbf{y}) = \int_{\mathbb{R}^N} p(f^*, \mathbf{f}|\mathbf{y})d\mathbf{f}, \tag{2.15}$$

where f^* is the function value of interest. Instead of a prior PDF over the weights, we have to specify a prior over function values and extend it over the unknown function value. This is where the RKHS and the so-called kernel trick come into action. Simply speaking, the kernel trick allows us to estimate similarities between different y_i based on their locations $\mathbf{x}_i$ only. This similarly translates into entries of the covariance matrix we need to define our probability density functions. Generally, the covariance of two continuous random variables $f(\mathbf{x}_1)$ and $f(\mathbf{x}_2)$—here we are alluding to the fact that our random variables are function values at different positions—is defined

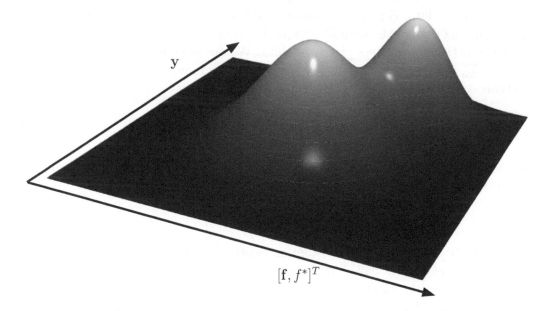

Figure 2.3 A conceptual visualization of the Bayesian perspective of function approximation from data. This function $p(f^*, \mathbf{f}, \mathbf{y})$ is, while inaccessible, the focus of the methodology. After marginalizing over the unknown f and conditioning on the data, the posterior PDF over f^* is obtained. That posterior is a key building block for autonomous experimentation since it comprises a "best guess" for the function at every $\mathbf{x}$ and an associated uncertainty, which can be used in an optimization (2.2) to find optimal next measurements.

as

$$Cov(f(\mathbf{x}_1), f(\mathbf{x}_2)) = \int_{\mathcal{F}} f(\mathbf{x}_1) f(\mathbf{x}_2) p(f) df \equiv k(\mathbf{x}_1, \mathbf{x}_2), \qquad (2.16)$$

which is a positive definite function and can therefore play the role of a kernel. That way, we can calculate $Cov(f^*, f(\mathbf{x}))$, without knowing f^*. This is one of the reasons why the Gaussian (stochastic) process is so popular: the covariance, and therefore the kernel, is really all we need in order to specify it. Following the steps outlined in the caption of Figure 2.3 for a Gaussian prior results in the posterior

$$\begin{aligned} p(f^*|\mathbf{y}) &= \int_{\mathbb{R}^N} p(f^*|\mathbf{f}, \mathbf{y}) \, p(\mathbf{f}|\mathbf{y}) \, df \\ &\propto \mathcal{N}(m^* + \boldsymbol{\kappa}^T (\mathbf{K} + \mathbf{V})^{-1} (\mathbf{y} - \mathbf{m}), \mathcal{K} - \boldsymbol{\kappa}^T (\mathbf{K} + \mathbf{V})^{-1} \boldsymbol{\kappa}). \end{aligned} \qquad (2.17)$$

The author hopes it came across how the connection between the RKHS and Bayesian inference led us to an efficient algorithm for stochastic function approximation. For general functions, we would not be able to compute the covariance; however, for functions that are elements of an RKHS—because it is reproducing—we can calculate covariances between function values we have never observed which we need to define joint probability distributions.

2.5 THE CONNECTION BETWEEN REGULARIZATION AND GP PERSPECTIVE

From the previous sections, it is now clear that there is a strong connection between the regularization perspective and Gaussian processes, induced by the kernel. In fact, comparing the posterior mean of a Gaussian process

$$\mu(\mathbf{x}) = \mathbf{m}^* + \boldsymbol{\kappa}^T (\mathbf{K} + \mathbf{V})^{-1} (\mathbf{y} - \mathbf{m}), \qquad (2.18)$$

where $\boldsymbol{\kappa} = k(\mathbf{x}_0, \mathbf{x}_j)$, and $\mathbf{m}$ is the prior mean, with the approximation we get from kernel ridge regression

$$r(\mathbf{x}) = \mathbf{y}^T (\mathbf{G} + \mu \mathbf{I})^{-1} \langle \boldsymbol{\Phi}(\mathbf{x}_i), \boldsymbol{\Phi}(\mathbf{x}) \rangle = \mathbf{y}^T (\mathbf{K} + \mu \mathbf{I})^{-1} \boldsymbol{\kappa}, \qquad (2.19)$$

we see that both are equivalent if we consider a zero-prior-mean GP ($\mathbf{m} = 0$). It is amazing to see how two methods that are derived from quite different initial assumptions lead to the same result.

2.6 ACQUISITION FUNCTIONALS

Having approximated a model function f, together with its uncertainties, it now becomes time to use this knowledge to our advantage in the autonomous loop via maximizing the acquisition functional. The choice of the right acquisition functional can significantly improve the performance of an autonomous experiment. New measurements will be placed where the acquisition functional is maximal; it can therefore be used to include the practitioner's expectations and knowledge in the experiment.

One of the most commonly used acquisition functionals—in this case a function—is the posterior variance of a Gaussian process

$$f_a(\mathbf{x}) = \sigma(\mathbf{x}). \tag{2.20}$$

This acquisition function will place new points where the uncertainty is high, and not necessarily where the most knowledge can be gained. For that, it is better to use information-theoretical quantities such as the entropy of a stochastic system. An optimal new measurement would maximize

$$f_a(\mathbf{x}) = H(p(\mathbf{f})) - H(p(\mathbf{f}|\mathbf{y})), \tag{2.21}$$

which is the difference in entropy before and after data was collected. This quantity is called the mutual information [283]. Since Shannon introduced the concept of entropy in 1948 [480], this quantity (2.21) is sometimes called the Shannon information gain. Both acquisition functions are tailored for pure exploration without any regard for features of interest. They can, however, be combined with the characteristics of the surrogate to focus on regions of interest. One of the most popular acquisition functions of this kind is the upper-confidence-bound (UCB) acquisition function

$$f_a(\mathbf{x}) = m(\mathbf{x}) + c\sigma(\mathbf{x}), \tag{2.22}$$

which will put priority on predicted maxima. Since in a GP framework, we obtain the posterior across the input domain, we can use probabilities in the acquisition functional—here the term "functional" really shines—

$$f_a(\mathbf{x}) = p(-\infty \leq f(\mathbf{x}) \leq b) = \frac{1}{\sigma\sqrt{2\pi}} \int_{-\infty}^{b} e^{\frac{f(\mathbf{x})-m(\mathbf{x})^2}{2\sigma^2}} df = 0.5(1 + \text{erf}(\frac{b - m(\mathbf{x})}{\sigma\sqrt{2}})), \tag{2.23}$$

which places priority on regions where the probability of $f(\mathbf{x}) \leq b$ is high. Overall, acquisition functionals should always be chosen with great care; not always do they have to make physical sense, as long as they are mathematically well-behaved.

Finding the maximum of the acquisition functional is no simple task. Most often we need feedback in an autonomous loop every few seconds or so. Because data point locations are associated with low uncertainties, the acquisition functional is prone to contain a large number of local optima, and even ridges and plateaus. Often, we can only hope to find local optima; however, local optimizers often perform poorly because of wide-spread, small gradients. Hybrid optimizers (local/global) can perform well.

2.7 SUMMARY, CLOSING THOUGHTS AND POINTERS TO RELATED CHAPTERS IN THIS BOOK

In this introduction, we have seen how autonomous experimentation can be interpreted as a function approximation problem and a subsequent optimization, organized in a loop. While many tools and methods exist for function approximation, kernel methods are among the most powerful due to the flexibility of defining the kernel and therefore, indirectly, the non-linear mapping $\mathbf{\Phi}(\mathbf{x})$ from the input space to a feature space. That motivates kernel ridge regression as a top contender. Even so, kernel ridge regression does not give us access to a measure of confidence. Bayesian methods, especially stochastic processes, do have this property and we have shown how there is an equivalence between a Gaussian (stochastic) process and kernel ridge regression for the quadratic loss function. In principle, this solves the function approximation problem. However, Gaussian processes come with the assumption of normal distributions which is sometimes a poor approximation. Furthermore, since we have to compute, store, and work with a covariance matrix, GPs don't scale well in general. More about these shortcomings and how to overcome some of them can be found in Chapter 4. In the "Applications" part of this book, you can see many example applications of GPs to autonomous experimentation.

CHAPTER 3

A Perspective on Machine Learning for Autonomous Experimentation

Joshua Schrier

Department of Chemistry, Fordham University, The Bronx, New York, USA

Alexander J. Norquist

Department of Chemistry, Haverford College, Haverford, Pennsylvania, USA

CONTENTS

T HIS CHAPTER introduces machine learning (ML) in the context of autonomous scientific experimentation. Our goal is to outline the types of roles that ML can play in supporting autonomous experimentation, and in doing so provide a language for discussing the applications presented in the following chapters. Although this will be most helpful for orienting new researchers to the vocabulary of concepts that can be used to frame their own research goals, it also serves as a framework for the work discussed in subsequent chapters. It is not our goal to present a practical how-to manual of general ML methods and programming environments; this is better served by other resources [7, 56, 336, 427, 544]. Nor is it our goal to discuss specific applications of ML for autonomous experimentation and their successes, which is the topic of the subsequent chapters in this book. Instead, we begin by defining ML and its relationship to the autonomous discovery process. We then define and contrast some of the dichotomies of ML algorithms. Finally, we discuss challenges and considerations in how best to apply ML to autonomous experiments.

DOI: 10.1201/9781003359593-3

3.1 WHAT DO WE MEAN BY MACHINE LEARNING (ML) AND HOW DOES IT SERVE THE AUTONOMOUS DISCOVERY PROCESS?

The most general definition of machine learning (ML) is as a method for creating functions (i.e., input-output relationships) not by explicit programming but instead by showing example data. "Function" here is used both in its mathematical sense, as in a relation that uniquely associates members of one set with members of another set, and in its computer programming sense, a sequence of computer instructions that perform a task. A learner attempts to generalize from experiences and make predictions about unseen examples. In contrast, a database merely allows for the storage and retrieval of a fixed set of previously encountered examples. This broad definition, formalized by Probably Approximately Correct computational learning theory, provides a framework for mathematical analysis of ML, which extends beyond computer programs to include processes such as evolution [540].

As famously noted by Breimann, the practice and epistemic philosophy of ML differs from the model-based parameter estimation traditionally practiced in statistics and the sciences [71]. The traditional approach assumes that one starts with a plausible model whose functional form is determined by some domain expertise and then uses data to fit some unknown, but meaningful, parameters contained in that model. Often, the experimenter desires to know the value of the parameters, rather than the models prediction. In contrast, machine learning takes a radically different approach. The underlying functional form has no necessary or predetermined relationship to a preconceived model of the world, and neither do the parameters have any claim to meaning. Only the quality of the input-output predictions made by the function matter. This is especially relevant to deep learning models, which are often highly over-parameterized compared to their data and yet remarkably effective in a wide variety of contexts. Whereas the traditional modeling approach attempts to reduce phenomena to a few essential variables and impose human knowledge about the relationships between the variables to constrain the prediction, modern ML practice tries to learn these relationships directly from the data. While this often strikes newcomers as counterproductive, as Richard Sutton has noted, the bitter lesson of the past 70 years of AI research has been that building domain knowledge into ML has at best offered short-term gains. The long-term advances have all come by discarding hand-coded specialized knowledge and replacing it with brute force computation and data [502].

One way to evaluate potential applications of ML is to define it as a technology that "makes predictions cheap" [9]. In situations where ML is inferior, or at best comparable, to a human evaluation, the advantage lies in making its predictions inexpensive and/or faster than would be possible by a human, increasing productivity, reducing cost, or making split-second decisions. In some situations, ML can exceed human performance. This is typical for problems in high dimensions or where it is necessary to apply rigorous statistical reasoning, where human predictions often suffer because of the use of innate heuristics that prioritize recency and a limited number of salient features when making predictions [525]. In either case, we need to accept

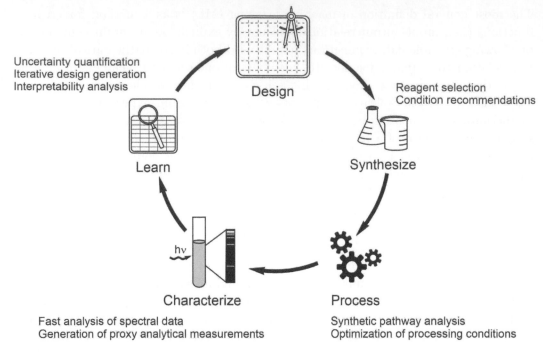

Figure 3.1 A hypothetical design/synthesize/process/characterize/learn cycle in an autonomous experiment, with examples of relevant machine learning tasks at each stage.

some probability of having a wrong answer, and so the best applications are where the cost of being wrong is small.

When posed in this way, it is clear that there are many possible applications for ML throughout the design-build-test-learn cycle of an autonomous experiment [592, 260]. (see Figure 3.1). For example, ML may be used as a fast/low-cost proxy for a slow/expensive computational or experimental process used to identify the most promising candidates, as we will see in many chapters of the Application part of this book. It could then be used to recommend possible synthesis procedures that are likely to achieve the desired target. It could be used to interpret experimental data (e.g., spectra or characterizations) to determine the presence or absence of desired properties, or again to substitute a slower characterization with a faster proxy measurement. In each of these scenarios, the resulting ML model may be used not only to obtain point estimates but also to quantify uncertainty, which can be used to guide the selection of subsequent iterations of the autonomous process. See Chapters 4 and 6.

Thinking about ML in terms of cheap but possibly wrong predictions can also be useful for guiding decisions about where to use them in an autonomous experiment. A purely ML-based approach is inappropriate if the cost of being wrong includes possible hazards to human safety or property damage, which could be avoided by a deterministic approach. On the other hand, being wrong is totally acceptable in the context of many scientific discovery problems, as a few wrong experiments are at worst innocuous and at best can be opportunities for serendipity. Indeed, for many scientific discovery problems, predictive accuracy within the domain of training problems is less important than being able to extrapolate to new regions of the search space.

3.2   DISTINCTIONS AND DICHOTOMIES FOR DISCUSSING ML FOR AUTONOMOUS EXPERIMENTS

To divide the broad definition of ML into smaller, more specialized subcategories, it is useful to describe ML approaches in terms of some general axes, depicted schematically in Figure 3.2. We will briefly note the extreme limits of each of these and some points along the continuum, with the goal of establishing a shared vocabulary for discussing ML algorithms.

Supervised vs. Unsupervised. Supervised ML problems use example input-output pairs (so-called labeled data) to learn a function that predicts the outputs for subsequent inputs. This may be further divided into regression (predict a numerical value) and classification (predict a discrete class). In contrast, unsupervised problems identify underlying structure within a set of unlabeled examples. For example,

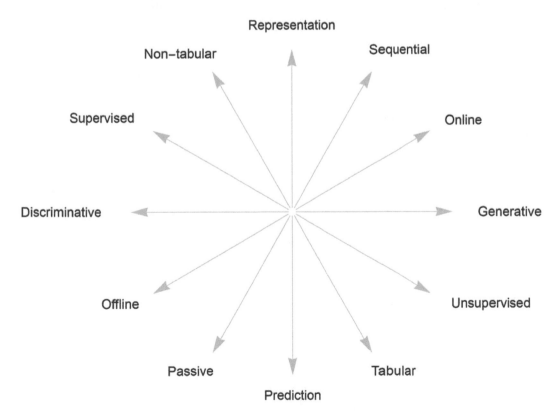

Figure 3.2   Dichotomies for discussing machine-learning methods

clustering attempts to divide the examples into a small number of groups of similar entries. Anomaly detection involves identifying data points that are significantly different from the majority of the data and are therefore potentially interesting or abnormal and worthy of closer investigation.

Within the supervised-unsupervised continuum are a vast array of other methods. The intuition is that obtaining labeled information is difficult, as it may be expensive, time-consuming, or require expert knowledge; in contrast, unlabeled data is easier to obtain. Semi-supervised methods assume a small pool of labeled data and a much larger pool of unlabeled data available to build a model. Transfer learning assumes that knowledge learned from one task (where we have a large pool of labeled data) is transferable to a related, yet distinct task (where we have only a smaller pool of labeled data). Meta-learning can be considered a form of transfer learning, in which a larger pool of labeled data on a collection of semi-related tasks is used to train models that can learn quickly when encountering a new task. One-shot (or few-shot) learning algorithms address the problem of trying to learn an appropriate distance function that can be used to classify new inputs as related to a small number of examples of that type of input. This is often achieved by using a pool of similar unlabeled (or differently labeled) data to learn the function (e.g., using a collection of faces to learn a face-to-face distance function, which then needs only a single example of a new persons face-to-identify that person in subsequent photographs).

Tabular vs. Non-tabular. Tabular machine learning refers to cases where the input data is structured in a table or spreadsheet-like format, with rows representing individual samples and columns representing the features or attributes of each sample. This might correspond to the settings and outcomes of an autonomous experimental apparatus. Non-tabular input data is quite diverse and might include images, videos, audio, text, or graphs, all of which occur in a variety of autonomous experiment types. In practice, tabular data problems with modest (circa 100 examples) are best treated with traditional machine-learning methods (especially tree-based methods such as Random Forests and XGBoost), without the need for deep learning [181]. In contrast, the challenge in non-tabular data is often learning an appropriate feature extraction or representation, and deep-learning methods have excelled in these areas. While tabular and non-tabular might seem like very distinct categories, they actually form a continuum. For example, a molecule might be represented as a non-tabular graph of atoms and bonds describing the underlying bonding patterns, or a tabular featurization of those attributes in the form of a fingerprint or property descriptors, or some combination of these types of features.

Offline vs. Online. Offline learning assumes that the algorithm has access to a data archive from which examples can be retrieved as needed, and in any desired order, during the learning process. This is often the case for many scientific experiments, where the datasets are modest and may even be loaded into memory directly, and where the underlying feature to be learned is time-invariant. In contrast, an online (or stream-based) learning problem assumes the algorithm must learn from a continuous stream of data presented in sequential order. This is often framed in terms of problems where the learner has a limited ability to store data for later use. This is less common in scientific applications but does appear in the context of even processing in high-energy physics experiments [123]. A more likely case is continual learning where the learner must adapt to the time-varying nature of the data, for example in autonomous beamline alignment optimization [325].

Discriminative vs. Generative. Discriminative models attempt to predict the output labels given the input features. This corresponds to the typical prediction that we think of (e.g., given a collection of pixels, is it a cat?) In the case of a classification problem, we can think about discriminative models as learning the boundary between different classes. The examples given in our discussion above have largely corresponded to discriminative models. Discriminative models can play many roles in an autonomous experiment, such as determining if a substance is present based on some type of spectral measurement. In contrast, a generative model aims to learn the underlying distribution of the data and generate new samples from the distribution (e.g., generate a collection of pixels that looks like a cat). In this case, the prediction is about what a new sample from the distribution would be. This can be made class-conditional to describe the distribution of data within a given class. In an autonomous experiment, this might correspond to generating an experiment plan or composition, perhaps conditioned on some property. This is particularly useful when the number of possible experimental candidates is too vast to be enumerated exhaustively. Instead, one can use a generative approach to sample relevant solutions.

These two extremes are often combined in practice. For example, a generative model might be used to model the distribution of the data and generate synthetic samples that can be used to augment the training data available for a discriminative model. In the reverse direction, the training of generative models often involves the use of discriminative models as a subcomponent. The classic example of this is in the architecture of generative adversarial networks (GANs), in which a discriminator network that distinguishes between generated and actual examples is incorporated.

Prediction vs. Representation. Many of the examples discussed above have focused on predicting the value of a function, but one might also consider the problem of learning a compact, lower-dimensional representation of the data. Such a representation might serve practical purposes, such as reducing the input dimensionality of the problem, increasing robustness to noise, or aiding in human interpretability, all of which have clear relevance to an autonomous experiment. Representation learning can be seen as a combination of unsupervised and supervised learning. In an unsupervised setting, the goal is to learn a representation—for example, a vector embedding—that captures the underlying structure without the use of labels. In a supervised setting, the goal is to learn a representation that is useful for a specific prediction task. An extreme form of this feature selection involves selecting a subset of features that are most relevant for a prediction task.

Passive vs. Sequential. The previous examples describe scenarios in which the learner has no choice about what data examples it can use it is provided the data and must do the best it can with what is available. In contrast, an autonomous experiment system enables a sequential learning process, in which the goal is to make iterative decisions that refine the model and its predictions. In sequential learning, the model selects the data it wants to be labeled, rather than relying upon some arbitrarily pre-defined sample. Specifying a policy function (or acquisition function) determines how to prioritize the acquisition of new samples. For example, it might be preferable to sample new points that explore the most uncertain model predictions, exploit the highest value predicted outcomes, or blend these in some way (a so-called explore-exploit tradeoff). Ultimately, this policy will depend on the application.

For example, in active learning, the goal is to learn the labeling function everywhere by sampling regions with the most uncertainty (i.e., an explore-only policy). This is a very natural approach—quantifying what is known and directing investigations toward unknowns which corresponds nicely with the scientific method. In practice, this often leads to more sampling near decision boundaries in classification (such as mapping phase boundaries in materials [111, 248, 270]), or in regions that are less smooth in regression. We shall see examples in the following chapters. Alternatively, if our goal is to optimize some function, this might be framed in terms of other policy functions, such as the upper confidence bound, probability of improvement, or expected improvement. In these optimization problems, machine learning is used to approximate the unknown function being optimized. A popular approach is to use Bayesian optimization along with Gaussian process regression models as the underlying function (see Chapter 6). In many laboratory settings, the order in which a series of decisions is made can give different results. Stated another way, taking the same action may have yielded a different function value depending on the state of the system, and the system's state is in turn determined by the sequence of actions that have been taken thus far, and the rewards may be cumulative or delayed. Reinforcement learning encompasses a variety of methods treating these types of problems [503], and we see examples in Chapter 9.

3.3   CHALLENGES FOR ML FOR AUTONOMOUS EXPERIMENTATION

Autonomous experimentation presents unique challenges and demands for ML algorithms. In this section, we discuss some of the general challenges and choices to be made when applying ML in autonomous experimentation. We offer these as questions to keep in mind when developing an ML plan for an autonomous experiment.

Data quantity and quality. Experimental scientific data is often small, wide, noisy, and biased. Physical experiments and detailed computer simulations both require limited resources that impose a budget on the number of examples that can be acquired. To some extent, this is partially mitigated by high-throughput methods, but it will always be desirable to have more examples. Thus, sample-efficient ML methods are clearly desirable. Experimental characterizations often result in a rich set of features (i.e., wide in the sense of a large number of tabular columns for each example row) and non-tabular data (e.g., microscopy, hyperspectral data, etc.) describing each sample. Noise arises from both systematic sources (such as instrument limitations) and unsystematic sources (e.g., environmental fluctuations, sample provenance, etc.). Uncertainty can be considered as arising from aleatoric sources (the inherent randomness of the experimental procedure or phenomenon being studied) and epistemic (the uncertainty of the model). In many scientific datasets, the distribution of inputs and outputs is often unrepresentative of the desired goal, and such biases must be accounted for by defining proper loss functions or data augmentation.

This is more likely to be a problem when making use of existing datasets, rather than in performing de novo autonomous experiments, but exogenous sources of bias (e.g., a selection of particular reagents and not others) can be harder to expunge.

Domain Knowledge. Scientific experiments are complex, and the datasets generated by these experiments often contain implicit domain-specific knowledge and understanding. Applying ML to these datasets without understanding these nuances may result in models that make trivial predictions or succumb to the limits of the available datasets discussed above. Experiences in synthetic biology, protein design, and materials science have indicated that the best approach is to create close, interactive collaboration between domain and ML experts to develop a common language for describing and solving problems [260]. A naive application of ML to the inputs and outputs of an autonomous experiment is limited to brute experimental facts. In contrast, incorporating domain knowledge provides access to additional information and constraints that can make the problem tractable. In its simplest form, domain experts can suggest relevant features (e.g., atomic properties describing a composition) to be provided. While in principle such features might be learned directly from a sufficiently large dataset, in practice, many autonomous problems have too limited of a dataset to do so, and so providing these features is highly advantageous. Physics-informed methods in deep learning can be used to impose constraints on the allowed functional forms and boundary conditions [227]. Knowledge about the domain, and in particular knowledge of ways to simulate the processes involved in the underlying problem can also be essential for devising effective data augmentation strategies. For example, reinforcement learning is often highly sample inefficient, and so one approach is the Sim2Real transfer learning method in which one first trains a reinforcement learner in a simulation environment and then fine-tunes the learned policy in the real experimental setting [604, 192, 134, 323, 209, 167].

Uncertainty quantification, generalization, and extrapolation. Scientific experiments are interested in finding unexpected and unforeseen relationships and outcomes. This places an emphasis on going beyond simply predictive accuracy but also to rigorously quantifying uncertainty in model predictions [3, 415]. Many applications of ML for autonomous experiments demand less predictive accuracy but more extrapolation outside the training set distribution. For example, an autonomous system particularly one with high experimental throughput can allow for testing many ML-generated predictions, and so long as one of them is successful the goal is accomplished. Additionally, if one is trying to discover new compositions, etc. then the ability to extrapolate beyond existing training data may be more important than predictive accuracy within the training set domain. Paradoxically, this might mean that relatively simple ML models, such as linear models and random forests, might be more appropriate for achieving the goals of autonomous experimentation [68].

Model interpretability and explainability. A common critique of machine learning for scientific problems is that it results in a black box function that does not provide insight or understanding. The description of autonomous experimentation in terms of self-driving laboratories may exacerbate this criticism, as it implies removing humans from the process entirely. Clearly, scientists want to be able to understand and interpret their results, and a field of interpretable machine learning or explainable AI (XAI) exists toward this end [209, 167]. Causal inference methods are one way to approach this problem, by incorporating additional information about the underlying possible causal relationships between variables and determining in a data-driven way which relationships are supported by the data [167]. It is useful to frame these not merely as something that occurs at the end of the process, but instead, as an opportunity to engage with human expertise throughout the experimental cycle [134, 323].

Selecting the next experiment. In many applications of autonomous experimentation, the goal is to determine some set of inputs (e.g., materials compositions and processing conditions) that result in a desirable output (e.g., materials property). How should one use ML to determine which inputs to examine in the laboratory? A direct design approach uses ML to accelerate predictions about the output of a given input. For example, we might use ML trained on a large database of calculated results to downselect or prioritize subsequent calculations and experiments. Such a direct design approach will only be applicable if the number of possible candidates can be computed in an exhaustive fashion. An optimization strategy might instead suggest that we use ML as a proxy for experimental optimization. Examples were discussed in the context of Bayesian Optimization/Gaussian Process Regression above and in many of the application chapters. A generative or inverse design strategy suggests learning the relationship between how the output properties condition the possible input space, allowing us to use our ML model to generate samples with the desired property. (For a recent discussion and review of these three approaches to experiment selection, in the context of materials, see Ref. [553].)

Taking advantage of relevant experimental scale and throughput. There are two extremes of thinking about autonomous experimentation. At one extreme, one might think of automating existing laboratory processes, an example of which is having a robotic arm move samples into existing equipment [78]. This roughly preserves the data scale and quality of existing laboratory processes (perhaps there is a less than an order of magnitude improvement by eliminating delays and allowing for continual operation). At the other extreme, one can think of using miniaturization (e.g., nanoreactors [508], ultra HTE [161], femtoliter microfluidics [398], etc.) to radically expand experimental throughput. This type of approach might greatly increase data size but requires different types of startup R&D efforts (beyond the scope of our considerations here). One might also expect that a new method might at least initially result in noisier data.

These two extremes may require very different types of ML algorithms to be successful. The low-throughput regime requires sample-efficient learning methods, and the sequential learning and optimization process may have many iterations, each comprised of a few individual samples. The high-throughput regime requires better handling of experimental noise and demands the sequential learning and optimization process to be performed with only a few iterations, but each iteration might have many samples in a batch. The same type of distinction is also relevant to characterization [592]. In one extreme, one uses ML to guide where to look during an autonomous experiment—i.e., when using an existing high-quality characterization, ML assists in the clever selection of experiments to perform, with the goal of performing as few experiments as possible to characterize the property of interest [248]. In the other extreme, ML can be used to change how to look at the sample during the autonomous experiment—i.e., one might consider using fast, inexpensive proxy-based characterizations and using ML to infer the desired property in lieu of the traditional characterization method [98].

II

Methods, Mathematics, and Algorithms

CHAPTER 4

Gaussian Processes

Marcus M. Noack

Applied Mathematics and Computational Research Division, Lawrence Berkeley National Laboratory, Berkeley, California, USA

CONTENTS

$\mathbf{A}$CCURATE function approximation and uncertainty quantification are vital for autonomous intelligent decision-making. The Gaussian Process is a particular type of stochastic process—sometimes called a random field—which is defined as a set

$$\{f(\mathbf{x}) : \mathbf{x} \in \mathcal{X}\} \tag{4.1}$$

of random variables $f(\mathbf{x})$ that are defined on an index set (often our input space) $\mathcal{X}$. Traditionally, the index set was time, and the stochastic fields were understood as temporally changing random variables. Random fields are seen as equivalent by most authors and have their origin in calculations on actual crop fields.

A Gaussian Process is a stochastic process with a particular choice of joint probability distribution, namely the normal distribution. While this limitation brings with it acceptable inaccuracies, it allows for exact Bayesian inference and fast computations. In this case, an equivalent concept is Kriging which originated from Geostatistics and the search for gold in South Africa [107].

DOI: 10.1201/9781003359593-4

4.1 THE BASIC THEORY

Given a dataset $\mathcal{D} = \{\mathbf{x}_i, y_i\}$, a normal (Gaussian) prior probability distribution over function values $f(\mathbf{x})$ is defined as

$$p(\mathbf{f}) = \frac{1}{\sqrt{(2\pi)^{\dim}|\mathbf{K}|}} \exp\left[-\frac{1}{2}(\mathbf{f} - \boldsymbol{\mu})^T \mathbf{K}^{-1}(\mathbf{f} - \boldsymbol{\mu})\right], \tag{4.2}$$

where $\mathbf{K}$ is the covariance matrix, calculated by applying the kernel $k(\mathbf{x}_1, \mathbf{x}_2)$ to the data-point positions, and $\boldsymbol{\mu}$ is the prior-mean vector, calculated by evaluating the prior-mean function at the data-point locations. We define the likelihood over data values $\mathbf{y}$ as

$$p(\mathbf{y}|\mathbf{f}) = \frac{1}{\sqrt{(2\pi)^{\dim}|\mathbf{V}|}} \exp\left[-\frac{1}{2}(\mathbf{y} - \mathbf{f})^T \mathbf{V}^{-1}(\mathbf{y} - \mathbf{f})\right], \tag{4.3}$$

where $\mathbf{V}$ is the matrix containing the noise—often non-i.i.d. noise but uncorrelated which translates into a diagonal $\mathbf{V}$. In the standard literature, often only i.i.d. noise is discussed, which is insufficient for many applications [360].

The most commonly employed kernel is the RBF (radial basis function) kernel [407]

$$k(\mathbf{x}_1, \mathbf{x}_2) = \sigma_s^2 \exp\left[-\frac{||\mathbf{x}_1 - \mathbf{x}_2||^2}{2l^2}\right], \tag{4.4}$$

where σ_s^2 is the, so-called, signal variance and l is the length scale. The length scale is often chosen to be isotropic but considering anisotropy can be beneficial. Axially anisotropic kernels are often used for what is called automated relevance determination (ARD) in machine learning. The signal variance and the length scale are the simplest examples of hyperparameters (ϕ) of the Gaussian process and can be found by solving the optimization problem

$$\arg\max_{\phi} \Big(\log(L(D, \phi)) =$$
$$-\frac{1}{2}(\mathbf{y} - \boldsymbol{\mu}(\phi))(\mathbf{K}(\phi) + \mathbf{V})^{-1}(\mathbf{y} - \boldsymbol{\mu}(\phi))$$
$$-\frac{1}{2}\log(|\mathbf{K}(\phi) + \mathbf{V}|) - \frac{\dim(\mathbf{y})}{2}\log(2\pi)\Big). \tag{4.5}$$

This optimization problem represents a detour around a fully Bayesian approach and has to be done with care to avoid over-fitting. The fully Bayesian alternative approximates the posterior probability distribution

$$p(\phi|\mathbf{y}) = \frac{p(\phi|\mathbf{y})p(\mathbf{y})}{\int p(\phi|\mathbf{y})p(\mathbf{y})d\mathbf{y}} \tag{4.6}$$

via MCMC (Chapter 5). A prior has to be specified which can be challenging; a flat prior reverts back to the optimization problem. Optimization algorithms offer some advantages compared to statistical techniques such as favorable scaling properties with dimensionality and a higher probability of finding close-to-global solutions. In the optimization case, a Laplace approximation around the found optima can help with over-fitting. A brief discussion with pointers to more information can be found in [570].

Once the hyperparameters are found, we condition the joint prior

$$p(\mathbf{f}, \mathbf{f}_0) = \frac{1}{\sqrt{(2\pi)^{\dim}|\boldsymbol{\Sigma}|}} \exp\left[-\frac{1}{2}\left(\begin{bmatrix} \mathbf{f} - \boldsymbol{\mu} \\ \mathbf{f}_0 - \boldsymbol{\mu}_0 \end{bmatrix}^T \boldsymbol{\Sigma}^{-1} \begin{bmatrix} \mathbf{f} - \boldsymbol{\mu} \\ \mathbf{f}_0 - \boldsymbol{\mu}_0 \end{bmatrix}\right)\right], \qquad (4.7)$$

where

$$\boldsymbol{\Sigma} = \begin{pmatrix} \mathbf{K} & \boldsymbol{\kappa} \\ \boldsymbol{\kappa}^T & \boldsymbol{\mathcal{K}} \end{pmatrix}, \qquad (4.8)$$

to obtain the posterior

$$\begin{aligned} p(\mathbf{f}_0|\mathbf{y}) &= \int_{\mathbb{R}^N} p(\mathbf{f}_0|\mathbf{f}, \mathbf{y})\, p(\mathbf{f}, \mathbf{y})\, d\mathbf{f} \\ &\propto \mathcal{N}(\boldsymbol{\mu}_0 + \boldsymbol{\kappa}^T\,(\mathbf{K} + \mathbf{V})^{-1}\,(\mathbf{y} - \boldsymbol{\mu}), \boldsymbol{\mathcal{K}} - \boldsymbol{\kappa}^T\,(\mathbf{K} + \mathbf{V})^{-1}\,\boldsymbol{\kappa}), \end{aligned} \qquad (4.9)$$

where $\boldsymbol{\kappa}_i = k(\mathbf{x}_0, \mathbf{x}_i, \phi)$, $\boldsymbol{\mathcal{K}} = k(\mathbf{x}_0, \mathbf{x}_0, \phi)$ and $\mathbf{K}_{ij} = k(\mathbf{x}_i, \mathbf{x}_j, \phi)$. $\mathbf{x}_0$ are the points where the posterior will be predicted. $\mathbf{f}_0$ are values of the latent function f at the points $\mathbf{x}_0$. The posterior includes the posterior mean $m(\mathbf{x})$ and the posterior variance $\sigma^2(\mathbf{x})$.

4.2 KERNEL THEORY

We are introducing kernels as an inner product in some favorable space where classification and interpolation problems become linear. As described in the mathematical perspective of autonomous experimentation (Chapter 2), many learning algorithms can be formulated so that positional information in the data only appears as inner products in the equations for linear interpolation. In order to maintain this nice property but add the capability for interpolating non-linear relationships, one applies the mapping $\Phi(\mathbf{x})$ to the input locations. A kernel computes the inner product of the images of such a map $k(\mathbf{x}_1, \mathbf{x}_2) = \langle \Phi(\mathbf{x}_1), \Phi(\mathbf{x}_2)\rangle$ and therefore has to be a symmetric positive semi-definite function of the form $\mathcal{X} \times \mathcal{X} \to \mathbb{R}$. If we think of the map Φ as a basis function centered at $\mathbf{x}_i$ that assigns every $\mathbf{x}$ a $\Phi(\mathbf{x}, \mathbf{x}_i)$ something interesting happens. Let's define a pre-Hilbert space of functions

$$\mathcal{H} = \{f(\mathbf{x}) : f(\mathbf{x}) = \sum_i^N \alpha_i \Phi(\mathbf{x}_i, \mathbf{x}), \forall \boldsymbol{\alpha} \in \mathbb{R}^N, \mathbf{x} \in \mathbb{R}^n\}, \qquad (4.10)$$

with symmetric basis functions $\Phi(\mathbf{x}_i, \mathbf{x})$ and coefficients α_i. Now we define an inner product on $\mathcal{H}$

$$\langle f, g\rangle_{\mathcal{H}} = \sum_i^N \sum_j^N \alpha_i \beta_j \Phi(\mathbf{x}_i, \mathbf{x}_j), \qquad (4.11)$$

which adopts the symmetry of Φ, is bilinear and strictly positive definite, and therefore a valid inner product. Realizing that $\Phi(\mathbf{x}_i, \mathbf{x})$ is also an element of this space means

$$\langle \Phi(\mathbf{x}_i, \mathbf{x}), \Phi(\mathbf{x}_j, \mathbf{x}) \rangle_{\mathcal{H}} = \Phi(\mathbf{x}_i, \mathbf{x}_j) = k(\mathbf{x}_i, \mathbf{x}_j). \tag{4.12}$$

The neat insight is that the kernel turns out to be a basis function and also the inner product!

For (4.10) to qualify as a Hilbert space we have to equip the space with the norm $||f||_{\mathcal{H}} = \sqrt{\langle f, f \rangle_{\mathcal{H}}}$ and add all limit points of sequences that converge in that norm. As a reminder, note that scalar functions over $\mathcal{X}$, e.g., $f(\mathbf{x})$, are vectors (bold typeface) in $\mathcal{H}$. After this intuitive definition of a kernel as a kind of basis function and inner product let's look at a more technical one:

Definition 4.1 *A kernel is a symmetric and positive semi-definite (p.s.d.) function* $k(\mathbf{x_1}, \mathbf{x_2})$, $\mathcal{X} \times \mathcal{X} \to \mathbb{R}$,
it therefore satisfies $\sum_i^N \sum_j^N c_i \, c_j \, k(\mathbf{x}_i, \mathbf{x}_j) \geq 0 \; \forall N, \; \mathbf{x} \in \mathcal{X}, \; \mathbf{c} \in \mathbb{R}^N$

This definition highlights the fact that the set of kernels is closed under addition, multiplication and linear transformation [166], which can be used to create new kernels without the need to prove positive definiteness. Given the definition of the pre-Hilbert space (Equation 4.10), it can be shown that for elements of $\mathcal{H}$

$$\langle k(\mathbf{x}_0, \mathbf{x}), f(\mathbf{x}) \rangle = f(\mathbf{x}_0), \tag{4.13}$$

which is the reason the completion of the space $\mathcal{H}$ is called Reproducing Kernel Hilbert Space. [1] (RKHS) [55]. Gaussian processes are based on prior probability distributions over the RKHSs. In this case, the kernels are understood as covariance functions

$$k(\mathbf{x}_1, \mathbf{x}_2) = \int_{\mathcal{H}} f(\mathbf{x}_1) f(\mathbf{x}_2) q(f) \, df, \tag{4.14}$$

where q is some density function.

4.3 ADVANCED KERNEL DESIGNS

In ordinary Gaussian process regression, we most often want to interpolate a function over the input space $\mathcal{X}$; distances are assumed to be isotropic and Euclidean, and we require first and second-order stationarity of the process. These requirements lead to the overwhelming use of stationary kernel functions

$$k(\mathbf{x}_1, \mathbf{x}_2) = k(||\mathbf{x}_1 - \mathbf{x}_2||; \sigma_s^2, l), \tag{4.15}$$

where l is the isotropic and constant length scale and σ_s^2 is the constant signal variance and a constant, often zero, prior-mean function. This gives rise to only two

---

[1] As an example of non-reproducing, we note that if $k = \delta(\mathbf{x}, \mathbf{x}_0)$, and the inner product is defined by $\langle f, g \rangle = \int fg \, dx$, the Hilbert space becomes L_2, which is not reproducing because the delta function is not $\in L_2$.

hyperparameters—the isotropic length scale and the signal variance—that have to be found by solving Equation 4.5. In most studies, the RBF kernel is used. See [407] for an insightful review. When several tasks are involved (multi-task learning), they are often assumed to be independent in the standard GP framework. If the tasks are assumed to be correlated, the used methods are either based on significant augmentations of the basic GP theory, on stationary separable kernels, or the task contributions to the covariance are assumed to be constant across the input space. [67].

While the standard approach described above yields an agnostic, widely applicable, and computation-friendly framework for regression, it has some considerable disadvantages.

1. Any kernel definition carries hidden **constraints on the underlying RKHS**. For instance, the exponential kernel does not allow for functions in the RKHS that are differentiable everywhere in the input domain; on the contrary, the RBF kernel will only allow functions in C^∞ to be elements of the RKHS. Most kernels are chosen for practical applications without these impositions in mind, leading to unrealistic approximations. However, what seems like a disadvantage can be turned into a useful asset. The natural occurrence of constraints on the RKHS can be used purposefully to inject domain awareness into the GP, significantly increasing the function approximation and uncertainty quantification quality.

2. The covariance, calculated by using stationary kernels can only depend on the distance between two data points, not on their respective locations in the input domain. In other words, two points on one side of the domain will have the same calculated covariance as two other points on the other side of the domain, as long as their respective distances are the same. Clearly, from a modeling perspective this is a significant simplification, and yet, used in the vast majority of kernel-based algorithms. **Non-stationary kernels** make no such simplification and should be preferred.

3. The assumption of stationarity in kernel designs also makes it difficult to learn similarities across tasks since there is no natural distance between arbitrarily indexed tasks. This has led to the development of many workarounds based either on significant simplifications—coregionalization for instance—which leads to poor capture of cross-task covariances, or need a set of additional tools to work, such as support vector regression [67]. Flexible non-stationary kernels can avoid the need for specialized tools for multi-output GPs and can encode accurate covariances across tasks. We will go into greater depth about **multi-task learning with GPs** in the next section.

4.3.1 Imposing Constraints on the RKHS via Stationary Kernels

To recap, stationary kernels are positive definite functions of the form

$$k(\mathbf{x}_1, \mathbf{x}_2) = k(||\mathbf{x}_1 - \mathbf{x}_2||), \tag{4.16}$$

where $|| \cdot ||$ is some—most often the Euclidean—norm. All functions $f(\mathbf{x}) = \sum_i^N \alpha_i k(\mathbf{x}, \mathbf{x}_i)$ adopt the differentiability properties of the chosen kernel $k(\mathbf{x}, \mathbf{x}_i)$. This forces us to constrain the differentiability of our model. A natural way of doing so is by utilizing Matérn kernels. The Matérn kernel class is defined as

$$k(\mathbf{x}_1, \mathbf{x}_2) = \frac{1}{2^{v-1}\Gamma(v)} \left(\frac{\sqrt{2v}}{l}r\right)^v B_v\left(\frac{\sqrt{2v}}{l}r\right), \tag{4.17}$$

where r is a valid distance measure in $\mathcal{X}$, B_v is the modified Bessel function and v is the parameter controlling differentiability. For $v = \frac{1}{2}$, we recover the exponential kernel $k = \sigma_s^2 \exp[-\frac{r}{2l}]$ which is non-differentiable at the origin; for $v \to \infty$, we recover the RBF kernel that is infinitely differentiable.

The knowledge of a latent function f being of the form $\sum_i g_i(x_i)$ significantly improves the extrapolation capabilities of the GP function approximation. Using the kernel

$$k(\mathbf{x}_1, \mathbf{x}_2) = \sum_i k_i(x_1^i, x_2^i). \tag{4.18}$$

gives the process the capability to propagate information far outside the regions covered by data. See Figure 4.1.

Constraining anisotropy—or allowing it—is another capability deeply embedded into stationary (and non-stationary) kernels. Anisotropy can be induced in several ways. One possibility is by using the product and additive kernels

$$k(\mathbf{x}_1, \mathbf{x}_2) = \exp\left[-\frac{|x_1^1 - x_2^1|}{l_1}\right] + \exp\left[-\frac{|x_1^2 - x_2^2|}{l_2}\right] \quad [Additive] \tag{4.19}$$

$$k(\mathbf{x}_1, \mathbf{x}_2) = \exp\left[-\frac{|x_1^1 - x_2^1|}{l_1}\right] \exp\left[-\frac{|x_1^2 - x_2^2|}{l_2}\right] \quad [Multiplicative]. \tag{4.20}$$

However, keep in mind that the additive kernel comes with additional properties presented before. A second way of achieving anisotropy is by altering the Euclidean distance metric in $\mathcal{X}$ such that

$$k(\mathbf{x}_1, \mathbf{x}_2) = k((\mathbf{x}_1 - \mathbf{x}_2)^T \mathbf{M} (\mathbf{x}_1 - \mathbf{x}_2)), \tag{4.21}$$

where $\mathbf{M}$ is any symmetric positive definite matrix [570, 360].

One of the most impressive capabilities of kernel methods (including GPs) is the ability to only allow functions with certain symmetries or periodicity properties to be considered [366]. Since kernels can be passed through any linear operator without losing their positive semi-definiteness [166], for any kernel $k(x_1, x_2)$, $L_{x_1}(L_{x_2}(k))$ is also a valid kernel function. For instance, for axial symmetry in two dimensions (see Figure 4.2), we can apply the linear operator

$$L(f(\mathbf{x})) = \frac{f([x^1, x^2]^T) + f([-x^1, x^2]^T) + f([x^1, -x^2]^T) + f([-x^1, -x^2]^T)}{4} \tag{4.22}$$

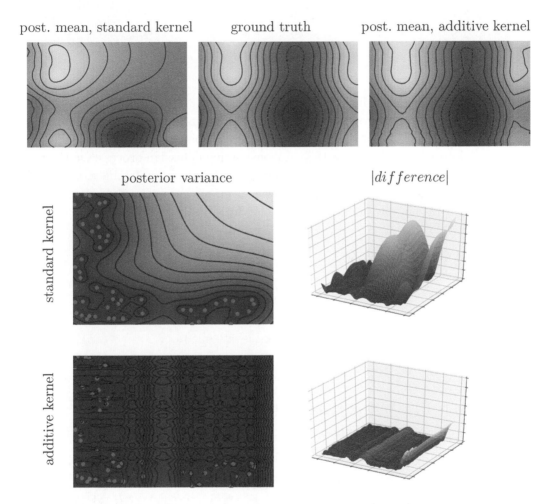

Figure 4.1 Exponential kernel $k(\mathbf{x}_1, \mathbf{x}_2) = 2 \exp\left[-\frac{|\mathbf{x}_1 - \mathbf{x}_2|}{0.5}\right]$ vs. additive kernel $k(\mathbf{x}_1, \mathbf{x}_2) = \exp\left[-\frac{|x_1^1 - x_2^1|}{0.5}\right] + \exp\left[-\frac{|x_1^2 - x_2^2|}{0.5}\right]$ to approximate a function of the form $\sum_i g_i(x_i)$. The additive kernel can propagate information to infinity even when data is only available in $(dim - 1)$-dimensional slices of the dim-dimensional input space. The posterior variance and error are significantly smaller compared to the use of the exponential kernel. Image adapted from [366].

standard kernel   ground truth   symmetric kernel

Figure 4.2 Axial-symmetry constraints on the RKHS underlying a Gaussian process. Displayed are Ackley's function (middle) and two GP posterior-mean functions. Left is the posterior mean calculated with an unconstrained Gaussian process. On the right is the posterior mean, calculated with imposed axial symmetry. In this example, the axial symmetry improves the uncertainty at a given number of measurements 4 fold and increases the computational speed 64 fold. Image courtesy of [366].

to any kernel function k to obtain

$$
\begin{aligned}
L_{\mathbf{x}_1}(k(\mathbf{x}_1, \mathbf{x}_2)) &= 1/4 \; (k(\mathbf{x}_1, \mathbf{x}_2) + k([-x_1^1, x_1^2]^T, \mathbf{x}_2) \\
&\quad + k([x_1^1, -x_1^2]^T, \mathbf{x}_2) + k([-x_1^1, -x_1^2]^T, \mathbf{x}_2)) \\
L_{\mathbf{x}_2}(k(\mathbf{x}_1, \mathbf{x}_2)) &= 1/4 \; (k(\mathbf{x}_1, \mathbf{x}_2) + k(\mathbf{x}_1, [-x_2^1, x_2^2]^T) \\
&\quad + k(\mathbf{x}_1, [x_2^1, -x_2^2]^T) + k(\mathbf{x}_1, [-x_2^1, -x_2^2]^T)) \\
&\Rightarrow
\end{aligned}
$$

$$
\begin{aligned}
L_{\mathbf{x}_2}(L_{\mathbf{x}_1}(k(\mathbf{x}_1, \mathbf{x}_2))) &= 1/16 \; (k(\mathbf{x}_1, \mathbf{x}_2) + k([-x_1^1, x_1^2]^T, \mathbf{x}_2) \\
&\quad + k([x_1^1, -x_1^2]^T, \mathbf{x}_2) + k([-x_1^1, -x_1^2]^T, \mathbf{x}_2) \\
&\quad + k(\mathbf{x}_1, [-x_2^1, x_2^2]^T) \\
&\quad + k(\mathbf{x}_1, [x_2^1, -x_2^2]^T) + k(\mathbf{x}_1, [-x_2^1, -x_2^2]^T) \\
&\quad + k([-x_1^1, x_1^2]^T, [-x_2^1, x_2^2]^T) + k([-x_1^1, x_1^2]^T, [x_2^1, -x_2^2]^T) \\
&\quad + k([-x_1^1, x_1^2]^T, [-x_2^1, -x_2^2]^T) + k([x_1^1, -x_1^2]^T, [-x_2^1, x_2^2]^T) \\
&\quad + k([x_1^1, -x_1^2]^T, [x_2^1, -x_2^2]^T) + k([x_1^1, -x_1^2]^T, [-x_2^1, -x_2^2]^T) \\
&\quad + k([-x_1^1, -x_1^2]^T, [-x_2^1, x_2^2]^T) + k([-x_1^1, -x_1^2]^T, [x_2^1, -x_2^2]^T) \\
&\quad + k([-x_1^1, -x_1^2]^T, [-x_2^1, -x_2^2]^T)).
\end{aligned} \tag{4.23}
$$

In the same way, rotational symmetries [366] can be enforced (see Figure 4.4) via

$$
k(\mathbf{x}_1, \mathbf{x}_2) = \frac{1}{36} \sum_{\phi \in p\pi/3} \sum_{\theta \in q\pi/3} \tilde{k}(\mathcal{R}_\phi \mathbf{x}_1, \mathcal{R}_\theta \mathbf{x}_2); \; p, q \in \{0, 1, 2, 3, 4, 5\}, \tag{4.24}
$$

Figure 4.3 Periodicity-constrained RKHS. Shown are the posterior mean functions after 50 data points have been collected randomly. Left: standard kernel. Middle: ground truth. Right: kernel as shown in 4.25. The function approximation accuracy is clearly superior in the right image. Image courtesy of [366].

where $\tilde{k}$ is any valid stationary kernel, ϕ and θ are angles, and $\mathcal{R}$ is a rotation matrix rotating the vector $\mathbf{x}$ by the specified angle. Periodicity (see Figure 4.3) can be enforced via

$$L_{\mathbf{x}_2}(L_{\mathbf{x}_1}(k(\mathbf{x}_1, \mathbf{x}_2))) =$$
$$1/9\ (k(\mathbf{x}_1, \mathbf{x}_2) + k(\mathbf{x}_1, [x_2^1, x_2^2 + p]^T) + k(\mathbf{x}_1, [x_2^1, x_2^2 - p]^T)$$
$$+ k([x_1^1, x_1^2 + p]^T, \mathbf{x}_2) + k([x_1^1, x_1^2 + p]^T, [x_2^1, x_2^2 + p]^T)$$
$$+ k([x_1^1, x_1^2 + p]^T, [x_2^1, x_2^2 - p]^T)$$
$$+ k([x_1^1, x_1^2 - p]^T, \mathbf{x}_2) + k([x_1^1, x_1^2 - p]^T, [x_2^1, x_2^2 + p]^T)$$
$$+ k([x_1^1, x_1^2 - p]^T, [x_2^1, x_2^2 - p]^T));\tag{4.25}$$

p is the period.

Figure 4.4 Six-fold symmetric test function (center) and GP function approximations. Left: approximation using the squared-exponential kernel. Right: GP approximation using kernel (4.24). Data points are implicitly used six times increasing the amount of information in the dataset six-fold. Image courtesy of [366].

Figure 4.5 Stationary (left) and non-stationary (right) kernel function over $\mathcal{X} \times \mathcal{X}$. Both functions are symmetric around the diagonal. However, the stationary kernel function is constant along the diagonals. The non-stationary kernel function has no such restricting property. This translates into flexible inner products in $\mathcal{H}$. Image courtesy of [366].

4.3.2 Non-Stationary Kernels

In contrast to stationary kernels that depend on some distance between an input-point pair $k(\mathbf{x}_1, \mathbf{x}_2) = k(||\mathbf{x}_1 - \mathbf{x}_2||)$, non-stationary kernels are more general symmetric p.s.d. functions of the form $k = k(\mathbf{x}_1, \mathbf{x}_2) \neq k(||\mathbf{x}_1 - \mathbf{x}_2||)$.

Stationary and non-stationary kernels are both symmetric positive-definite functions since they induce inner products in $\mathcal{H}$. The difference between the two classes can be illustrated visually when the input space is one-dimensional (see Figure 4.5). The key takeaway is that stationary kernel functions are constant along diagonals, unlike non-stationary kernels, which translates into potentially highly flexible inner products. This allows them to effectively encode similarities across the input and the output space.

When designing non-stationary kernels, we have to ensure positive semi-definiteness, just like in the stationary case. It is difficult to prove positive semi-definiteness in closed form for general functions, so kernels are usually created by building them from other kernels and using their basic definition.

The kernel $f(\mathbf{x}_1)f(\mathbf{x}_2)k(\mathbf{x}_1, \mathbf{x}_2)$ can easily be shown to be a valid kernel and is simple, yet powerful since f can be any arbitrary function. Figure 4.6 illustrates the behavior of this particular kernel for a linear function f in one dimension. It shows how the length scale of a stationary kernel is averaged across the domain, leading to under and over-estimated variances. The two-dimensional equivalent can be seen in Figure 4.7 with

$$f(\mathbf{x}) = (\phi_1 (\sqrt{50} - ||\mathbf{x}||)) + \phi_2. \tag{4.26}$$

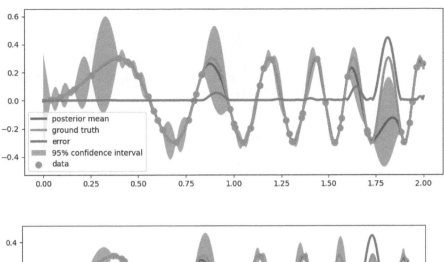

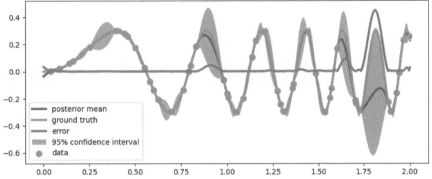

Figure 4.6 GP function approximation of a latent function that has a varying length scale. The approximation on top was accomplished using a Matérn kernel. The approximation at the bottom was achieved by using the kernel $k(x_1, x_2) = x_1 x_2 k_{matern}$. The stationary-kernel GP overestimates uncertainties on the left and underestimates them on the right. The GP that uses a non-stationary kernel makes no such mistakes and therefore increases the accuracy of the function approximation and uncertainty quantification. Figure courtesy of [366].

4.4 MULTI-TASK LEARNING WITH GPS

So far we considered the case in which a scalar function $f(\mathbf{x})$ is being learned. It is natural to ask if and how we can learn a vector-valued function $\mathbf{f}(\mathbf{x})$, where the components of $\mathbf{f}$ are correlated, using a Gaussian process. This is referred to by many different terms in the literature, e.g., multi-task learning, transfer learning, multi-output learning, multi-target learning, multi-response learning, and in the case of a GP, multivariate GPs. In this short overview, we follow the in-depth review by [19]. The transition from single-task GPs to multi-task GPs can entirely be absorbed by the kernel definition. We learned that, for approximating a scalar function, the kernel is a symmetric positive semi-definite function $k : \mathcal{X} \times \mathcal{X} \to \mathbb{R}$. For multi-task GPs, this definition changes to $k : \mathcal{X} \times \mathcal{X} \to \mathbb{R}^T \times \mathbb{R}^T$, where T is the number for tasks $t \in \mathcal{X}_o$. The methods for multi-task learning vary in the way they specify the kernel.

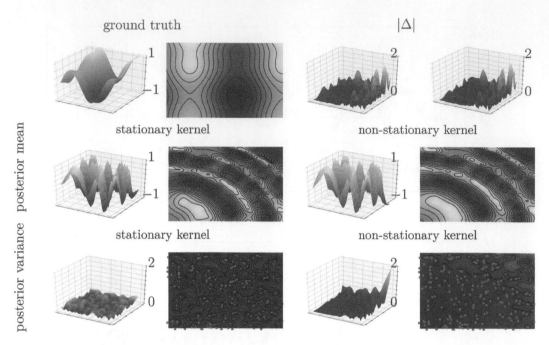

Figure 4.7 GP approximation of a function in two dimensions using stationary and non-stationary kernels. Similarly to Figure 4.6, the variance is overestimated in the region adjacent to the origin. In regions farther away from the origin, the stationary kernel causes the GP to underestimate the posterior variances. Image adapted from [366].

4.4.1 Separable Kernels

One conveniently simple way to learn multiple tasks is to separate the total covariance contribution into covariances stemming from distances in the input space $\mathcal{X}_i$ from contributions in the output space $\mathcal{X}_o$. Then

$$k(\mathbf{x}_1, \mathbf{x}_2, t_1, t_2) = k_{\mathcal{X}_i}(\mathbf{x}_1, \mathbf{x}_2) k_{\mathcal{X}_o}(t_1, t_2), \tag{4.27}$$

where $k_{\mathcal{X}_i}$ and $k_{\mathcal{X}_o}$ are both scalar kernels. Equation 4.27 can be rewritten as

$$K(\mathbf{x}_1, \mathbf{x}_2) = k_{\mathcal{X}_i}(\mathbf{x}_1, \mathbf{x}_2) \mathbf{B}. \tag{4.28}$$

Because $\mathbf{B}$ acts as a kernel it has to be symmetric and positive semi-definite. From the definition of kernels, we know that we can sum kernels to create new ones. Therefore,

$$\sum_l k_{\mathcal{X}_i}^l(\mathbf{x}_1, \mathbf{x}_2) \mathbf{B}_l \tag{4.29}$$

is also a valid kernel. Kernels of this kind are commonly referred to as Sum-of-Separable (SoS) kernels. Many methods for multi-task kernel learning differ in how they calculate the $\mathbf{B}_l$; the two most common are regularization methods and coregionalization models. For more information on both approaches see [19, 315].

4.4.2 Coregionalization Models

Coregionalization models were championed by geoscientists under the name of co-kriging. It is based on the basic assumption that the components f_t can be written as a linear combination of some underlying latent functions

$$f_t(\mathbf{x}) = \sum_q^Q c_d^q u_q(\mathbf{x}), \qquad (4.30)$$

where c_d^q are scalar coefficients and u_q have zero mean and covariance $cov[u_q, u_{q'}] = k_q(\mathbf{x}_1, \mathbf{x}_2)$ if $q = q'$ and 0 otherwise. Using a grouping of functions into groups that share the same covariance, we can rewrite 4.30 so that

$$f_t(\mathbf{x}) = \sum_q^Q \sum_i^G c_d^{qi} u_q^i(\mathbf{x}). \qquad (4.31)$$

The covariance between two components f_t and $f_{t'}$ is then given by

$$cov[f_t(\mathbf{x}), f_{t'}\mathbf{x}'] = \sum_q^Q \sum_{q'}^Q \sum_i^G \sum_{i'}^G c_d^{qi} c_{d'}^{q'i'} \ cov[u_q^i(\mathbf{x}), u_{q'}^{i'}(\mathbf{x}')], \qquad (4.32)$$

which reduces to

$$cov[f_t(\mathbf{x}), f_{t'}\mathbf{x}'] = \sum_q^Q \sum_i^G c_d^{qi} c_{d'}^{q'i'} k_q(\mathbf{x}, \mathbf{x}). \qquad (4.33)$$

This shows that the coregionalization model is a way to find a valid kernel $k_{\mathcal{X}_o}(t_1, t_2) = \sum_i^G c_d^{qi} c_{d'}^{q'i'}$.

4.4.3 Non-Separable Kernels

Of course, one can consider non-separable kernels in the first place. Flexible non-stationary kernels can be used to calculate covariances across the input and the output space. For instance $k(\mathbf{x}_1, \mathbf{x}_2) = f(\mathbf{x}_1)f(\mathbf{x}_2)$ is a valid kernel, independent of the structure of $\mathcal{X}$. Therefore, $k(\mathbf{x}_1, t_1, \mathbf{x}_2, t_2) = f(\mathbf{x}_1, t_1)f(\mathbf{x}_2, t_2)$ is a valid kernel and can be used to calculate covariances across the input space and the tasks. For more on this approach see [366]. The main drawback of this method is the number of hyperparameters that have to be found through the training of the GP.

4.5 GPS FOR LARGE DATASETS

Gaussian processes have certain advantages compared to mainstream ML. For instance, GPs scale better with the dimensionality of the problem than neural networks and are more exact in approximating the latent function [302]. Even more importantly, GPs provide the desired robust Bayesian uncertainty quantification. Even so, due to their unfavorable scaling of $O(N^3)$ in computation and $O(N^2)$ in storage [570], the applicability of GPs has largely been limited to small datasets (small

N). Datasets commonly encountered in the materials sciences, earth, environmental, climate, biosciences, and engineering are often out of reach of a GP. The reason for the unfavorable numerical complexity of GPs is the inversion – or solving a linear system—involving a typically dense covariance matrix [570]. Several methods to reduce the GP's scaling burden exist and are mostly based on approximations. The so-called local-GP-experts method divides the dataset into several subsets. Each subset serves as an input into separate GPs and the predictions are combined [104, 157]. This is equivalent to defining one GP with a block-diagonal covariance matrix. It is customary, but not mandatory, to perform the division into subsets by locality—which leads to the name "local GP experts". Another approximation is the inducing-points method. As the name suggests, inducing-points methods attempt to find a substitute set of data points of length M, in order to achieve a favorable data structure that translates into sparsity of the covariance matrix. Some examples include fixed-rank Kriging [108], KISS-GP [572], and the predictive process [42, 146]. Generally speaking, inducing-points methods are highly dependent on the kernel definition and therefore limit flexibility in the kernel design. This is a major drawback since recent applications are increasingly taking advantage of flexible non-stationary kernel functions [442]. Often, the inducing points are chosen on a grid, a process only viable in low-dimensional spaces. For highly nonlinear functions, M approaches N, and the advantage of the method disappears. Structure-exploiting methods are a special kind of inducing-points method in which pseudo-points are placed on grids—note the mentioned connection to inducing points. The resulting covariance matrix has Toeplitz algebra, which leads to fast linear algebra calculations needed to train and condition the GP. The success of those methods is not agnostic to the kernel definition and, as mentioned above, its use is limited to low-dimensional spaces. A technique originating in the statistics communities, the Vecchia approximation [547, 229], picks a subset of the data points to condition the prior on, instead of calculating the full posterior. The success of this method is kernel-dependent and has largely been applied to stationary kernels. Methods for better scaling of exact GPs have also been developed but generally receive less attention. It is, for instance, possible to let a flexible non-stationary kernel discover if a problem is naturally sparse. For this, the kernel has to be able to return zero, which means it has to be compactly supported [362].

4.6 GP TRAINING

The vast majority of studies implementing GPs for specific application areas use stationary kernels. As a reminder, stationary kernels are of the form $k(||\mathbf{x}_1 - \mathbf{x}_2||)$. The most popular example is the radial-basis-function (RBF) kernel, also called the squared exponential kernel

$$k(\mathbf{x}_1, \mathbf{x}_2) = \sigma_s^2 \exp[-\frac{||\mathbf{x}_1 - \mathbf{x}_2||^2}{2l^2}], \tag{4.34}$$

where $|| \cdot ||$ is some norm, most often the Euclidean norm, and σ_s^2 and l are the signal variance and the length scale respectively. In this simplest of all cases, we only have to infer the values of the two hyperparameters σ_s^2 and l from the data. Stationary kernels like the RBF kernel have, with few exceptions, the advantage of simple training due to a low number of hyperparameters. The use of non-stationary kernels requires finding many hyperparameters which is a highly involved process due to their sheer quantity; even in low-dimensional spaces, some 100 or even 1000 hyperparameters are not uncommon. Even worse, their impact on the training objective function is not independent giving rise to an ill-posed optimization problem. In any case, we need robust procedures to find the hyperparameters, which this section is all about. For more detail, we refer the reader to [569].

4.6.1 Cross Validation

The idea of cross-validation is to divide the dataset into a number n of subsets (as few as $n = 2$) and then train on all but one and use that one as the validation set. Any number of loss functions can be used, with the squared-error loss function being the most common one. It is recommended to set $n > 2$ and to repeat the training for different permutations of training and validation sets. The drawback of this method is that we have to repeat the training a number of times which slows down the process considerably. In the extreme case, all data points but one are in the training set; the remaining point is used for validation. This methodology is often referred to as leave-one-out cross-validation and leads to N training rounds—N is the number of data points—which is often prohibitive.

4.6.2 Log Marginal Likelihood optimization

The arguably most common method of training a GP is to maximize the log marginal likelihood (Equation 4.5). We restate the optimization here for convenience.

$$\arg\max_{\phi} \Big(\log(L(D, \phi)) =$$

$$-\frac{1}{2}(\mathbf{y} - \boldsymbol{\mu}(\phi))(\mathbf{K}(\phi) + \mathbf{V})^{-1}(\mathbf{y} - \boldsymbol{\mu}(\phi))$$

$$-\frac{1}{2}\log(|\mathbf{K}(\phi) + \mathbf{V}|) - \frac{\dim(\mathbf{y})}{2}\log(2\pi) \Big). \tag{4.35}$$

The first of the three terms in the log marginal likelihood is for the data fit; the second can be interpreted as a complexity penalty and the third is the normalization term. The complexity penalty naturally protects against over-fitting the model which is a strength of this methodology. Another strength of this method is that numerical optimization tools are powerful and widely available, even for HPC hardware; many are able to overcome challenges of solution non-uniqueness [361, 371].

4.6.3 Bayesian Hyperparameter Training

The fully Bayesian approach to finding the hyperparameters is by approximating

$$p(\phi|\mathbf{y}) = \frac{p(\mathbf{y}|\phi)p(\phi)}{\int p(\mathbf{y}|\phi)p(\phi)d\phi}. \tag{4.36}$$

The posterior cannot be calculated directly because the denominator is not analytically tractable. The term $p(\mathbf{y}|\phi)$ is the marginal likelihood, which we use in the log marginal likelihood optimization. $p(\phi|\mathbf{y})$ can be approximated using a Markov Chain Monte Carlo algorithm.

4.7 A NOTE ON GPS ON NON-LINEAR SPACES

It is natural to ask if GPs can be defined on non-linear spaces. In the geosciences, this problem has a rich history, since most calculations are done on the surface of the earth—an inherently non-linear space. The issue with GPs (and other kernel methods) on non-linear spaces is that kernels that are positive semi-definite functions on linear spaces are not necessarily p.s.d. on non-linear spaces, given we use geodesics as our distance measure. Solutions to this problem discovered so far can be divided into two categories. The first is based on manifold-intrinsic distances derived from stochastic partial differential equations. The second method uses embeddings and extrinsic distances [273]. This is a relatively new research area and will surely advance rapidly in the near future.

CHAPTER 5

A High-Level Introduction to Uncertainty Quantification

Mark D. Risser

Climate and Ecosystem Sciences Division, Lawrence Berkeley National Laboratory, Berkeley, California, USA

Marcus M. Noack

Applied Mathematics and Computational Research Division, Lawrence Berkeley National Laboratory, Berkeley, California, USA

CONTENTS

U NCERTAINTY quantification (UQ) as a term has propagated from a set of well-defined mathematical and statistical tools to somewhat of a buzzword in recent decades. This chapter attempts to skip the hype and give a practical introduction to the fundamental methods of UQ. For the statistician, UQ most often means rigorous Frequentist or Bayesian quantification of uncertainties, while in the applied sciences

DOI: 10.1201/9781003359593-5

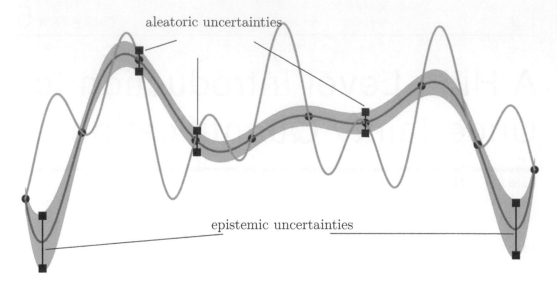

aleatoric uncertainties

epistemic uncertainties

Figure 5.1 Uncertainty quantification (UQ) in a function-approximation context. Noise in the data can be classified as aleatoric uncertainties, epistemic uncertainties are induced by missing data and therefore knowledge. Of course, the gray envelope in this figure is the result of both kinds of uncertainties.

uncertainties are often quantified by ensembles. Quantifying uncertainties without carefully studying the properties of each and every member of the ensemble can be dangerous because discrepancies might be introduced not by inherent stochasticity but rather by an inherent disability for a particular member to converge to a particular solution. In this chapter, we focus on the traditional definition of UQ, not ensemble methods. We largely follow the structure in Dekking et al. [124].

Sources of uncertainties are manifold and can broadly be divided into two classes: epistemic uncertainty—originating from lack of knowledge—and aleatoric uncertainties—originating from the intrinsic uncertainty in the system. In the context of stochastic function approximation and autonomous experimentation, aleatoric uncertainties are the noise in the data and epistemic uncertainties stem from places in parameter space that have not been measured yet (see Figure 5.1).

This chapter takes a general and agnostic approach to the basic building blocks of uncertainty quantification but we will make connections to function approximation and Autonomous Experimentation regularly.

5.1 THE PROBABILITY SPACE

To allow mathematical formality in UQ, we have to first define the sets we are working with. The triple (Ω, Σ, P) is called a probability space. Ω is the sample space; a set of all possible outcomes of an experiment. These can be categorical, for instance, heads and tails for tossing a coin, or numerical; if they are categorical, we have to map them to numbers and this is where random variables come into play. We will

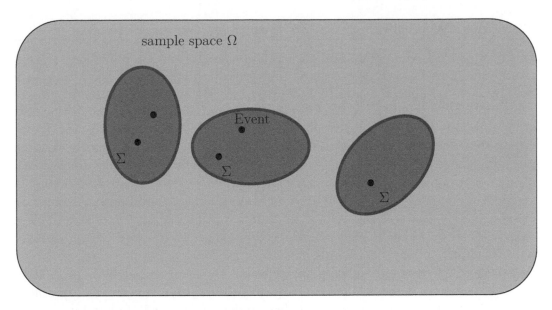

Figure 5.2 Visualization of the probability space, also called the probability triple (Ω, Σ, P). The sample space Ω contains all possible outcomes. The set Σ is a collection of all subsets of the sample space; that means it also contains Ω. The events, or outcomes, are particular elements of the sample space that have been observed. Each event is then assigned a probability by the probability measure P.

discuss those in the next section. Σ is a set of events. P is a probability measure defined as $P : \Sigma \to [0, 1]$. For instance, Ω could be $[-2, 2]$, the position along an interval of the real line, Σ is any combination of open intervals on Ω, and $P = 0.25(b - a); a, b \in [-2, 2], a < b$. This would be the probability of a position being within the interval $[a, b]$. See Figure 5.2 for a visualization of the probability space. The three elements of the probability space have to satisfy some axioms. Ω cannot be an empty set, so it has to contain at least one outcome. Σ has to be a so-called σ-algebra on Ω, meaning $\Sigma \subseteq 2^{\Omega}$ so that Σ may contain Ω, is closed under complements, countable unions and, by extension, countable intersections. For any collection of pairwise disjoint sets A_i, we have $P(\bigcup A_i) = \sum P(A_i)$. The probability measure on the entire sample space $P(\Omega) = 1$. To make the concept a bit more tangible, let's consider one more example: the roll of a dice. In this case $\Omega = \{1, 2, 3, 4, 5, 6\}$, $\Sigma = 2^{\Omega}$ contains $2^6 = 64$ events, for example $\{1\}, \{2\}, \{1, 2\}, \{1, 3\}, \ldots$. The probability measure $P(\{1\}) = P(\{2\}) = P(\{3\}) = P(\{4\}) = P(\{5\}) = P(\{6\}) = \frac{1}{6}$, $P(\{\}) = 0$, $P(\{1, 2, 3, 4, 5, 6\}) = 1$. Although the definition of the probability space seems overkill to tackle simple problems like rolling dice once, it becomes vitally important for more-complicated scenarios. Often, events that are elements of the sample space are not numbers but categorical outcomes such as heads or tails for the toss of a coin. To perform quantitative probabilistic calculations, we have to introduce a map that takes events and returns numbers. This map is called a random variable.

5.2 RANDOM VARIABLES

One of the biggest conceptual changes when moving from a deterministic worldview to a probabilistic one is that what used to be variables that take on scalar values are now random variables that are characterized by distributions: probability mass or probability density functions. The formal definition of a random variable is that it is a function from our sample space to a measurable space, i.e., $X : \Omega \to S$, and $x \in S$. Often—but certainly not always—$S \subset \mathbb{R}$. If we speak of random variables we often mean the image of the function X, i.e., the function values. If S is multi-dimensional, a random variable turns into a multivariate random variable, also called a random vector $\mathbf{X}$. Lowercase x or $\mathbf{x}$ are observed outcomes. Random variables and vectors are classified by the characteristics of the set S. If S is countable, the random variable is called a discrete random variable whose distribution is characterized by a probability mass function (e.g., the roll of a dice, or the toss of a coin). If S is uncountably infinite, we speak of a continuous random variable that is characterized by a probability density function. Random variables are arguably the most fundamental building block of probability theory, allowing concepts like random graphs and random functions to be defined by extension. A stochastic process, for instance, can be defined as a set of random variables, most often interpreted as function values.

5.3 PROBABILITY MASS FUNCTIONS, PROBABILITY DENSITY FUNCTIONS, AND PROBABILITY DISTRIBUTIONS

The distribution of random variables is defined by probability mass or probability density functions, in the discrete and the continuous case respectively. Both are used to define probabilities of events. The probability mass function is defined as the probability of a discrete event, i.e.,

$$f(x) = P[X = x] \text{ or } f(\mathbf{x}) = P[\mathbf{X} = \mathbf{x}]. \tag{5.1}$$

Since we know that some event will occur when the experiment is performed, the probability mass function has to satisfy $\sum_{x \in S} f(x) = 1$, and probabilities cannot be negative, which means $f(x) > 0 \forall x \in S$. Remember, S is the measurable equivalent of Ω. An example probability mass function for the roll of two dice can be seen in Figure 5.3. We see something very fundamental about probability theory and statistics in that figure; when we observe several events, we "spawn" extra dimensions and define, so-called, joint distributions (probability mass or density functions) over that new higher-dimensional space. Probability mass functions are often referred to as histograms in the statistics literature—in that case, they are most often defined over a one-dimensional sample space and are not normalized.

Probability density functions are the equivalent of probability mass functions for continuous random variables. A probability density function is an integrable function $f(x)$ such that

$$P[a < x \le b] = \int_a^b f(x)dx. \tag{5.2}$$

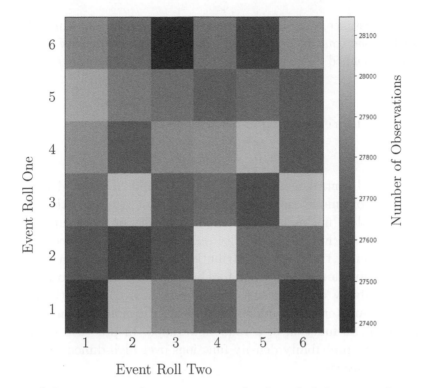

Figure 5.3 A histogram—a.k.a. an unnormalized probability mass function for the roll of two dice. The events are the tuples of the two outcomes of each roll. The plot shows the number of times each of the 36 events ($\{1, 1\}, \{1, 2\}, \{1, 3\}, ...$) occurred after 10^6 experiments. Unsurprisingly, the events occur in very similar quantities.

Following the logic for probability mass functions, probability density functions have to be greater than 0 everywhere and satisfy

$$\int_{-\infty}^{\infty} f(x)dx = 1. \tag{5.3}$$

There is some ambiguity when it comes to the term probability distribution. Sometimes the term is used interchangeably with probability density or mass functions, often to avoid some of the restrictions of density functions (for instance densities are difficult to define in infinite-dimensional spaces). Other times, the term refers to the cumulative distribution function defined by

$$F(t) = \int_{-\infty}^{t} f(x)dx = P[x < t]. \tag{5.4}$$

and equivalently by a sum for probability mass functions.

Probability mass and probability density functions can be characterized by a handful of values, called moments. The most-commonly used moments—ordered by decreasing relevance—are the mean, the variance, the skewness, and the kurtosis. Most work in statistics is done by looking at mean and variance only. The normal distribution, for instance, only has non-zero first and second moments. The nth moment of a real-valued continuous function—most probability density functions fall in this category—is defined by

$$\mu_n = \int_{-\infty}^{\infty} x^n f(x) dx. \tag{5.5}$$

Second and higher moments are often taken about the mean

$$\tilde{\mu}_n = \int_{-\infty}^{\infty} (x - \mu_1)^n f(x) dx \tag{5.6}$$

and are then called central moments. While probability mass or density functions have only one set of unique moments, the same is not true the other way around; a set of moments does not assign a probability density or mass function uniquely. It is worth noting that not all probability mass or density functions have moments that exist, for instance, the Cauchy distribution does not have even a first moment. Moments are an important computation tool because they allow us to store even high-dimensional probability distributions economically in a parameterized way. Other popular ways of characterizing random variables are the moment-generating function and the characteristic function of a random variable. In practical calculations, we most often use joint probability density functions over high-dimensional, commonly Euclidean, sample spaces.

5.4   THE NORMAL DISTRIBUTION AND THE CENTRAL LIMIT THEOREM

While we will dive into a variety of different probability density and mass functions in a later section, we want to highlight the normal distribution here because of its theoretical and practical impact. The normal distribution is defined in N dimensions as

$$f(\mathbf{x}) = \mathcal{N}(\boldsymbol{\mu}, \boldsymbol{\Sigma}) = \frac{1}{\sqrt{(2\pi)^N |\boldsymbol{\Sigma}|}} \exp[-\frac{1}{2}(\mathbf{x} - \boldsymbol{\mu})^T \boldsymbol{\Sigma}^{-1}(\mathbf{x} - \boldsymbol{\mu})], \tag{5.7}$$

where $\boldsymbol{\mu} \in \mathbb{R}^N$ is the distribution's mean and $\boldsymbol{\Sigma}$ is the $N \times N$ covariance matrix—don't confuse this with the non-bold Σ from earlier. See Figure 5.4 for a plot of the normal distribution over a two-dimensional sample space. Why is the normal distribution so important? The reason is the Central Limit Theorem, which states that when adding up (infinitely) many random variables—even if they themselves are not normally distributed—the resulting distribution will be normal. Of course, we never add up infinitely many random variables but the theorem can still be used for an approximation when many (perhaps >30) random variables are added. It now just so happens that many random events we observe are actually a weighted sum of other random variables (at least approximately). As an example, think of the superposition of waves in earthquake simulations or the sum of environmental conditions for reservoir modeling. As long as a sum of many random variables is used, the resulting distribution of the sum is approximately Gaussian! Figure 5.5 shows the effect of the Central Limit Theorem in an impressive way; while a roll of a dice is a probability mass function with probability $P[X = x_i] = f(x_i) = 1/6$ for each outcome x_i, the sum of many tosses follows a normal distribution.

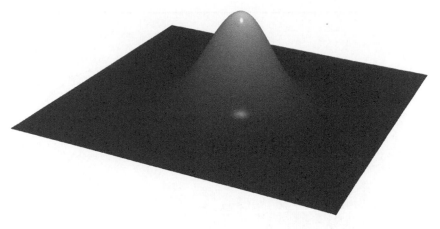

Figure 5.4   A joint normal distribution over a two-dimensional sample space.

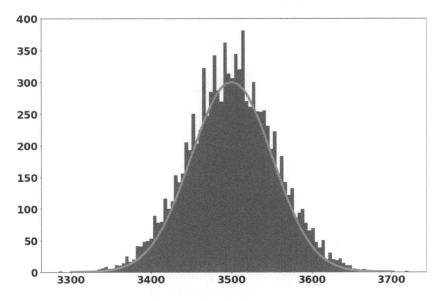

Figure 5.5   The probability mass function of the sum of outcomes of 10,000 rolls of a dice. While the outcome of one roll is certainly not normally distributed, the sum clearly is well-approximated by a normal distribution (orange).

5.5   JOINT, MARGINAL, AND CONDITIONAL PROBABILITIES

We have already discussed joint probability distributions (density functions) earlier but want to define them here more rigorously. Given a multivariate random variable (a random vector) defined on the same probability space, a **joint probability density function** assigns a probability measure to every element of the sample space, and via the integral, a probability to every subset of possible outcomes. A tangible example is the random vector of body height and shoe size of an example population. Given the joint probability distribution, we can, for instance, calculate the probability of someone being between 180 and 182 cm tall and having a shoe size between 32 and 34 cm. Joint probability distributions are often denoted $f(x, y, ...)$, where x and y are

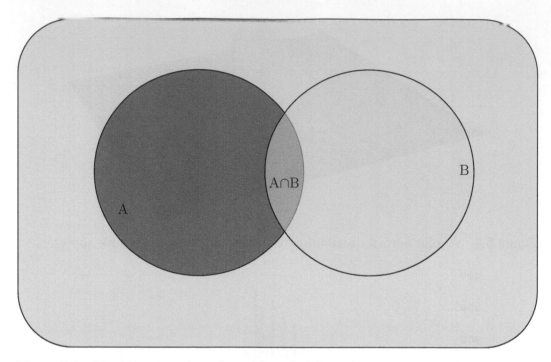

Figure 5.6 Visualization of conditional probability. If one is interested in the probability $P(x \in A)$ but has already observed that $x \in B$, one can ask for the probability of x being within A, given that it is also in B, this is written as $P(x \in A | x \in B) = \frac{P(x \in A \cap B)}{P(x \in B)}$. It is important to know that this is often simply written as $P(A|B) = \frac{P(A \cap B)}{P(B)}$.

particular outcomes of the random variables X and Y, or simply $f(\mathbf{x})$. An important property of a set of random variables or vectors is the extent to which they are interrelated or "dependent". When two random variables or vectors are completely unrelated or "independent", their joint distribution can be written as the product of their individual distributions, e.g., $f(x, y) = f(x)f(y)$.

When two random variables or vectors are dependent, the concept of a **conditional probability density function** becomes important. Conditional probability distributions describe the behavior of one or more random variables given some information regarding the outcome(s) of (an)other random variable(s) (see Figure 5.6). For example, what if we are now interested in the probability distribution of body heights given that a shoe size of 35 cm was observed already? The conditional probability distribution

$$f(x|y) = f(x|Y = y) = \frac{f(x, y)}{f(y)}. \tag{5.8}$$

describes the probability distribution of the random variable X if we have observed that event $Y = y$ happened. Note the important relationship between conditional and joint probabilities with regard to independence: rearranging Equation 5.8 and comparing with the definition of independence given above, we see that X and Y are independent if and only if $f(x|Y = y) = f(x)$. In other words, X and Y are independent if probabilities regarding X are the same whether or not we have specific information about the outcome of Y. This concept is easily transferrable to higher-dimensional random vectors $\mathbf{X}$ and $\mathbf{Y}$. In a different scenario, one might calculate the joint probability distribution from measurements; however, later the interest is only on the distribution of one random vector or variable, ignoring the other. In that case, the **marginal probability distribution** is needed. Its definition changes slightly depending on the underlying sample space. For probability mass functions (countable sample space), the marginal probability distribution is defined as

$$f(X = x_j) = \sum_i f(X = x_j, y_i). \tag{5.9}$$

In the uncountable infinite sample space case (probability density functions) the marginal distribution is defined as

$$f(x) = \int_{-\infty}^{\infty} p(x, y) dy. \tag{5.10}$$

In almost all practical situations, the conditional probability cannot be calculated via 5.8 because the joint probability density function is not known. However, marginal probabilities and other conditional probabilities are often known or can be estimated. Bayes' theorem [137, 507, 220] is a famous result that describes how to derive conditional distributions for one random variable given marginal and conditional distributions of another random variable.

5.6 BAYES' THEOREM

Bayes' theorem describes how to derive conditional distributions for one random variable given marginal and conditional distributions of another random variable. For example, consider the conditional distribution given in Equation 5.8: the order of conditioning is irrelevant, such that we can write $f(x, y) = f(y)f(x|y) = f(x)f(y|x) = f(y, x)$. In the case when it is natural to define conditional distributions only for one of the random variables, e.g., X given Y, we can obtain the conditional distribution over x given $Y = y$ as follows:

$$f(x|Y = y) = \frac{f(Y = y|x)f(x)}{f(y)}. \tag{5.11}$$

Bayes' theorem provides the theoretical underpinning for the so-called "Bayesian" philosophy of statistics, which supposes that all unknown quantities in statistical analysis (often called "parameters" or "hyperparameters") are themselves random variables. In contrast to the Bayesian philosophy is the so-called "Frequentist" perspective, which supposes that hyperparameters have a single, fixed (but unknown) value. Hence, from a Bayesian perspective the phrase "the true average height of all human males lies between 180 and 182 cm with probability 0.23" is well defined, while in a Frequentist sense this phrase is incorrect and meaningless (since from that perspective the true average height has a deterministic yet unknown value). In the Bayesian approach, Bayes' theorem is used to calculate the conditional distribution (the so-called "posterior" distribution) of unknown statistical parameters (here denoted with X) given data (here denoted Y).

5.7 PROPAGATING UNCERTAINTIES

In many applied-science fields—earthquake and weather simulations, imaging, and flow simulations among many others—operators, often in the form of partial differential equations are used to model real-world phenomena. The question is often: How does uncertainty in the inputs translate into uncertainties in the output? In the case of linear operators, the new mean and covariance matrix can be computed analytically since

$$E[aX] = \sum_i x_i p_i a = a \sum_i x_i p_i = aE[X]$$
$$Var[aX] = E[(aX - E(aX))(aX - E(aX))] =$$
$$a^2 E[(X - E(X))(X - E(X))] = a^2 Var[X]. \tag{5.12}$$

For non-linear operators, we have basically three options: First, a local linear approximation via a Taylor expansion. Second, sampling methods like Monte Carlo or Markov-Chain Monte Carlo methods, and third, polynomial chaos expansion [397]. As a related technique, we also want to introduce stochastic processes in this section.

5.7.1 Local Linear Approximation

In situations in which, in a neighborhood supported by the given uncertainties, the operator in question is not too far from linear, a local Taylor approximation can yield good results. In practice, mostly first-order Taylor approximations (here around the point $\mathbf{x}^0$) are considered

$$f(x_1, x_2) \approx f(x_1^0, x_2^0) + \frac{\partial f}{\partial x_1}(x_1^0 - x_1) + \frac{\partial f}{\partial x_2}(x_2^0 - x_2)$$
$$= f(x_1^0, x_2^0) + \frac{\partial f}{\partial x_1}dx_1 + \frac{\partial f}{\partial x_2}dx_2. \tag{5.13}$$

The variance of the function at the new point is then given by

$$Var[f] \approx |\frac{\partial f}{\partial x_1}|^2 \sigma_{dx_1}^2 + |\frac{\partial f}{\partial x_2}|^2 \sigma_{dx_2}^2 + 2\frac{\partial f}{\partial x_1}\frac{\partial f}{\partial x_2}Cov[dx_1, dx_2] \tag{5.14}$$

Of course, for more accurate uncertainty propagation higher-order Taylor approximations could be considered; higher-order derivatives are also linear operators.

5.7.2 Markov Chain Monte Carlo

MCMC is a simple algorithm with many applications that is interpreted differently by the various communities that use it. The core idea behind the Markov Chain Monte Carlo (MCMC) algorithm is to approximate some property of a function—the mean function value, an integral, or the function itself—by repeated function evaluations organized sequentially as a chain so that a "jump" to a new evaluation position only depends on the last position and a proposal distribution, not on any

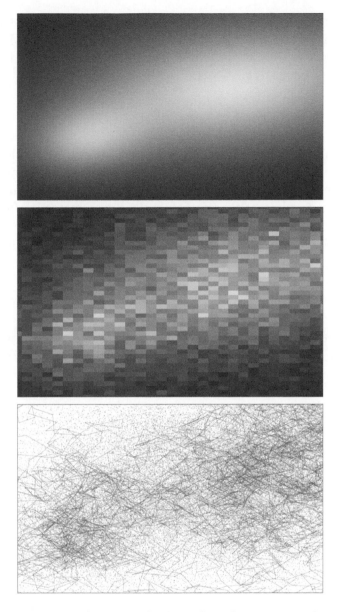

Figure 5.7 Illustration of the MCMC algorithm. From top to bottom. First, the distribution—or simply a function—to be approximated. Second, a two-dimensional histogram of all points that were collected by the MCMC. The key takeaway is that density resembles the function itself. Third, a subset of the paths the MCMC took. There are many more jumps in regions of high function values.

other past positions that were evaluated (Figure 5.7). Given the last position, a new position is drawn from

$$\mathbf{X}_t \sim p(\mathbf{x}_t|\mathbf{x}_{t-1}). \tag{5.15}$$

There are many adaptations and customizations of the core MCMC algorithms, such as the Metropolis-Hasting, the Hamilton-MCMC, and the scMCMC algorithms.

These adaptations are mostly about choosing the right proposal distribution. Within probability theory, uncertainty quantification, and statistics, MCMC is used to approximate posteriors. To calculate a posterior analytically, Bayes' theorem suggests we have to derive the posterior

$$p(x|y) = \frac{p(y|x)p(x)}{p(y)} = \frac{p(y|x)p(x)}{\int p(y|x)p(x)dx}. \tag{5.16}$$

Often, neither can the numerator be calculated analytically as any closed-form probability density function nor is the denominator available in closed form. The numerator can, however, be evaluated point-wise, i.e., choose an x and evaluate the likelihood, the prior, and the product of the two, while the denominator is often impossible to derive or evaluate. The pseudocode 5.1 suggests that the function we are interested in only appears in the fraction to calculate the acceptance probability

$$\alpha = f(x')/f(x) = \frac{\frac{p(y|x)p(x)}{p(y)}}{\frac{p(y|x')p(x')}{p(y)}} = \frac{p(y|x)p(x)}{p(y|x')p(x')} \tag{5.17}$$

which does not contain the integral. This means to approximate the posterior, we only have to evaluate $p(y|x)p(x)$ repeatedly. See Figure 5.7 for a graphical illustration.

```
1  pick a random starting x_0
2  calculate f(x_0)
3  for i in range(1,N):
4      x_proposal ~ p(x_i|x_{i-1})
5      calculate f_proposal = f(x_proposal)
6      α = f_proposal/f(x_{i-1})
7      u ~ U(0,1)
8      if u ≤ α: x_i = x_proposal
9      else: x_i = x_{i-1}
```

Listing 5.1: A simple Markov Chain Monte Carlo algorithm, called the Metropolis algorithm, to estimate a posterior distribution.

5.7.3 Polynomial Chaos

Polynomial chaos expansion (PCE) was first introduced by Norbert Wiener [566] in 1938. Today it is a widely used technique for creating surrogate models in many scientific disciplines and machine learning [79]. For PCE—much like for stochastic processes—we postulate a true but unknown model function $f(\mathbf{x}, \boldsymbol{\omega})$, where $\mathbf{x}$ is the position in some input space and $\boldsymbol{\omega}$ is a vector of some uncertain input parameters. PCE seeks to find a surrogate for the model function f. The main idea is to define an approximation [79] for f so that

$$f(\mathbf{x}, \boldsymbol{\omega}) \approx \sum_i^M c_i(\mathbf{x}) \Psi_i(\boldsymbol{\omega}), \tag{5.18}$$

where $c_i(\mathbf{x})$ are coefficients, and $\Psi_i(\boldsymbol{\omega})$ are multivariate orthonormal polynomial basis functions defined by multiplying univariate polynomials

$$\Psi_\alpha(\boldsymbol{\omega}) = \prod_j^N \phi_j^{\alpha_j}(\omega), \tag{5.19}$$

where α is a multivariate index containing combinatoric information on how to enumerate all products of individual univariate orthonormal polynomials. Having chosen the polynomials, the solution comes down to finding the coefficients c_i. Polynomial chaos expansion is a vast topic and more information can be found in [79, 566, 303].

5.7.4 Stochastic Processes and Random Fields

The concept of stochastic processes is driven by the idea that a set of function evaluations $\{f(\mathbf{x}_1), f(\mathbf{x}_2), f(\mathbf{x}_3), f(\mathbf{x}_4), ...\}$ is understood as a random vector $\mathbf{f}$. This is sometimes referred to as a random field. There is another way to understand stochastic processes, and this is by calling any process that has a stochastic component (such as an MCMC) a stochastic process. Here we will use the former definition. Understanding function values as jointly distributed random variables has some interesting consequences. We can place a probability density function over the function evaluations and tune it such that the likelihood of observations is maximized. Combined with the idea of kernels to calculate covariances between unobserved point-pairs, this allows us to predict posterior distributions for function values in unobserved regions, which makes the stochastic process a powerful tool for uncertainty quantification and Autonomous Experimentation. See Chapter 4 for more on this topic.

5.8 EXAMPLE DISTRIBUTIONS

Most calculations in uncertainty quantification are done on analytical distributions due to the friendly and efficient computations they provide. What follows is by no means a comprehensive list but it presents a starting point. More importantly, it represents the various interesting aspects of distributions one wants to use or study.

5.8.1 The Normal Distribution

The normal distribution, also called the Gaussian distribution, is by far the most commonly used distribution on this list. This has to do with the Central Limit Theorem which states that the distribution of all sums of random variables approaches a normal distribution. Many practical things in our world happen to be the sum of some underlying effects. The normal distribution is a continuous probability distribution, defined in one dimension by

$$f(x) = \frac{1}{\sqrt{2\pi}\sigma} \exp[-\frac{1}{2}\left(\frac{x-\mu}{\sigma}\right)^2], \quad -\infty \leq x \leq \infty, \tag{5.20}$$

where μ is the mean and σ is the standard deviation. The variance of the distribution is σ^2. Its multi-variate equivalent is defined by

$$f(\mathbf{x}) = \frac{1}{\sqrt{(2\pi)^k |\Sigma|}} \exp[-\frac{1}{2}(\mathbf{x} - \boldsymbol{\mu})^T \Sigma^{-1} \mathbf{x} - \boldsymbol{\mu})], \qquad (5.21)$$

where Σ is the covariance matrix, $\boldsymbol{\mu}$ is the multivariate prior mean, and $\mathbf{x} \in \mathbb{R}^k$. The normal distribution is often denoted by $\mathcal{N}(\boldsymbol{\mu}, \Sigma)$. The cumulative distribution of the normal distribution is mostly used in one dimension and is then given by

$$\int_{-\infty}^{x} f(x)dx = \frac{1}{2}[1 + \mathrm{erf}(\frac{x - \mu}{\sqrt{2}\sigma})] \qquad (5.22)$$

5.8.2 The Poisson Distribution

The Poisson distribution is a discrete probability distribution—defined over countable sample spaces—that is used for the number of occurrences of an event (see Figure 5.8). It is defined by

$$p(x) = \frac{\lambda^k e^{-\lambda}}{x!}, x = 0, 1, 2, \ldots, \qquad (5.23)$$

where λ is a positive real number that is the distribution's mean and variance. The cumulative distribution function is given by

$$F(x) = \sum_{i}^{x} \frac{e^{-\lambda}\lambda^i}{i!} \qquad (5.24)$$

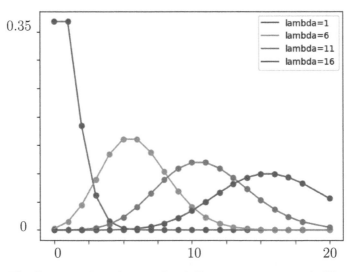

Figure 5.8 The Poisson distribution for different parameters λ. The Poisson distribution is a discrete probability distribution that is used for random variables describing the number of occurrences of events.

5.8.3 The Student's t-Distribution

The Student's t-distribution—or simply t-distribution—is a continuous probability distribution arising from drawing small samples from normal distributions. Its probability density function in one dimension is given by

$$f(x) = \frac{\Gamma(\nu+1)}{\sqrt{\nu\pi}\Gamma(\frac{\nu}{2})}\left(1 + \frac{x^2}{\nu}\right)^{-\frac{\nu+1}{2}}, \quad -\infty \leq x \leq \infty, \tag{5.25}$$

where Γ is the gamma-function and ν is the so-called degree of freedom of the distribution. The explicit form of the cumulative distribution function of the Student's t-Distribution depends on the value for ν [590]. If $\nu = \infty$ the distribution is Gaussian. The t-distribution is bell-shaped but has heavier polynomial tails compared to the normal distribution with its exponential tails. The multivariate equivalent of the probability density function is given by

$$f(\mathbf{x}) = \frac{\Gamma(\frac{\nu+p}{2})}{\Gamma(\frac{\nu}{2})\nu^{p/2}\pi^{p/2}\sqrt{|\mathbf{\Sigma}|}}\left(1 + \frac{1}{\nu}(\mathbf{x} - \boldsymbol{\mu})^T \mathbf{\Sigma}^{-1}(\mathbf{x} - \boldsymbol{\mu})\right)^{-\frac{\nu+p}{2}} \tag{5.26}$$

5.8.4 The Binomial Distribution

The binomial distribution is a discrete probability distribution over the number of successful outcomes of independent experiments, each with binary results (0/1) and associated probabilities p and $1 - p$. The probability mass function (see Figure 5.9) is defined as

$$f(x, n, p) = \frac{n!}{x!(n-x)!}p^x(1 - p)^{n-x}, x = 0, 1, 2, \ldots, \tag{5.27}$$

for x successes in n independent binary experiments. The associated cumulative distribution function is defined as

$$Pr(X \leq x) = \sum_{i=0}^{x} \frac{n!}{i!(n-i)!}p^i(1 - p)^{n-i} \tag{5.28}$$

Because of the Central Limit Theorem, the distribution approaches a normal distribution with an increasing number of experiments.

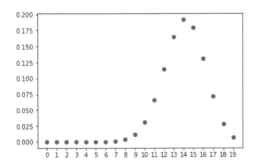 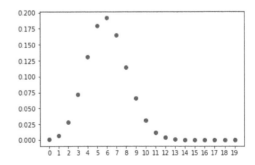

Figure 5.9 The binomial distribution for $n = 20$ and $p = 0.7$ (left), and $p = 0.3$ (right).

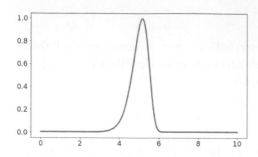 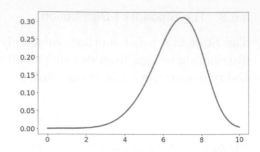

Figure 5.10 The Weibull distribution for (left) $\gamma = 14, \alpha = 5.2, \mu = 0$, and (right) $\gamma = 6, \alpha = 7.2, \mu = 0$.

5.8.5 The Weibull Distribution

The Weibull distribution (see Figure 5.10) is defined for continuous random variables and arises from failure (or life data) analysis. Its density function in one dimension is given by

$$f(x, \lambda, k) = \frac{\gamma}{\alpha} \left(\frac{x - \mu}{\alpha} \right)^{\gamma - 1} e^{-(\frac{x-\mu}{\alpha})^\gamma}, x > 0 \tag{5.29}$$

5.9 UNCERTAINTY QUANTIFICATION AND AUTONOMOUS EXPERIMENTATION

Autonomous Experimentation can have many aspects that are based on probabilities and statistics. Gaussian processes, for instance, are characterized by joint probability density functions over function values. The rules of probability theory make it possible to infer knowledge and compute posterior probability distributions that can be used for decision-making; after all, measurements will be needed where posterior uncertainties are high. But the impact of probability theory is much greater. Many data-analysis algorithms are based on Bayesian statistics, and other statistical tools such as bootstrapping, cluster analysis, and dimensionality reductions (PCA). The key takeaway from this section is that a good working knowledge of uncertainty quantification is necessary to understand many aspects of autonomous experimentation and machine learning.

CHAPTER 6

Surrogate Model Guided Optimization Algorithms and Their Potential Use in Autonomous Experimentation

Juliane Mueller

Computational Science Center, National Renewable Energy Laboratory, Golden, Colorado, USA

CONTENTS

IN this chapter, we will review the basics of derivative-free optimization methods based on surrogate models, and we outline how these methods can straightforwardly be applied to autonomously steering experimentation. The underlying assumption is that the fitness function to be optimized and constraints that may be present cannot be expressed analytically because their values depend on conducting experiments in either a laboratory setting or on high-performance computing

DOI: 10.1201/9781003359593-6

systems by carrying out black-box simulations. Thus, gradient information cannot be exploited. In this chapter, we will first summarize general solution approaches that use surrogate models and active learning. Then we dig into various adaptations of these approaches that have been developed to accommodate constraints, integer variables, and multi-objective problems. We will provide recommendations regarding the suitability of these approaches for specific problem classes. Note that the area of derivative-free optimization is quite large and we do not claim to provide an exhaustive review here. We focus on global surrogate modeling strategies and some main classes of problems that often arise in science and engineering for which these approaches are efficient and effective. We will close the chapter by outlining future research avenues that should target the specific challenges of autonomous experimentation.

6.1 INTRODUCTION TO SURROGATE MODELING AND ACTIVE LEARNING FOR OPTIMIZATION

Increasing computing power has allowed us to develop increasingly complex simulation models that better approximate physical phenomena in the real world. These simulations give rise to related optimization problems in science and engineering applications, ranging from wind turbine wake control [17], to quantum computing [330], hyperparameter tuning of machine learning models [331, 39], environmental applications [527, 96], and many more. Mathematically, an optimization problem is written as

$$\min_{\mathbf{x} \in \Omega} f(\mathbf{x}) \tag{6.1}$$

$$\text{s.t. } g_j(\mathbf{x}) \leq 0, j = 1, \ldots, J \tag{6.2}$$

where $\Omega \subset \mathbb{R}^d$ is the d-dimensional search space, f is the objective function to be minimized, and g_j are constraints. In this chapter, we are focusing on problems where the evaluation of f (and g_j if present) is highly time-consuming (minutes or more per evaluation) and black box, i.e., an analytic description is not available. For this class of problems, there is usually also no gradient information available, necessitating the use of derivative-free optimization methods.

These types of problems can come with multiple different characteristics, such as multiple conflicting objective functions that must be minimized simultaneously [328], integer constraints on some or all parameters [333, 334], computationally expensive or fast to evaluate constraints g_j (or both) [437, 435, 329, 335], and uncertainty arising from evaluating the objective and/or constraint functions [562]. These special characteristics should be taken into account when developing solution algorithms.

In the following, we discuss how surrogate models and active learning strategies can be used creatively to tackle these types of problems and how surrogate model approaches presented in the literature can straightforwardly be adapted to guiding experiments in the laboratory.

A surrogate model $m(\mathbf{x})$ is a computationally inexpensive approximation of a function: $f(\mathbf{x}) = m(\mathbf{x}) + e(\mathbf{x})$, where e denotes the difference between the true function

Input: Number of function evaluation budget; number of initial points;
Step 1: Create an initial experimental design, e.g., using Latin hypercube sampling, and evaluate the expensive function.
Step 2: While function evaluation budget not reached:
    Step 2a: Build / Update the surrogate model using all input-output pairs.
    Step 2b: Solve an auxiliary optimization problem using the surrogate model to select the next evaluation point.
    Step 2c: Evaluate the expensive function.
Return: Best solution found.

Figure 6.1   Pseudocode for a surrogate model assisted optimization algorithm. Each step must be adapted depending on the characteristics of the optimization problem.

f and the surrogate m. During the optimization routine, the surrogate model is used to guide the search rather than the true function [66]. This significantly reduces the number of queries that must be made to the costly function and thus speeds up the time to solution of the optimization problem. In the literature, surrogate models are widely used when f is computationally expensive. However, since m is trained only on input-output pairs, $\{(\mathbf{x}, f(\mathbf{x}))\}_{i=1}^{n}$, it is indifferent with respect to how the value of f at an input $\mathbf{x}$ is obtained, and thus the surrogate model concept can straightforwardly be applied to cases where f is the outcome from a lab experiment.

Surrogate modeling is often combined with active learning strategies, where in each iteration of the optimization algorithm, the surrogate model is used to identify which new inputs $\mathbf{x}_{\text{new}}$ should be evaluated next and given the new input-output pair $(\mathbf{x}_{\text{new}}, f(\mathbf{x}_{\text{new}}))$, the surrogate model is updated. The pseudocode of a general surrogate model guided optimization algorithm is shown in Figure 6.1.

The initial experimental design strategy may be based on Latin hypercube designs, optimal design strategies, or it could be augmented with user-specified points that are known to perform well. The type of surrogate model employed could be, e.g., radial basis functions (RBFs) [412, 411], Gaussian Processes (GPs) [457, 217], or multivariate adaptive regression splines [152] to name a few, and the problem characteristics will determine which model is the most suitable.

Step 2b is the implementation of the active learning strategy, and it depends on the problem characteristics and the surrogate model [332]. Sampling methods should balance exploration (global search) and exploitation (local search) and can be set up such that one or more new evaluation points can be chosen in each iteration of the algorithm. The surrogate model is updated each time new input-output data are available and it thus actively learns the true function's landscape in search for the location of the optimum. Surrogate model algorithms usually stop after a budget of function evaluations or total compute time has been used up and they return the best solution encountered during the search. Note that, due to the black-box nature of the problems, it cannot be guaranteed that the returned solution is globally optimal. Most surrogate model algorithms only offer convergence in probability.

Research in optimization algorithm development using surrogate models focuses mostly on Step 2b, i.e., modeling and solving an auxiliary problem that enables one to optimally select the next sample point(s). In this step, additional information from the optimization problem such as the presence of (computationally cheap or expensive) constraints g_j or integer constraints, as well as information from multiple conflicting objective functions and unsuccessful function evaluations should be directly taken into account when generating sample points in order to help the search. Taking this information into account during sampling has the benefit of potentially reducing the search space. For example, if some parameters have integer constraints, it would be a waste of resources to evaluate a solution that does not satisfy this constraint as it will never be a solution to the problem. Rather the search space could be reduced to the integer values only, thus yielding a finite number of values for integer-constrained parameters.

6.2 SURROGATE MODELS

Different types of surrogate models with different functionalities have been developed in the literature and have been used as computationally cheap approximations in black-box expensive optimization tasks. Here we provide a few examples of surrogate models that are widely used. Ensembles of different surrogate models have also been explored in the literature [168].

6.2.1 Gaussian Process Models

Gaussian process models (GPs) have their roots in geostatistics [307]. A GP treats the function it approximates like the realization of a stochastic process:

$$m_{\text{GP}}(\mathbf{x}) = \mu + Z(\mathbf{x}) \tag{6.3}$$

where μ is the mean of the stochastic process and $Z \sim \mathcal{N}(0, \sigma^2)$ represents the deviation from the underlying mean. It is assumed that the correlation between the errors depends on the distance between points. The correlation between two random variables $Z(\mathbf{x}_k)$ and $Z(\mathbf{x}_l)$ is described as

$$corr(Z(\mathbf{x}_k), Z(\mathbf{x}_l)) = \exp\left(-\sum_{i=1}^{d} \theta_i |\mathbf{x}_k^{(i)} - \mathbf{x}_l^{(i)}|^{\gamma_i}\right). \tag{6.4}$$

Here, θ_i determines how quickly the correlation decreases in the ith dimension, the superscript $^{(i)}$ denotes the ith component of the parameter vector, and γ_i represents the smoothness of the function in the ith dimension. Often, γ_i are fixed, whereas θ_i, μ, and σ are determined my maximum likelihood estimation.

The GP prediction at a new point $\mathbf{x}_{\text{new}}$ is then

$$m_{\text{GP}}(\mathbf{x}_{\text{new}}) = \hat{\mu} + \mathbf{r}^T \mathbf{R}^{-1} (\mathbf{f} - \mathbf{1}\hat{\mu}), \tag{6.5}$$

where

$$\mathbf{r} = \begin{bmatrix} corr(Z(\mathbf{x}_{\text{new}}), Z(\mathbf{x}_1)) \\ \vdots \\ corr(Z(\mathbf{x}_{\text{new}}), Z(\mathbf{x}_n)) \end{bmatrix}, \tag{6.6}$$

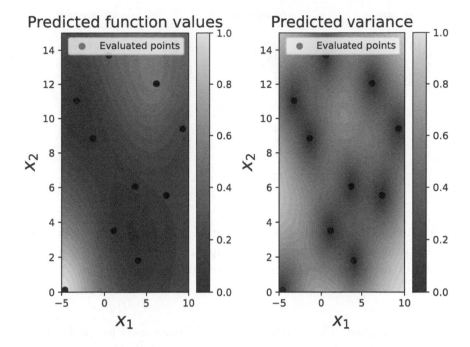

Figure 6.2 A GP is trained on 10 samples (red points) and used for making predictions across the full space. Left: Predicted function values. Right: Associated prediction uncertainty.

$\mathbf{R}$ is the $n \times n$ correlation matrix whose (k, l)-th element is defined as in (6.4), n is the number of training data points (input-output pairs), $\mathbf{1}$ is a vector of ones of appropriate dimension, and $\mathbf{f}$ is a vector of objective function values. The mean squared error corresponding to the GP prediction is

$$s^2(\mathbf{x}_{\text{new}}) = \hat{\sigma}^2 \left(1 - \mathbf{r}^T \mathbf{R}^{-1} \mathbf{r} + \frac{(1 - \mathbf{1}^T \mathbf{R}^{-1} \mathbf{r})^2}{\mathbf{1}^T \mathbf{R}^{-1} \mathbf{1}} \right), \tag{6.7}$$

and

$$\hat{\mu} = \frac{\mathbf{1}^T \mathbf{R}^{-1} \mathbf{f}}{\mathbf{1}^T \mathbf{R}^{-1} \mathbf{1}}, \quad \hat{\sigma} = \frac{(\mathbf{f} - \mathbf{1}\hat{\mu})^T \mathbf{R}^{-1} (\mathbf{f} - \mathbf{1}\hat{\mu})}{n}. \tag{6.8}$$

The advantage of using GP models is that they provide an estimate of uncertainty together with the function value prediction (see Figure 6.2). When using a GP for interpolation, this uncertainty is zero at points where the function has been evaluated (for zero measurement/simulation noise) and everywhere else it is larger than zero. This uncertainty estimate can help us identify where in the parameter space additional data should be collected to reduce the uncertainty of the model.

One of the challenges in fitting GP models is that they scale with $\mathcal{O}(n^3)$ with n the number of evaluated points in the training dataset. This can lead to computational bottlenecks when GPs are used for problems where more than a few hundred data

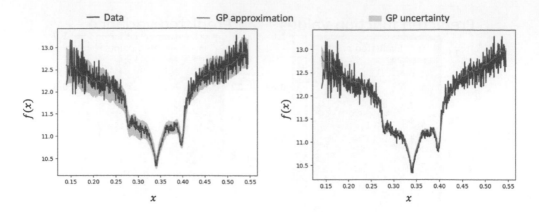

Figure 6.3 Left: GP with heteroskedastic kernel; Right: GP with homoskedastic kernel. The GP using a heteroskedastic kernel captures the different variance magnitudes in the data better. Image credits: Iméne Goumiri [177].

points can be obtained. Thus, GPs have mostly been used for problems with very few parameters and very limited dataset size, see e.g., [509]. However, in recent years, attempts have been made to reduce the complexity associated with training GP models for large-scale applications [158, 337, 476, 140, 363], which may enable the efficient application of GPs more broadly (see also Chapter 4).

Different kernels can be used in GP models and the one provided in (6.4) is just one example, namely a squared exponential kernel for $\gamma_i = 2$. Other kernels can be used such as the Matérn kernel, white noise kernel, a constant kernel, as well as composite kernels. Usually, it is a priori unknown which kernel is the best for a given problem, and often multiple GPs with different kernels are trained to understand which one performs best. Research is also ongoing in the development of kernels that depart from the assumption of independent and identically distributed noise across the space and isotropy [441, 363, 355]. For example, heteroskedastic kernels are able to better represent the different variance magnitudes across the space (see Figure 6.3), and thus provide us with more accurate uncertainty estimates.

6.2.2 Radial Basis Functions

Radial basis functions (RBFs) are also widely used as surrogate models in derivative-free optimization [438, 437, 435, 333, 329]. The functional form is

$$m_{\mathrm{RBF}}(\mathbf{x}) = \sum_{i=1}^{n} \lambda_i \varphi(\|\mathbf{x} - \mathbf{x}_i\|_2) + p(\mathbf{x}), \tag{6.9}$$

where $\|\cdot\|_2$ denotes the Euclidean norm, φ is the kernel, and p is a polynomial tail whose order depends on the kernel choice (see Table 6.1).

TABLE 6.1   Widely used kernels for RBF models and polynomial orders.

Kernel name	$\varphi(r)$	Polynomial order
Cubic	r^3	2
Thin plate spline	$r^2 \log(r), r > 0$	2
Gaussian	$\exp(-r^2/\rho^2)$	0
Multiquadric	$\sqrt{r^2 + \rho^2}$	2
Inverse mnultiquadric	$(r^2 + \rho^2)^{-1/2}$	0

In order to determine the parameters $\boldsymbol{\lambda} = [\lambda_1, \ldots, \lambda_n]^T$, and the coefficients of the polynomial tail β, a linear system of equations must be solved:

$$\begin{bmatrix} \boldsymbol{\Phi} & \mathbf{P} \\ \mathbf{P}^T & \mathbf{0} \end{bmatrix} \begin{bmatrix} \boldsymbol{\lambda} \\ \boldsymbol{\beta} \end{bmatrix} = \begin{bmatrix} \mathbf{f} \\ \mathbf{0} \end{bmatrix}, \text{ where } \mathbf{P} = \begin{bmatrix} \mathbf{x}_1^T & 1 \\ \mathbf{x}_2^T & 1 \\ \vdots \\ \mathbf{x}_n^T & 1 \end{bmatrix}, \tag{6.10}$$

$\boldsymbol{\Phi}$ is an $n \times n$ matrix whose (i, j)-th element is $\varphi(\|\mathbf{x}_i - \mathbf{x}_j\|_2)$, $\mathbf{f}$ is the vector of objective function values, and $\mathbf{0}$ is a matrix / vector of 0s of appropriate dimensions. Fitting an RBF to a set of input-output data pairs is computationally faster than fitting a GP since no hyperparameters must be tuned by maximum likelihood. However, note that RBFs do not provide uncertainty estimates with the function value predictions.

6.2.3   Multivariate Adaptive Regression Splines

Multivariate adaptive regression splines (MARS) [152] are non-parametric models and are of the general form

$$m_{\text{MARS}}(\mathbf{x}) = \sum_{i=1}^{m} a_i B_i(\mathbf{x}), \tag{6.11}$$

where B_i are basis functions and a_i are constant coefficients. The basis functions can either be constant, a hinge function (e.g., $\max\{0, x - const\}$ or $\max\{0, const - x\}$), or a product of hinge functions. MARS models are built in two phases, a forward pass where terms in the form of basis functions are added and a backward pass where terms that have little influence are removed. MARS models have also been for global optimization tasks, e.g., [304, 22].

6.3   SAMPLING APPROACHES

Here we discuss a few different adaptive sampling approaches. Adaptive sampling is also referred to as active learning and model management in the literature. These methods are based on formulating an auxiliary optimization problem that depends on the characteristics of the optimization problem to be solved as well as the surrogate models used for approximation. Sampling strategies should be able to balance exploitation (local search) and exploration (global search) in order to find local improvements of the best solution found so far but also not get stuck in such local optima and search globally for other regions of attraction. In the following, we divide sampling approaches based on the types of parameters in the problem (continuous vs. integers); the type of constraints (computationally cheap, expensive, or hidden); and the number of objective functions to be optimized. The list of problem classes is not exhaustive, and neither are the references provided. Our goal is to give the reader an idea of the versatility of application problems that can be solved with surrogate model-based optimization and a starting point for a literature search.

6.3.1 Problems with Continuous and/or Integer Variables

Here we consider unconstrained optimization problems of the form

$$\min f(\mathbf{z}) \tag{6.12}$$

$$\text{s.t.} \ -\infty < x_{i_1}^l \le x_{i_1} \le x_{i_1}^h < \infty, i_1 = 1, \ldots, k_1 \tag{6.13}$$

$$u_{i_2} \in \{u_{i_2}^l, u_{i_2}^l + 1, \ldots u_{i_2}^h\}, -\infty < u_{i_2}^l, u_{i_2}^h < \infty, i_2 = 1, \ldots, k_2 \tag{6.14}$$

$$\mathbf{x} \in \mathbb{R}^{k_1}, \mathbf{u} \in \mathbb{Z}^{k_2}, \mathbf{z}^T = (\mathbf{x}^T, \mathbf{u}^T), \tag{6.15}$$

where $\mathbf{x}$ denote the continuous variables and $\mathbf{u}$ are integer variables (if present).

Sampling in the continuous space Traditionally, surrogate model-based sampling strategies have been developed for problems in which all parameters are continuous. In conjunction with GPs, the expected improvement (EI) sampling has been widely used (see e.g., [217]). Denote $f^{\text{best}} = f(\mathbf{x}^{\text{best}})$ the best function value found so far. Then, for minimization problems, an improvement I is due to a value f^* that is less than f^{best}: $I = f^{\text{best}} - f^*$. The expected improvement at any unsampled point $\mathbf{x}$ in the search space is

$$\mathbb{E}(I)(\mathbf{x}) = s(\mathbf{x}) \left(\nu \Phi(\nu) + \phi(\nu) \right), \quad \nu = \frac{f^{\text{best}} - m_{\text{GP}}(\mathbf{x})}{s(\mathbf{x})} \tag{6.16}$$

where Φ and ϕ are the normal cumulative distribution and density functions, respectively, and $s(\mathbf{x}) = \sqrt{s^2(\mathbf{x})}$ in (6.7). With this definition, we are able to balance local and global search since both the predicted improvement and the uncertainty of the predictions is taken into account in the EI function. The EI function is maximized in order to select a new sample point $\mathbf{x}^{\text{new}}$. Figure 6.4 shows a 2-D illustration of the GP predictions (left) and the EI (right) across the problem domain. We can see that the EI is zero at the points where the function has already been evaluated (red dots) and it is non-negative everywhere else. This leads to a highly multimodal landscape with many flat regions that has to be optimized with a global optimization algorithm in order to find the point with maximal EI. This in turn becomes increasingly difficult as more samples are acquired and the solution to the auxiliary optimization problem heavily depends on the starting guess. Multi-start methods are required and many of the local searches quickly become stuck in flat spots of the EI surface, thus not achieving the goals of the sampling strategy.

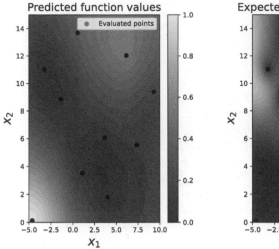
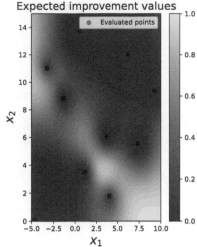

Figure 6.4 Left: GP predictions across the space given 10 sample points; Right: Corresponding EI surface. We can see that the EI function already has many flat regions which are not a solution to (6.16).

Other sampling methods that can be used in conjunction with a GP [216] are maximizing the probability of improvement

$$\mathbb{P}(I)(\mathbf{x}) = \Phi\left(\frac{f^{\mathrm{best}} - m_{\mathrm{GP}}(\mathbf{x})}{s(\mathbf{x})}\right), \tag{6.17}$$

which focuses on searching locally for improvements. Sampling decisions can also be based on minimizing the lower confidence bound

$$LCB(\mathbf{x}) = m_{\mathrm{GP}}(\mathbf{x}) - \kappa s(\mathbf{x}), \tag{6.18}$$

for some $\kappa \geq 0$. For $\kappa > 0$, this allows us to take the GP's prediction uncertainty into account when sampling. However, κ must be adjusted by the user and the results depend on the choice for κ.

In order to avoid some of the drawbacks arising from using GPs and EI, sampling strategies that depend on distance metrics and RBF surrogate model predictions have been introduced. For example, in the stochastic RBF method [438], a large number of candidate points is generated around the best point found so far ($\mathbf{x}^{\mathrm{best}}$) by adding random perturbations (see Figure 6.5). Then two scores are computed for each candidate point, one based on the RBF model's prediction and one based on the distance of the candidate point to the set of already sampled points. A weighted sum of both scores is calculated and the candidate with the best score is selected as a new sample point. The weights associated with either score allow prioritizing the RBF prediction (local search) or the distance to already sampled points (global search), respectively, and by cycling through a range of weights, a transition between local and global search can be achieved. This method does not require the optimization of

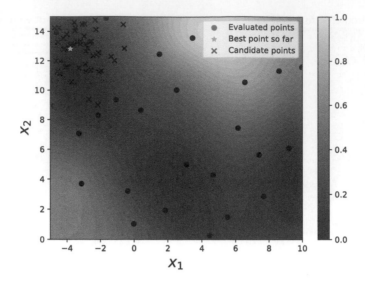

Figure 6.5 Candidate points (green crosses) are created around the best point sampled so far (yellow star). The contours were created based on the sampled points (red).

an auxiliary problem such as expected improvement and therefore has less computational overhead. One disadvantage is that for large dimensional problems, even small perturbations added to all values of $\mathbf{x}^{\text{best}}$ may lead to candidate points that are far away from $\mathbf{x}^{\text{best}}$. Thus, in a follow on paper [439], the authors proposed a dynamic way to create candidate points by perturbing each value of $\mathbf{x}^{\text{best}}$ with a probability that decreases as more sample points are collected. This allows to focus of the search more and more locally around $\mathbf{x}^{\text{best}}$ as the search progresses and preserves the local search aspect of the sampling. Also, this sampling strategy may become trapped in local optima. Thus, ideas that employ restart of the algorithm from scratch have been implemented as well.

Sampling in the integer space When integer variables are present in the optimization problem, it is often a waste of time to try and evaluate the expensive objective function at points that violate the integer constraints. On the one hand, a point that violates the integer constraints will not be a solution to the problem and would be useless. Rounding the solution can easily lead to a non-optimal solution and may violate other constraints (if present), and in some cases, the black-box simulation may not even evaluate successfully if the integer constraints are violated (e.g., if one tried to train a neural network with a non-integer number of nodes per layer). Thus, taking the integer constraints into account when creating new sample points can eliminate the above drawbacks that arise from trying to solve the relaxed problem.

The stochastic RBF approach as well as the dynamic perturbation approach can be easily adapted [333, 327]. Instead of adding random perturbations drawn from a normal distribution to the values of $\mathbf{x}^{\text{best}}$, one can perturb integer-constrained

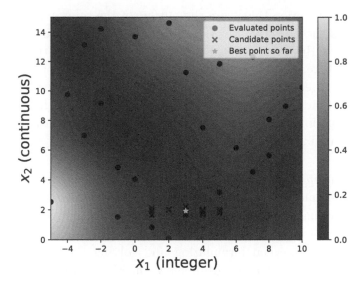

Figure 6.6 Candidate points (green crosses) *with integer constraints on x_1* are created around the best point sampled so far (yellow star). The contours were created based on the sampled points (red). Note that the function is only defined for points where x_1 is an integer, but we illustrate the full space for better visualization and because the approximation model is not restricted to integers.

parameters by integer increments and thus create a large set of candidate points that satisfy the integer constraints by definition (see Figure 6.6). Then, the same metrics (distance and RBF model prediction) can be employed as discussed above to select the best candidate point.

Similar to the perturbation approach, other sampling routines can be adapted to respect the integer constraints. For example, if we use EI, we can use a mixed-integer optimization algorithm to maximize (6.16) subject to integer constraints. Note however that although from a practical perspective, a GP can be fit to data with integer constraints, the GP predictions (and thus the EI values) may be uninformative if the samples are too far from each other and the hyperparameters for the GP length scales come out too small after tuning. The authors of [114] have developed a GP-based algorithm for mixed-integer optimization and we refer the reader to this publication.

6.3.2 Optimization Problems with Constraints

In this section, we discuss the handling of three types of constraints, namely computationally cheap constraints, computationally expensive black-box constraints, and hidden constraints (see [255] for a taxonomy on constraints in simulation-based optimization). Again, well-designed sampling methods can directly take these constraints into account and arrive at better solutions faster than the widely used approach of lumping constraints into the objective function with a penalty term.

If the constraint function values are the output of the same expensive simulation that also computes the objective function, new input-output pairs are obtained for all functions at once. However, in some cases, the constraint functions are computed by a simulation different from the objective function, or the constraint values are computationally inexpensive to obtain. In such cases, it makes sense to first compute the constraint function values, and only if the evaluated point is feasible, should the costly objective function value be calculated. Thus, the computational expense can be kept at a minimum.

Computationally cheap constraints Constraints that are computationally cheap to evaluate can and should be taken into account when selecting sample points. The idea is to never evaluate the computationally expensive function at a point that violates the constraints because those solutions will not be useful and it saves us computation time. Thus, when formulating and solving the auxiliary problem in Step 2b of the surrogate model algorithm shown in Figure 6.1, the computationally cheap constraints should be directly included. For example, if $g_j^c(\mathbf{x}) \leq, j = 1, 2, \ldots, J$ denote the computationally cheap constraints, then, in the case of using a GP with expected improvement sampling, an auxiliary problem could be formulated as

$$\min_{\mathbf{x} \in \Omega} \mathbb{E}(I)(\mathbf{x}) \tag{6.19}$$

$$\text{s.t. } g_j^c(\mathbf{x}) \leq 0 \ \forall j, \tag{6.20}$$

where Ω denotes the bounding box of the search space. Similarly, if we were to use an RBF and sample with the candidate point approach, we can simply remove candidates that violate the constraints from consideration, and then choose the best of the remaining feasible candidates as a new point. A flowchart of the surrogate model optimization algorithm when computationally cheap constraints are present is shown in Figure 6.7. We can see that if the constraints are not satisfied for a sample point, we bypass the objective function evaluation.

Computationally expensive constraints When constraints are computationally expensive, we distinguish two cases: (1) The constraint function values can be obtained independently of the objective function values (e.g., different simulations compute constraints and objective, respectively); (2) The constraint and objective function values cannot be obtained independently (e.g., the same simulation returns all constraint and objective function values). In both cases, denoting the constraints by $g_j^e(\mathbf{x}) \leq 0 \ \forall j$, they can be treated similarly to expensive objective functions, namely by approximating them with surrogate models $m_{g_j^e}$, i.e., $g_j^e(\mathbf{x}) = m_{g_j^e}(\mathbf{x}) + \epsilon_{g_j^e}(\mathbf{x})$, where $\epsilon_{g_j^e}$ is the difference between the true constraint function value and the surrogate model approximation [335, 437]. Note that we prefer to use a separate surrogate model for each constraint (rather than one surrogate model for a lump sum of constraints) as this increases overall accuracy.

In the first case, we want to check for the feasibility of a point by evaluating the constraints and evaluate the objective only if all constraints are satisfied. Depending on the setup, this can save compute time, e.g., when objective function evaluations

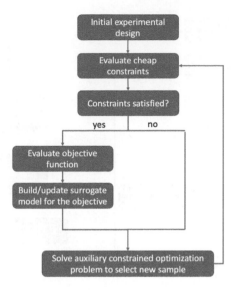

Figure 6.7 Flowchart of a surrogate model algorithm for problems with *computationally cheap* constraints. The constraints should be directly used in the auxiliary problem to find new samples. The algorithm stops once the computational budget has been exhausted.

take significantly longer than constraint evaluations and thus the benefits of parallelism would only be limited (see also the flowchart on the left in Figure 6.8). Thus, the surrogate models for the constraints are usually computed based on more input-output data points than the surrogate for the objective function. Regardless of feasibility, the surrogate models for the constraints are updated in each iteration of the optimization algorithm, while the surrogate model for the objective function is only updated when a feasible point has been found.

The second case is algorithmically somewhat simpler because all functions are evaluated at all sample points (see the flowchart on the right in Figure 6.8). Note that building RBFs for multiple functions that use the same sample site matrix allows us to simply swap out the right-hand side of (6.10) and thus reduce the computational complexity of building several surrogate models.

Similarly to the case of computationally cheap constraints, the surrogate models of the constraints should be incorporated into the definition of the auxiliary optimization problem that is solved to select new sample points. For example, problem (6.19)-(6.20) can be adapted:

$$\min_{\mathbf{x}\in\Omega} \mathbb{E}(I)(\mathbf{x}) \tag{6.21}$$

$$\text{s.t. } m_{g_j^e}(\mathbf{x}) \le 0 \ \forall j, \tag{6.22}$$

or in the case of candidate point sampling, candidates that are *predicted* to be infeasible can be excluded. Note that the surrogate models are only approximations of the true constraints and the feasible region defined by Ω and (6.22) may in fact be empty. In such a case, in order to find a first feasible point, one could, e.g., sample

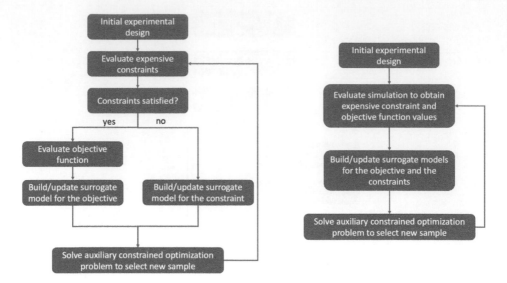

Figure 6.8 Flowchart of a surrogate model algorithm for problems with computationally expensive constraints. Left: Case 1, objective and constraint functions can be evaluated independently. The objective is only evaluated if all constraints are satisfied. Right: Case 2, objective and constraint functions cannot be evaluated independently. In both cases, surrogate models are used to approximate the objective and constraints and they are then used in the auxiliary optimization problem formulation. The algorithm stops once the computational budget has been exhausted.

at a point that violates the smallest number of constraints or that has the smallest constraint violation defined as

$$v(\mathbf{x}) = \sum_{j=1}^{J} \max\{0, m_{g_j^e}(\mathbf{x})\}. \tag{6.23}$$

One could also minimize the surrogate models for the constraint functions simultaneously in a multi-objective sense to potentially discover feasible points. However, these approaches do not guarantee that a feasible point will eventually be found, especially for problems with very small feasible regions. Thus, any knowledge of feasible points should be provided to the algorithm, e.g., as part of the initial experimental design, in order to avoid wasting function evaluations on infeasible points. Since surrogate models only approximate the true constraint functions, it may also make sense to allow sampling at points with a small predicted constraint violation in order to better learn the boundary of the feasible region.

Hidden constraints   A special class of problems arises when constraints are not explicitly known, but rather they appear at function evaluation time. For example, a simulation may not finish due to an underlying solver not converging [259, 329]. In these cases, an objective function value cannot be obtained. However, when using surrogate model-based optimization, the lack of the objective function value will not

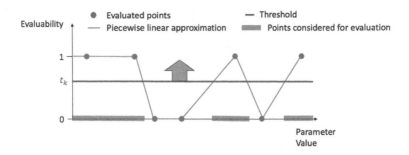

Figure 6.9 Cartoon of the piecewise linear approximation of the evaluability function. Only the parameter values shaded in blue will be considered for evaluation because their predicted evaluability lies above the threshold t_k.

allow us to update the surrogate model in each iteration of the algorithm, and thus the iterative sampling will not advance. Thus, we have to exploit any knowledge about the hidden constraint when making iterative sampling decisions. To this end, in [329], the authors defined a function to represent whether a point was evaluated successfully:

$$g(\mathbf{x}_e) = \begin{cases} 1 & \text{if } \mathbf{x} \text{ evaluated successfully} \\ 0, & \text{otherwise} \end{cases} \tag{6.24}$$

where $\mathbf{x}_e$ are points whose evaluation has been attempted. An RBF surrogate, $s_g(\mathbf{x})$, with a linear kernel (using $\varphi(r) = r$ in (6.9)) was then used to predict the evaluability of any point in the search space, and thus each point has a probability between 0 and 1 to be evaluable (see also the sketch in Figure 6.9). Then the auxiliary optimization problem to be solved in each iteration k of the optimization becomes

$$\min_{\mathbf{x} \in \Omega} s_f(\mathbf{x}) \tag{6.25}$$

$$\text{s.t. } s_g(\mathbf{x}) \geq t_k \tag{6.26}$$

where s_f is the surrogate model of the objective function that is trained only on successfully evaluated points, and t_k is a dynamically adjusted threshold that increases with the number of iterations k. Thus, as the algorithm proceeds and approaches the maximum number of function evaluations that can be attempted, only points that are increasingly likely to evaluate successfully will be considered for evaluation. This shows that even the lack of objective function values can still provide useful information for guiding the adaptive search. A flowchart of the algorithm for hidden constraints is shown in Figure 6.10.

6.3.3 Multi-objective Optimization

Multi-objective optimization problems arise when multiple conflicting objective functions $(f_1, f_2, \ldots, f_l)$ must be optimized simultaneously and a single solution that optimizes all objectives does not exist:

$$\min_{\mathbf{x} \in \Omega} [f_1(\mathbf{x}), f_2(\mathbf{x}), \ldots, f_l(\mathbf{x})]^T. \tag{6.27}$$

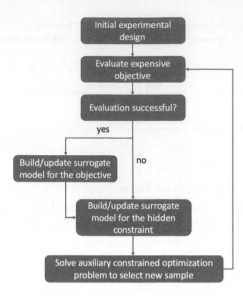

Figure 6.10 Flowchart of the hidden constraints algorithm. If the objective function does not evaluate successfully, we cannot update its surrogate model. However, we can still exploit the information to better model the hidden constraint.

The solution to a multi-objective optimization problem is a set of trade-off (Pareto-optimal) solutions, i.e., solutions that are not comparable since improving one objective function will lead to deteriorating at least one other objective function.

Commonly used approaches to multi-objective optimization are to minimize a weighted sum of the objectives where the weights represent the importance of each objective. Also, an ϵ-constraint approach can be used, where one objective function is minimized and the others act as constraints whose values may not exceed some limit ϵ. However, when dealing with black-box expensive objective functions, the assumptions underlying these solution approaches may not be fulfilled. In this case, the computationally expensive objective functions can be approximated each with a separate surrogate model: $f_i(\mathbf{x}) = m_i(\mathbf{x}) + e_i(\mathbf{x})$, where e_i represents the difference between the true function and the surrogate model m_k. The surrogate approximation optimization problem can then be stated as

$$\min_{\mathbf{x}\in\Omega}[m_1(\mathbf{x}), m_2(\mathbf{x}), \ldots, m_l(\mathbf{x})]^T \qquad (6.28)$$

This problem could be solved directly with evolutionary methods such as NSGA-II [121] or III [122], or different sampling strategies can be developed to exploit the information provided by each surrogate. For example [328] introduced an algorithm that uses several different sampling strategies that aim at, e.g., improving the approximate Pareto front locally, filling large gaps in the Pareto front, and exploring the extrema of each objective function. Other sampling strategies can be used, and if one or more of the objective functions are computationally cheap or given analytically, then surrogate models should not be used for these functions. Rather, their easily available function values should be exploited first when deciding where in the search space to evaluate the expensive objectives next. Multi-objective optimization problems may also come with additional constraints, which can be computationally cheap or expensive. In these cases, similar strategies as described in Section 6.3.2 can be employed (see also [436]).

6.4   HOW GOOD IS MY OPTIMIZATION ALGORITHM?

Since surrogate models and active learning methods are intended to solve computationally expensive black-box optimization problems, most do not have convergence guarantees or only convergence in probability. Most sampling methods guarantee that no point is evaluated more than once (duplicate sampling for deterministic problems is a waste of computing time and causes problems when fitting the surrogate models). If sampling indefinitely, they therefore will eventually sample the space densely and find the global optimum [487, 434]. In practice, the computational budget is very restricted and therefore it cannot be guaranteed that the final solution is an optimum or even a stationary point. However, compared to the best solutions known to domain experts for their specific applications, the solutions returned by surrogate model algorithms are usually significantly better.

When developing surrogate model-based optimization algorithms, it is important to compare their performance to both state-of-the-art numerical optimization approaches and approaches used in the specific application in order to analyze sampling efficiency and effectiveness as well as demonstrate the potential benefits that advanced optimization methods can have. In computationally expensive optimization, the goal is to find the best solutions within the least number of function evaluations, which motivates the development of sophisticated sampling methods. However, in many applications, researchers rely on random and/or grid sampling. As the name indicates, in random sampling, points that are randomly chosen in the search space are evaluated with the objective function, and the best-evaluated point is declared the optimal solution. In grid sampling, each dimension of the parameter space is subdivided into intervals and the objective function is evaluated at each grid point. The point with the best objective function value is declared the optimal solution. However, neither approach allows to hone in on interesting regions of the parameter space. No local refinement is carried out in search of improvements. Moreover, these approaches tend to over-sample uninteresting regions of the search space, thus leading to a waste of computing resources. Due to the lack of any adaptive sampling in these methods, one can assume that more sophisticated optimizers will easily outperform them and a numerical comparison to random and grid sampling is usually not sufficient or informative in regards to an algorithm's efficiency. As research in numerical optimization algorithms progresses, it is therefore important to raise the bar and compare to the most successful algorithms that exist. For example, one successful optimization method is mesh adaptive direct search [29] which has seen lots of extensions to increase the types of problems that can be solved including problems with categorical variables and multiple objective functions [30, 6]. NOMAD [256] is an excellent implementation of the MADS algorithm.

Other widely used methods for problems that do not offer derivative information include evolutionary methods [484]. However, evolutionary algorithms generally
require a very large number of function evaluations. This is often impractical when
function evaluations are time-consuming. However, also these methods have been
paired with surrogate models in order to alleviate the computational burden [112].

Generally, we recommend clarifying a few questions before deciding about the
type of solution algorithm to employ. These include

- How many dimensions does my optimization problem have?

- How many function evaluations can I afford?

- How many objective functions do I have? Are they computationally expensive?

- How many constraints do I have? Are they computationally expensive?

- Are integer constraints present?

- Is there uncertainty in the objective or constraint function evaluations?

Answers to these questions and appropriate mathematical modeling of the optimization problem are necessary to decide which solution algorithms are most likely to be
successful. For example, surrogate model-based methods as described in the previous
sections are often not the most efficient for solving high-dimensional problems. An
initial experimental design is needed to fit a first surrogate model, and in rare cases,
this design can already exhaust the function evaluation budget. In these cases, a local
search around a solution that is known to perform well may be the best bet for finding
improvements. If a large number of constraints is present and/or the feasible region is
known to be small, it is advised to include any known feasible solutions in the initial
experimental design because otherwise, the surrogate model algorithms may require
too many evaluations to find the feasible region before any improvement attempts
can happen. Ideally one would make the upper and lower bounds of the variables as
tight as possible, which can help accelerate the sampling. As alluded to previously,
the more problem-specific information a surrogate model algorithm can be equipped
with, the better.

6.5 SURROGATE MODELING IN AUTONOMOUS EXPERIMENTATION

The methods described in the previous section are extremely generalizable and have
been used to address problems in climate science, quantum computing, high energy
physics, materials science, design tasks, etc. Although mostly focused on problems
where simulation models are used to evaluate the objective and constraint functions,
there are opportunities to apply the methods to autonomous laboratory experimentation.

Differences between simulation and experimental applications (besides how data
is generated, computer vs. experiments) lie in the type of errors and uncertainties
as well as the amount of data that can be generated. In simulations, errors and uncertainty arise, for example, due to assumptions being made in modeling, resolution,

and simplification of physics but also the ground truth data (and its measurement errors) used to calibrate the simulations. In experiments, measurement errors and uncertainties as well as deviations from adjusting experimental conditions exactly are predominant. There is often spatial and temporal heterogeneity between the errors and uncertainties, which must be accounted for when using surrogate models for optimization and uncertainty quantification. Thus, in order to carry the ideas from surrogate-based optimization for simulations to autonomous experimentation, one can start by adapting the surrogate models. For example, the use of heteroskedastic kernels in GPs may be fruitful (see also the illustration in Figure 6.3). Alternatively, if the variances of the errors are known across the search space, one may explore avenues in which a dedicated surrogate model is used to approximate and predict these errors. The sampling approaches for different types of problems (integer, constrained, multi-objective, etc.) can then be used in conjunction with the new surrogate models.

Alternatively, new sampling methods can be devised that directly take into account the goals of the specific experiment. For example, one could sample such that the uncertainty is minimized across the search space or the information gain achieved with the new experiments is maximized. In contrast to the data-starved regimes in which the surrogate modeling strategies described in the previous sections operate, some lab experiments allow for high-throughput settings, i.e., it is possible to collect tens to hundreds of new data points in each iteration of the algorithm and the total number of experiments that can be conducted is in the thousands. Even in these settings, active learning and surrogate modeling are likely more successful than widely used trial and error or random sampling methods. Due to the lack of an analytic description that maps experimental settings to the observed quantities, we still need an approximation function. While "standard GPs" and RBFs may not be computationally the most efficient maps, advances in machine learning (ML) can be exploited. The ML model can take as inputs the various experimental configurations and quickly predict the quantities of interest everywhere in the search space (the ML model acts as the surrogate in Step 2a in Figure 6.1). The sampling method (Step 2b in Figure 6.1) must be adjusted to allow for the collection of a large number of new experimental configurations. The set of new sample points must not only aim at optimizing some target function but also at gaining as much information as possible in each iteration of the algorithm. Figures 6.11 and 6.12 illustrate how different approaches to batch sampling can lead to different amounts of information gain and improvements in the surrogate model's global accuracy.

One of the challenges to be addressed in autonomous experimentation is the seamless integration of experiment and computation, without having to involve a human in the loop who carries samples from the lab bench to characterization and then manually moves data from characterization tools to a computer. Ideally, labs can be built such that these transitions happen automatically. This may require new lab spaces, and new in situ characterization techniques in order to remove the sequential nature of the process and thus accelerate the scientific discovery, and it requires new processes specifically developed for the application under consideration. On the other hand, we expect that surrogate modeling and active sampling techniques will remain

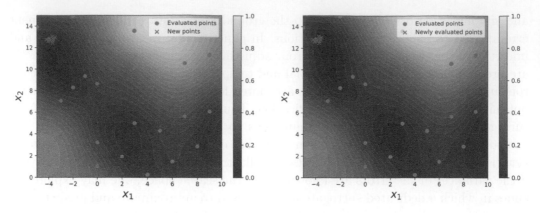

Figure 6.11 Batch sampling does not maximize information gain across the space. The surrogate model (contours) is barely updated after the new points (green crosses in the left figure) were evaluated and added to the surrogate model (compare contours on the left vs. contours on the right).

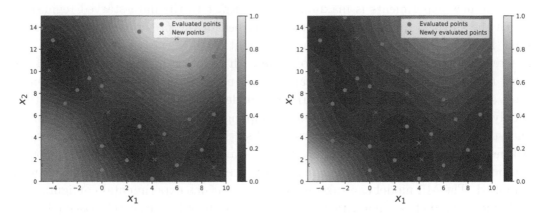

Figure 6.12 Batch sampling maximizes information gain across the space. The surrogate model (contours) changes significantly after the new points (green crosses in the left figure) were evaluated and added to the surrogate model. Compare the contours on the left vs. the contours on the right. Also compare the contours on the right to the contours in the right figure of Figure 6.11.

generally applicable across domains with low needs for algorithmic adaptation and changes.

6.6 FINAL REMARKS

Although the ideas of surrogate modeling have been around since the 1950s [69] (introduced as response surface methodology), research in the area is still ongoing and the types of problems that can be successfully addressed with surrogate modeling and active learning keep increasing. There has been a lot of research for solving continuous problems, but less so for integer or categorical variables. Similarly, problems with multiple levels of simulation (or experimental) fidelity are currently finding a lot of attraction, which may be motivated by the accessibility to diverse computing resources. While increasing computation power has allowed us to improve and refine simulations by modeling more physics and using finer resolutions, tuning the parameters remains a challenging task, and lower-fidelity models that run quicker can provide us with insights into the behavior of the high-fidelity models.

More research is also needed in optimization under uncertainty where uncertainty can arise from different sources, including the data used for calibration tasks or stochastic dynamics of the simulations. Using a fully Bayesian approach for solving the inverse problem becomes computationally intractable when function evaluations are expensive and the number of parameters to be inferred is large. While some research is underway [598], new sampling approaches are needed that adaptively acquire good estimates of the objective function values and that take into account specific problem characteristics.

ACKNOWLEDGMENTS

This work was authored by the National Renewable Energy Laboratory, operated by Alliance for Sustainable Energy, LLC, for the U.S. Department of Energy (DOE) under Contract No. DE-AC36-08GO28308. Funding was provided by the Office of Science, Office of Advanced Scientific Computing Research, Scientific Discovery through Advanced Computing (SciDAC) program through the FASTMath Institute. The views expressed in the book chapter do not necessarily represent the views of the DOE or the U.S. Government. The U.S. Government retains and the publisher, by accepting the article for publication, acknowledges that the U.S. Government retains a nonexclusive, paid-up, irrevocable, worldwide license to publish or reproduce the published form of this work, or allow others to do so, for U.S. Government purposes.

CHAPTER 7

Artificial Neural Networks

Daniela Ushizima

Applied Mathematics and Computational Research Division, Lawrence Berkeley National Laboratory, Berkeley, California, USA

CONTENTS

A RTIFICIAL Neural Networks (ANNs) were invented to mimic the structure and function of the human brain and its ability to learn, recognize patterns, and make decisions. In the late 1940s, researchers explored the idea of creating mathematical models that could learn and perform tasks inspired by the way the brain works. The result was a model consisting of a weighted sum of inputs passed through a non-linear function to generate an output. Among these researchers, there were two of the early pioneers of ANNs: Warren McCulloch, a neurophysiologist, and Walter Pitts, a logician, who proposed a model of artificial neurons in 1943 called the McCulloch-Pitts neuron [310] that could take binary inputs and produce a binary output. Based on this model, but instead capable of working on real-valued inputs, Frank Rosenblatt [450] designed the perceptron for supervised learning of binary classifiers, in which supervision refers to the process of feeding labeled data to an algorithm. Together with backpropagation algorithms [454], these methods served as the foundation for later developments in neural network research, discussed in Section 7.1-7.4.

DOI: 10.1201/9781003359593-7

Machine learning techniques such as ANNs involve two phases: **training** to construct a data-driven model and **inference** to use these models for prediction, with the training phase being the more computationally expensive. While ANN models require training (see Section 7.3), they are not restricted to supervised tasks; this means that ANNs can be used for clustering, for example with Kohonen maps [238] or Autoencoders [8], which are two classic ANN architectures reliant on creating a lower dimensional representation of the input data for clustering.

ANNs have proven to be particularly useful in tasks that involve large amounts of data and complex relationships between variables. Over the years, ANNs have been used in a wide variety of applications, including pattern detection, image and speech recognition, natural language processing, and robotics [218]. While the accuracy of these data-driven models is highly dependent on the size and quality of the datasets, other aspects such as architecture complexity, training and optimization protocols, and metrics for performance evaluation are essential to create and select the right model for certain applications, as described in Section 7.2-7.5. Section 7.7 discusses the current challenges in applying ANN to surveillance using systems centered in computer vision.

Today, ANNs continue to be an active area of research, particularly for deep learning and in the design of new architectures and algorithms that can improve accuracy, speed, and/or enable new applications. This chapter aims to summarize key points about ANN as well as describe use-cases in which ANNs play a major role in the automation of experimental data analysis.

7.1 BUILDING BLOCKS OF ANNS

The mathematical neurons are the basic building blocks of ANNs [211, 320], with each neuron receiving inputs from other neurons, then computing a weighted sum of n inputs, x_j for $j = 1, 2, .., n$. This will produce an output which can equal to 1, if the sum is above a certain threshold w_0 or 0 otherwise, being this range when using activation functions such as sigmoid or softmax:

$$\tilde{y} = \theta \left(\sum_{j=1}^{n} w_j x_j - w_0 \right) \tag{7.1}$$

where w_j is the synapse weight associated with the jth input, and $\theta(.)$ is the activation function. This equation only holds for a single layer, and defines an ANN architecture that often works for simple examples, while a stack with many layers on top of each other is usually considered for complex ANN applications. As illustrated in Figure 7.1, ANNs can be viewed as a weighted directed graph, with nodes corresponding to neurons and edges to weights. ANNs are composed of interconnected layers of artificial neurons, following eight main concepts [426, 320]:

- Input Layer: This is the layer that receives the input data x_j as in Equation 7.1, which can be in the form of text, images, audio, or other types of data.

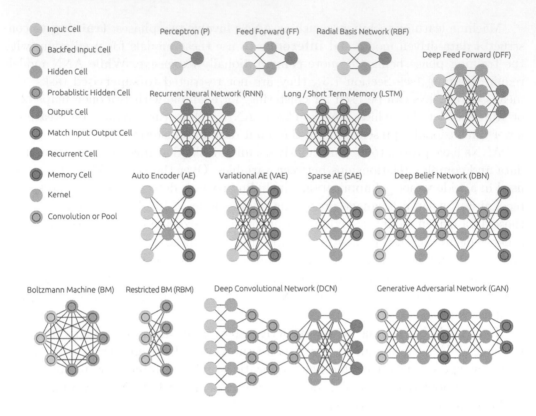

Figure 7.1 Diagram of main ANN architectures and their specialized nodes: modified from van Veen and Lejnen [543].

- Hidden Layers: These are the intermediate layers between the input and output layers. Each hidden layer consists of multiple neurons that perform computations on the input data.

- Output Layer: This is the layer that produces the output or prediction based on the input data. The number of neurons in the output layer depends on the type of problem being solved. For instance, the output layer may have a single neuron in a regression problem, or a single neuron with an activation function in a binary classification, or multiple neurons, one for each class, and a softmax function to predict a multinomial probability distribution.

- Activation Functions: These are functions that determine the output of a neuron based on the weighted sum of the inputs. In Equation 7.1, it is represented by the θ function. Common activation functions include the sigmoid, softmax that have an output range between 0 and 1. Notice that the Rectified Linear Unit (ReLU) activation function, which is commonly used in neural networks, has an output range from 0 to positive infinity. The Leaky ReLU and Parametric ReLU (PReLU) variants of ReLU allow for negative values as well. Other activation functions like the Hyperbolic Tangent (tanh) function have a range between −1 and 1. The non-linearity in artificial neural networks (ANNs) arises from the activation functions applied to the neurons within the network. Activation functions introduce non-linear transformations to the input data or the output of the previous layer, allowing the network to model complex relationships and capture non-linear patterns in the data.

- Loss Function: It measures the difference between the predicted output and the target output. The loss function is used to guide the optimization process during training [426]. Frequently, the use of the term loss function refers to the error for a single training example, and cost function refers to the average of the loss functions over the entire training dataset.

- Weights: These are the parameters that determine the strength of the connections between neurons. The weights w_j in Equation 7.1 are adjusted during the training process to minimize the error between the predicted output $\hat{y}$ and the target output. This means that weights are chosen to minimize the loss function and biases act as a thresholding for the activation function.

- Biases: These are values act as an activation threshold, and they are constant within each layer and added to the weighted sum of the inputs to each neuron - see w_0 in Equation 7.1. The bias allows the neuron to activate even when all the input values are zero.

- Optimization Algorithm: It adjusts the weights and biases of the neurons to minimize the loss function. Common optimization algorithms include gradient descent, Adam, and RMSProp, which are further discussed in Section 7.3.

These main parts of an ANN can be adjusted by changing its structure, such as number of neurons in a layer, and tuning parameters, such as the learning rate, to optimize its performance for a specific task. Such variations can give rise to different types of ANNs, some with special architectures and characteristics, as categorized in the next section.

7.2 NEURAL NETWORK ZOO

The diversity of ANNs has been referred to as a neural network zoo due to the rate in which new architectures and new acronyms are created. Over the past years, members of the Asimov Institute designed a series of charts [543] in an attempt to lay down common notations and create a taxonomy that represents the avalanche of models. Adapted from one of their diagrams, Figure 7.1 summarizes some of the main architectures, represented as a combination of key nodes, which can be input, output, hidden, probabilistic hidden, recurrent, memory, kernel, convolutional or pool cells. Additionally, this section presents a description of each of these components, except input, output, and hidden layers, presented in the previous section.

- Probabilistic Hidden Layer: This layer introduces stochasticity in the network by incorporating probabilistic models, such as Restricted Boltzmann Machines (RBMs) or Gaussian Belief Networks. These models allow for more flexibility in capturing complex dependencies and learning latent variables in the data.

- Recurrent Layer: These layers have connections that allow information to flow in a loop, enabling the network to retain and utilize information from previous time steps or iterations. Recurrent Neural Networks are commonly used for sequential data processing tasks like natural language processing and speech recognition.

- Memory Cells: Memory cells, such as Long Short-Term Memory cells or Gated Recurrent Units (GRUs), are specialized recurrent layers that have the ability to store and retrieve information over longer sequences. Memory cells are particularly useful in tasks that require modeling long-term dependencies or handling sequential data.

- Kernel/Convolutional Cells: Kernel or convolutional cells are specific to Convolutional Neural Networks (CNNs) used for image or signal processing. These cells use convolutional operations to extract local patterns and features from the input data, enabling the network to learn spatial hierarchies and translation-invariant representations.

- Pooling Cells: Also known as pooling layers, they are commonly used in CNNs to reduce the spatial dimensions of the data. Pooling operations, such as max pooling or average pooling, aggregate information from local regions to create a more compact representation. This helps in reducing the computational load and extracting robust features from the input.

Figure 7.1 emphasizes the graphs with cells that perform some computation over the inputs; it simplifies the architectures in terms of its key nodes, and understandably, it omits important interactions between nodes. For completion, six broadly used ANN architectures are further described below:

Feed-forward Neural Networks (FF): These are the most basic type of ANNs still being applied to regression and classification problems. Multi-Layer Perceptrons (MLPs) are a specific type of FNN that has at least one hidden layer. In FFs, the information flows in a forward direction through the network, as opposed to having cyclic feedbacks through neurons, such as recurrent connections (see RNN and LSTM in Figure 7.1), which enables modeling dynamical systems with memory. The FF output is calculated based on the input data and the weights between the neurons. If such an ANN has more than one hidden layer, it can be referred as being a deep neural network [426], however the term is often applied to more complex architectures as those described as follows.

Convolutional Neural Networks (CNN): Also known as convnets, these ANNs are commonly used for image recognition and processing tasks. CNNs are made up of multiple layers of neurons, including convolutional layers, pooling layers, and fully connected layers. CNNs can learn spatial hierarchies of features, particularly due to its ability of tuning a set of learnable filters that slide over the input data, typically an image. The filters detect local features such as edges, corners, and shapes in the image, and then pass them to the next layer in the network. As a result of an iterative training procedure, the hidden layers gain the ability to represent important

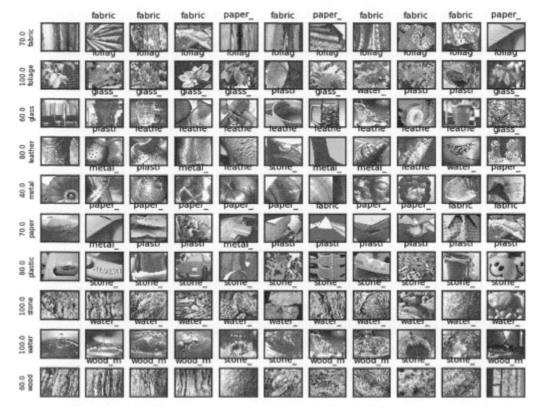

Figure 7.2   Classification using CNN featurization for image recommendation given queries (images in first column) to a large database with images from different materials [533, 24].

features of the task domain. The Figures 7.2 to 7.7 show applications of CNN to image classification and segmentation, with design details discussed in Section 7.7.

Recurrent Neural Networks (RNN): These have been used for sequence processing tasks, such as natural language processing and speech recognition. RNNs [139] use feedback connections to allow information to be passed between neurons in previous time steps. One of the main criticisms is that RNNs can suffer from vanishing gradients, meaning that the network can fail to capture long-term dependencies in the sequence.

Long Short-Term Memory Networks (LSTM): Similarly to RNNs, these ANN are particularly useful for processing long sequences of data. The advantage of LSTMs [191] over the more general RNNs is the ability to solve the vanishing gradient problem. LSTMs have "memory cells" and "gates" that allow the network to remember past inputs and selectively update their memory based on new inputs.

Autoencoder Neural Networks (AE): These are used for unsupervised learning tasks to address dimensionality reduction, feature extraction, and even data augmentation because they can create efficient encodings from high-dimensional data. Autoencoders [170, 14] learn to compress input data into a lower-dimensional representation known as the "latent space", and then reconstruct the original data from this representation.

Generative Adversarial Networks (GAN): Just like AE, these ANNs are used for unsupervised learning, however, they are designed for generative tasks, such as image and video synthesis. GANs [171] consist of two neural networks that compete with each other: a generator network that creates new samples, and a discriminator network that tries to distinguish between real and fake samples.

There are many other architectures and variations beyond those illustrated in Figure 7.1, for example, an architecture called **Transformer** [546] that was introduced in 2017 and quickly became the most popular algorithm for natural language processing (NLP) tasks, such as language translation and text classification, outperforming previous ANN architectures. Each type of ANN, including Transformers, is designed for specific tasks and can often be optimized (see Section 7.3) and customized (see Section 7.4) for different applications. For example, new implementations of Transformers have been created to solve computer vision problems, giving rise to vision Transformers [132], further discussed in Section 7.5.

7.3 TRAINING AND OPTIMIZATION

Training an ANN requires choosing an optimization function to improve the ANN performance. The key concept is the minimization or maximization of some loss function $f(a)$ by changing a, which can be represented by $a^* = arg\ min\ f(a)$, given a starting value a. The derivative, $f'(a)$ or $\frac{df}{da}$ denotes the slope of f at the point a, which specifies how to scale a change in the input x to calculate the corresponding change in the output $\hat{y}$. For example, the derivative indicates how to change a to minimize $f(a)$, and a technique called **gradient descent** will reduce $f(a)$ by moving a in small steps with opposite sign of the derivative [170].

Most NN frameworks provide efficient optimization modules to search for the parameters that minimize a cost function. Thanks to the efficiency of current methods allied to the easy-to-use modules, developers can quickly optimize a neural network with multiple layers by running backpropagation [454] (a.k.a. backprop) algorithms that compute gradients of the loss function with respect to the weights of the ANN for an input-output example. The backpropagation algorithm is usually implemented using matrix operations to efficiently compute the gradients and update the weights.

A pseudocode for backpropagation is described in Algorithm 1. Notice that the choice of loss function depends on the type of problem being solved, e.g., mean squared error for regression, cross-entropy for classification as further described in Section 7.5. The gradient of each layer with respect to its inputs and weights can be computed using the chain rule of calculus and the derivative of the activation function used in each layer.

There are many optimization algorithms to choose from, but other factors such as the learning rate can also be tuned to optimize the convergence and accuracy of the ANN. Here are some of the strategies used to optimize ANN [60, 170]:

Stochastic gradient descent (SGD): This is a popular optimization technique used in developing ANNs. It involves calculating the gradient of the loss function with

---

Algorithm 1 ANN training with backpropagation

Require:

1: $X \leftarrow$ Training data with n examples
2: $Y \leftarrow$ Labels for examples in X
3: $W \leftarrow$ Weights for layers, initialized with random values
4: $l \leftarrow$ Number of layers
5: $f \leftarrow$ Loss function
6: $\lambda \leftarrow$ Learning rate hyperparameter
7: $e \leftarrow$ Number of epochs
8: **for** each training example (x, y) **do**:
9:     Feed the x through ANN and compute $\hat{y}$
10:     Calculate the error between $\hat{y}$ and y using f
11:     Compute ∇f with respect to the weights using backpropagation
12:     Update w using an optimization algorithm (e.g., gradient descent):
13: **end for**
14: Update W
15: Repeat for e iterations or until convergence

---

respect to the weights of the network using a random subset of the training data (a mini-batch), and then updating the weights in the opposite direction of the gradient. SGD can be extended with momentum, which accumulates past gradients to move more smoothly toward the optimum.

Adaptive learning rate methods: Adaptive learning rate methods, such as Adagrad, Adadelta, Adam, and RMSProp [586], adjust the learning rate during training based on the magnitude of the gradients. These methods can improve the convergence speed and stability of the optimization process, especially for deep neural networks.

Batch normalization: This is a technique that normalizes the activations of each layer to ensure that the mean activation is close to zero and the standard deviation is close to one. Batch normalization can improve the convergence speed and generalization performance of neural networks.

Weight regularization: This method adds a penalty term to the loss function that discourages the weights from taking large values. This can prevent overfitting and improve the generalization performance of the neural network. L1 and L2 regularization are the most commonly used types of weight regularization.

Dropout: This technique randomly drops out some of the neurons during training. This can prevent overfitting and improve the generalization performance of the neural network.

Early stopping: This algorithmic option stops the training process before the neural network overfits the training data. This is done by monitoring the validation loss and stopping the training process when the validation loss stops decreasing.

The choice of optimization technique depends on the specific problem and the type of ANN architecture under consideration. While the appropriate optimization technique can improve the performance and stability of the ANN, their settings are as relevant, so fine-tuning the hyperparameters of the optimization technique often determine the best performance.

7.4 METRICS OF ANN'S PERFORMANCE: REGRESSION VS. CLASSIFICATION

There are different ways to measure the ANN's performance, for example, computational speed during training, energy consumption during computation, number of GPUs and/or CPUs during parallelization, and others. Prediction accuracy is a broadly accepted ANN performance metric that depends on several factors, such as the continuity of outcomes. If so, then regression algorithms should be selected, otherwise classification methods are considered for the generation of a discrete set of possible values. This section describes performance metrics involved in the solution of regression problems and classification.

In **Regression Problems** [170, 211, 581], the output of the neural network is a continuous value, therefore different metrics are used to evaluate the regression performance: **(a) mean squared error (MSE):** the average of the squared differences between the predicted and actual values. **(b) mean absolute error (MAE):** the average of the absolute differences between the predicted and actual values. **(c) R^2:** a statistical measure that represents the proportion of the variance in the dependent variable that is explained by the independent variables.

In **Classification Performance** [170, 230, 537], the ANN categorizes examples into one of a discrete set of possible outcomes, such as different types of materials [24], as illustrated in Figure 7.2. Performance evaluation often includes one or more of the following metrics to select proposed methods: **(a) accuracy:** the proportion of correct predictions over the total number of predictions. **(b) precision:** the proportion of true positives over the total number of positive predictions. **(c) recall:** the proportion of true positives over the total number of actual positives. **(d) F1-score:** the harmonic mean of precision and recall. **(e) confusion matrix:** a table that shows the number of true positives, true negatives, false positives, and false negatives for each class [24].

For other types of problems, such as sequence or image generation and topic modeling [534], different metrics may be used to evaluate performance, such as perplexity or visual inspection. In general, one chooses an appropriate evaluation metric that reflects the goals of the task (see metrics in the next section designed for segmentation tasks) and the nature of the data. Additionally, multiple metrics are often used to get a more comprehensive evaluation of the neural network's performance.

7.5 ANNS FOR SEMANTIC SEGMENTATION

Semantic segmentation is a computer vision task that involves labeling each pixel in an image with a class label, such as object categories [580] or scene semantics. The evolution of ANN for semantic segmentation can be roughly categorized into three phases, further described as follows:

MLP Era (1960s-1980s): The first ANNs were based on the Multilayer Perceptron (MLP) architecture [450], which is a feed-forward neural network with multiple layers of nodes. MLPs were first introduced in the 1960s and were popularized in the 1980s. MLPs are primarily used for supervised learning tasks such as classification and regression. MLPs have limitations when it comes to processing images, as they require flattening the image into a vector, which results in the loss of spatial information.

CNN Era (1990s-2010s): Convolutional neural networks (CNNs) [170] were introduced in the early 1990s, and they revolutionized image recognition tasks, particularly taking advantage of high-speed hardware processing. CNNs consist of several layers of convolutional and pooling layers, followed by fully connected layers. Convolutional layers help preserve spatial information, making them ideal for image recognition tasks. The use of CNNs in image recognition has led to significant advancements in computer vision applications, such as object detection, face recognition, and image segmentation.

Vision Transformer Era (2010s-Present): In recent years, researchers have introduced a new type of neural network architecture called Vision Transformers (ViTs) [132]. ViTs replace the convolutional layers with self-attention layers that enable the network to capture long-range dependencies between image patches. ViTs use a transformer architecture that was first introduced in the field of NLP. ViTs have been shown to outperform CNNs in some image recognition tasks.

While MLPs were the starting point for neural networks, CNNs revolutionized the application of ANN to become one of the most popular deep learning models used for almost every complex (e.g., non-bimodal distribution between foreground and background) computer vision tasks. Both MLP and CNN architectures are widely used in computer vision applications, and more recently, ViTs have emerged as a new architecture for image recognition tasks at scale, with promising results for large image datasets [132, 416].

Each of these architectures has its own strengths and weaknesses, and selecting the latest model does not necessary means that it is the best choice for a segmentation problem. Therefore, it is important that researchers consider metrics to assess segmentation results, for example, by choosing different **loss functions to improve performance of segmentation tasks**. Here are some common loss functions [170, 320] used in semantic segmentation:

Binary cross-entropy loss: This is a common loss function used for binary segmentation problems, where each pixel is either labeled as foreground or background. The binary cross-entropy loss measures the difference between the ground truth label (y) and the predicted probability ($\hat{y}$) for each pixel [119, 416] as illustrated in Equation 7.2. The binary cross-entropy is often calculated as the average cross-entropy across all data examples.

$$\mathcal{L}(y, \hat{y}) = -y \log(\hat{y}) - (1 - y)log(1 - \hat{y}) \tag{7.2}$$

Categorical cross-entropy loss: This is a common loss function used for multi-class segmentation problems, where each pixel is labeled with one of several possible class labels. The categorical cross-entropy loss measures the difference between the ground truth label y and the predicted class probability distribution outcome $\hat{y}$ for each pixel [338] as in Equation 7.3, where m denotes the number of classes, $y_{i,c}$ is a binary indicator equal to 1 if c is the correct class for the observation i, and $\hat{y}_{i,c}$ is the probability of observation i for class c.

$$\mathcal{L}(y, \hat{y}) = -\sum_{c=1}^{m} y_{i,c} \log(\hat{y}_{i,c}) \tag{7.3}$$

Dice loss: Also know as F1-score (Equation 7.4), this is a loss function that measures the overlap between the predicted segmentation mask and the ground truth mask. The Dice loss is based on the Dice coefficient, which is a similarity measure between two sets. The Dice loss can be useful for imbalanced datasets, where one class dominates the other [25].

$$\mathcal{L}(y, \hat{y}) = 1 - \frac{2 * \sum_{i=1}^{n} y_i \hat{y}_i}{1 + \sum_{i=1}^{n} y_i^2 + \sum_{i=1}^{n} \hat{y}_i^2} \tag{7.4}$$

Jaccard loss: This is a loss function that measures (Equation 7.5) the intersection over union (IoU) between the predicted segmentation mask $\hat{y}$ and the ground truth mask y. The Jaccard loss is based on the Jaccard index, which is another similarity measure between two sets. The Jaccard loss is similar to the Dice loss and can also be useful for imbalanced datasets [32].

$$\mathcal{L}(y, \hat{y}) = 1 - \frac{\sum_{i=1}^{n} y_i \hat{y}_i}{1 + \sum_{i=1}^{n} y_i^2 + \sum_{i=1}^{n} \hat{y}_i^2 - \sum_{i=1}^{n} y_i \hat{y}_i} \tag{7.5}$$

It is critical to choose a loss function that reflects the goals of the task and the nature of the data, and to evaluate the performance of the neural network using appropriate metrics and datasets. Another critical choice in the design of ANN is to select an adequate machine learning platform, such as those discussed in the next section.

7.6 MAIN TOOLKITS TO CREATE ANNS

There have been numerous toolkits proposed to create ANNs over the years, with some of the most widely used deep learning platforms being: TensorFlow, PyTorch, Keras, Caffe and MXNet. TensorFlow is an open-source library for machine learning and deep learning developed by Google. It supports both CPU and GPU computations and offers a high-level interface for building and training neural networks. Another open-source machine learning library is PyTorch, under development by Facebook that provides dynamic computational graphs that make it easier to debug and visualize neural networks during development.

There is a community perception that PyTorch is easier than TensorFlow, which is subjective and can vary based on individual preferences and prior experience. Here are a few reasons why some people find PyTorch more user-friendly:

- Dynamic Computational Graph: PyTorch uses a dynamic computational graph, which allows for more flexibility and intuitive debugging. Developers can define and modify the computation graph on-the-fly, making it easier to experiment and iterate during model development.

- Pythonic Syntax: PyTorch is built and designed to seamlessly integrate with Python. Its syntax closely resembles standard Python programming, making it easier for developers to grasp and use. This familiarity can lower the learning curve and facilitate faster prototyping.

- Easier Debugging: With PyTorch, developers can easily access and inspect intermediate values within the network during the debugging process. This feature enables efficient troubleshooting and error diagnosis.

- Strong Community Support: PyTorch has gained popularity in the research community, and it has a vibrant and active user community. The availability of comprehensive documentation, tutorials, and resources, along with active community support, can make it easier for newcomers to get started and find solutions to common problems.

While the choice between PyTorch and TensorFlow depends on the specific requirements of the project and the individual's familiarity and preferences, there are some reported advantages of TensorFlow over PyTorch such as:

- Scalability: TensorFlow is designed to scale efficiently to larger datasets and distributed computing environments, making it a good choice for large-scale projects.

- Pre-trained models: TensorFlow has a wide range of pre-trained models available, which can be used for transfer learning and as a starting point for custom models.

- Production deployment: TensorFlow provides production-ready deployment tools such as TensorFlow Serving and TensorFlow Lite, which make it easier to deploy trained models to production environments.

- Industry support: TensorFlow is backed by Google, which provides strong support and resources for users and developers. This might come at a cost as most developers consider Pytorch to possess a more intuitive and Pythonic API.

Instead, Keras introduces a high-level neural networks API written in Python that runs on top of TensorFlow, Theano (now Aesara), and CNTK. Keras provides a flexible and user-friendly interface for building and training deep learning models as it combines the readability of Python with fast prototyping capabilities, making it competitive in the deep learning space. The oldest of these frameworks is Caffe, a deep learning framework developed by the Berkeley Vision and Learning Center (BVLC) and community contributors, which was released in April 2014. It is optimized for image classification and supports both CPU and GPU computation.

Developed by the Apache Software Foundation in 2015, MXNet continues to be an open-source project, and it has been one of the deep learning toolkits adopted and supported by Amazon, particularly when using Amazon SageMaker. MXNet enables multiple programming languages and can run on CPUs, GPUs, and distributed clusters. For those with a background in MATLAB or Python, Caffe might be a good choice, however MXNet has been considered more flexible because it allows users to write code in Python, R, Julia, and more. It is likely that new deep learning platforms will continue to emerge and gain popularity as the field of deep learning evolves.

The next section shows applications using either Tensorflow or PyTorch to implement deep learning models toward deploying computer vision modules to enable systems to perform tasks more autonomously.

7.7   APPLICATIONS OF ANNS: FROM CHEMICAL COMPOUNDS TO CELLS

Laboratories worldwide have increasingly sought to automate steps in the process of investigating the structure and function of different molecules and materials. To do so, large amounts of data and analysis algorithms have turned raw data into information [130]. By combining ANN with lab automation and robotics, self-driving labs have become the new frontier in accelerating the pace for exploration of sample spaces, for example, in chemistry and material sciences [5]. The next subsections describe examples of current efforts on developing ANN as part of pipelines for self-driving labs.

7.7.1   X-ray Serial Crystallography Classification

In "A Convolutional Neural Network-Based Screening Tool for X-ray Serial Crystallography" [230], a new new deep learning-based tool is designed to screen large amounts of images containing X-ray serial crystallography (XSC) experimental records. The researchers developed a CNN that can analyze diffraction images from crystalline samples and identify which ones are suitable for high-resolution structural analysis. An AlexNet-variant is trained on a large dataset of diffraction images to distinguish between high-quality images and those with poor diffraction quality. The tool can be used to automate the screening process and will help researchers save time and resources in selecting the most suitable samples for XSC experiments. The researchers show the effectiveness of their screening tool by applying it to experimental data (Figure 7.3) from XSC experiments on diverse proteins, imaged at two facilities: the Spring-8 Angstrom Compact free electron LAser (SACLA) in Japan, and the Linac Coherent Light Source (LCLS) at SLAC National Accelerator. The CNN-based tool showed high accuracy (92% for data from Rayonix detector) and efficiency in identifying high-quality diffraction images, demonstrating its potential as a valuable resource for crystallography researchers.

7.7.2   X-ray Scattering Classification

The article "Convolutional neural networks for grazing incidence x-ray scattering patterns: thin film structure identification" [279] presents a new method for using CNNs

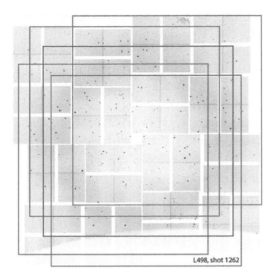

Figure 7.3 Example of X-ray serial crystallography image and augmentation technique designed to train our AlexNet-variant architecture [230].

to identify thin film structures from grazing incidence small angle X-ray scattering (GISAXS) patterns. The researchers developed an algorithm to turn scattering patterns images into high-dimensional vectors, and a model capable of predicting the structure of the thin film based on these new representations. Figure 7.4 shows the

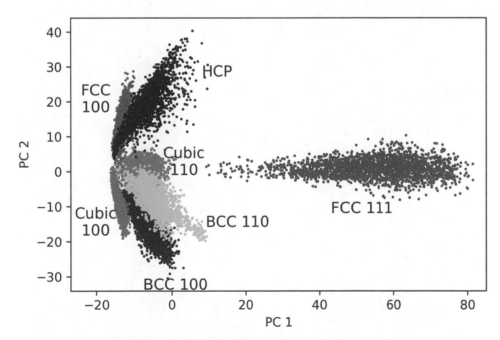

Figure 7.4 Classification of scattering patterns from GISAXS: principal component analysis from CNN-based featurization emphasizes difference between 111 unit cells and other clusters, as well as overlap between Simple Cubic 110 and BCC 110 clusters [279].

principal components of the representations generated by the proposed CNN-based model that was trained on a large dataset of simulated patterns to be able to identify the abstract structures in GISAXS patterns. The researchers demonstrated the advantages of their model by applying it to experimental GISAXS data, and showed high accuracy (98.12%) and efficiency in identifying the structures, making it a valuable resource for studying thin films. The approach has potential to save significant time and resources in thin film structure identification, and could contribute to the development of new materials and devices with desired properties.

7.7.3   X-ray Tomography Semantic Segmentation

Figure 7.5 illustrates one of the fiber detection results discussed in "A reusable neural network pipeline for unidirectional fiber segmentation" [119] paper that proposes a new ANN pipeline for the segmentation of unidirectional fibers from synchrotron-based microtomography with 3D images of ceramic matrix composites. The pipeline

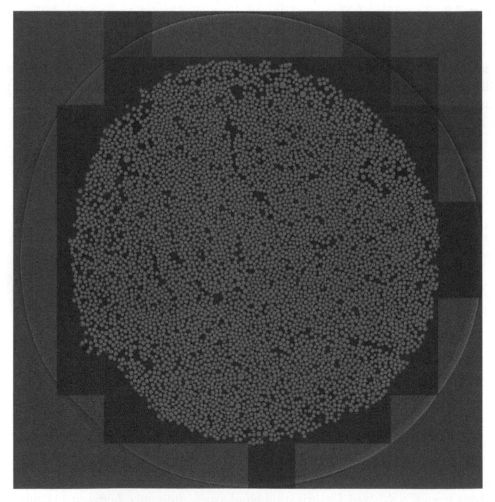

Figure 7.5   Segmentation of fibers from 3D images of ceramic matrix composites acquired with synchrotron-based microtomography [119].

consists of three stages: image preprocessing, fiber segmentation using both 2D and 3D versions of U-net, Tiramisu, and post-processing of the segmented fibers. The researchers used a dataset of high-resolution microscopy images of unidirectional fiber composites to train the CNN, which can accurately segment the fibers from the background in the images. The post-processing step further improves the accuracy of the segmentation by removing false positives and connecting broken fiber segments. The resulting segmented fibers can be used for further strength analysis and characterization of the fiber composites regarding different tensile forces. The researchers demonstrated the usability of their pipeline, e.g., model ability to generalize, by applying it to various fiber composite samples and comparing the results with manual segmentation performed by material scientists. The CNN-based pipelines showed high accuracy with Dice and Matthews coefficients greater than $92.28 \pm 9.65\%$, reaching up to $98.42 \pm 0.03\%$, showing that these automated approaches can match human-supervised methods in segmenting the fibers, making it a valuable resource for materials science researchers studying fiber composites.

7.7.4 Medical X-ray: Radiography Segmentation

Figure 7.6 shows the segmentation and classification results using one of the images in a very large dataset described in "In validating deep learning inference during chest Xray classification for COVID19 screening" [458, 459], which discusses the relevance of chest X-ray (CXR) imaging in the early diagnosis and treatment planning for patients with suspected or confirmed infection due to the high fatality rate of the virus. With the increase in covid cases and limited testing capabilities, new methods for lung screening using deep learning emerged quickly. However, quality assurance discussions have lagged behind. The authors proposed a set of protocols to validate deep learning algorithms, including their ROI Hide-and-Seek protocol, which assesses the classification performance for anomaly detection and its correlation to the presence/absence of radiological signatures. Through systematic testing, results demonstrated the weaknesses of current techniques that use a single image representation to design classification schemes. This work also offered perspectives on the limitations of automated radiography analysis when using heterogeneous data sources, showing that many works during the pandemic were just detecting different data sources [554, 2], not signs of COVID-19.

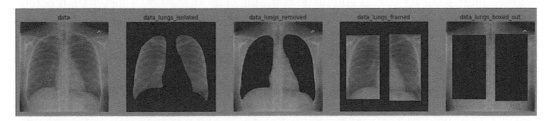

Figure 7.6 Proposed Hide-and-Seek protocol [458]: lung segmentation using U-net over raw image (leftmost) enables creation of 4 extra inputs by determining different region of interest (ROI) for classification tests using ResNet-50.

Input   **Ground**   **Prediction**

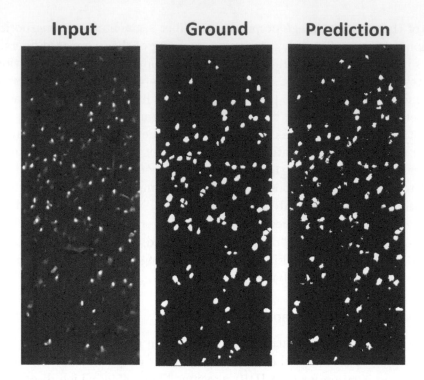

Figure 7.7   Detecting neuron cell immunofluorescence automatically using U-Net-variants. Neurons detected with an accuracy of 97%, recall of 0.81, and 80% faster than previous manual approach [318, 319].

7.7.5 Microscopy for Cell Segmentation

The article "Accelerating Quantitative Microscopy with U-Net-based Cell Counting" [318] presents a new approach to expedite the analysis of microscopy images using a U-Net-based deep learning model (Figure 7.7). The model was trained on a dataset of fluorescently labeled cells and it can accurately identify and count individual neuron cells in images. The success of their approach was measured by testing it against several microscopy images of brain tissue and comparing the results with manual counting previously performed by pathologists. The U-Net-based model showed high accuracy (97%) and made counting cells at least 80% faster than before, outperforming other cell counting methods adopted previously. The study suggests that deep learning can be a tool for accelerating quantitative microscopy and have contributed to advances in fields such as neuroscience [535] and cancer research [338].

7.8 ANNS, PITFALLS, AND THE FUTURE

Artificial intelligence has both enchanted and haunted humanity since the mid-20th century, but it was more of a distant research project or science fiction than a large set of technologies exponentially growing and permeating the fabric of our society at high-speed. Controversial or not, the rise of self-driving labs using ANNs for chemical

Figure 7.8   DALL-E system automatically generates illustrations based on prompts, such as "robot takes over the world" [383].

and materials sciences [5] is undeniable and could be a powerful tool in autonomous experimentation. However, the current landscape compels researchers to detect and mitigate the potential pitfalls that should be considered before deploying predictive and generative models (Figure 7.8) in substitution for human labor.

For example, many models present limited generalization because ANNs are prone to overfitting, which means they can become too specialized to the training samples and fail to generalize to new, unseen data. This can be a problem in autonomous decision making where the ANN is expected to make decisions in a dynamic and changing environment. One of the common practices against overfitting is to assign responsibility to the data quality and the need to "bring more data" as ANNs do require large amounts of high-quality data to be trained effectively. In autonomous experimentation, it can be challenging to obtain sufficient amounts of high-quality data that accurately represent the diversity and complexity of the environment, particularly when samples under investigation are rare or unique.

The potential for limited generalization is oftentimes worsened by scrapping data from non-curated public sources, which can lead to a high chance of skewing the analysis inadvertently. Automated models fed with low-quality data, and in the absence of accountability for model creation can induce serious issues, with ANNs making mistakes and producing unexpected results, which will be dangerous in autonomous experimentation settings. Ensuring the safety and reliability of ANN systems will require careful data selection, model training, testing and validation, and will require humans in the loop.

Another criticism has been the interpretability of ANNs, often deployed as "black-boxes" that can be difficult to interpret since it is hard to understand why the system is making certain decisions. This can be a problem in scientific experimentation when it is mandatory to recognize the reasoning behind the system's actions. To make matters worse, the computational complexity of ANNs can be expensive, requiring days to train and run, which can be an impediment to their usage in real-time autonomous experimentation where decisions need to be made quickly and on the spot.

Finally, there are several ethical and social concerns about the use of ANNs, such as privacy, security, and fairness, that need to be carefully considered. As this chapter is written, an open letter has been published calling for all AI labs to immediately pause for at least 6 months the training of AI systems more powerful than GPT-4 [597]. While high-tech companies dispute being the leader in AI solutions, AI systems are now competing with humans over increasingly more specialized tasks, with one dire disadvantage: AI systems are not held accountable for errors in the same way that humans are, if at all.

Overall, ANNs can be a powerful tool for autonomous experimentation, but their use requires careful examination of these inherent pitfalls and their impact on the laboratory performance, regulations, safety, and latent ethical implications.

ACKNOWLEDGMENTS

This chapter acknowledges the contributions of all those who participated in this study, particularly Lawrence Berkeley National laboratory (LBNL) CAMERA and MLExchange research teams and former students who are listed in alphabetical order: F. Araujo, A. Badran, E. Chagnon, D. Diniz, Y. Huang, J. Li, C. Liu, S. Miramontes, D. Perlmutter, J. Quenum, R. Sadre, R. Silva, A. Siqueira, M. Trevisan, P. Viseshchitra, and K. Xu. This work was supported by the Office of Science, of the U.S. Department of Energy (DOE) through the Advanced Scientific Computing Research and Basic Energy Sciences program under Contract No. DE-AC02- 05CH11231, LBNL Bridges program and the LBNL Workforce Development & Education program. Any opinion, findings, and conclusions or recommendations expressed in this material are those of the authors and do not necessarily reflect the views of DOE or the University of California.

CHAPTER 8

Artificial Intelligence Driven Experiments at User Facilities

Phillip M. Maffettone, Daniel B. Allan, Andi Barbour, Thomas A. Caswell, Dmitri Gavrilov, Marcus D. Hanwell, Thomas Morris, Daniel Olds, Max Rakitin, and Stuart I. Campbell

National Synchrotron Light Source II, Brookhaven National Laboratory, Upton, New York, USA

Bruce Ravel

Material Measurement Lab, National Institute of Standards and Technology, Gaithersburg, Maryland, USA

CONTENTS

DOI: 10.1201/9781003359593-8

U SER FACILITIES such as the National Synchrotron Light Source II (NSLS-II) at Brookhaven National Laboratory (BNL) are state-of-the-art centers that lead the world in advanced scientific instrumentation. Due to this technological capacity, astronomical data production rates, and combined advances in artificial intelligence (AI) and automation, user facilities serve as a hotbed for materials acceleration platforms (MAPs) [479, 294]. Also known as "self-driving laboratories" (SDLs), these AI-imbued experimental platforms are having an exceptional impact at scientific user facilities that service the broader research community. While NSLS-II came online in 2015 as the brightest light source in the world [44], it is hardly alone in this capacity. With imminent upgrades at the Advance Photon Source, Advances Light Source, and Linear Coherent Light Source and advances in detector technology, the data generation rates at the US Department of Energy (DOE) X-Ray light sources are skyrocketing. These coming upgrades only begin to encapsulate the landscape of user facilities. Broadly, these include other DOE centers in nanotechnology and neutron science, international government-sponsored user facilities, and commercial contract research organizations (CROs) [294]. The DOE light sources alone will be producing data in the exabyte (1 billion gigabytes) scale over the next decade [239]. The need to integrate varying degrees of autonomy into this data collection and analysis is thus clear and compounded by the diversity of experiments and science that a user facility must service.

In the following chapter, we will describe the contemporary efforts for autonomous experimentation at user facilities through the lens of recent achievements at NSLS-II. While we will address particular advancements in the science, we will focus deeper on how a modern scientific user facility can adapt to meet the needs of its user community and make optimal use of its resources—which in this case is a light source 3 times as bright as the sun that can only be utilized for so many hours in a year. First, we will outline the changes in infrastructure that underpins the capacity for developing MAPs and scaling autonomous experimentation platforms. This includes details of internet technologies (IT), networking, data storage, and computational resources. Next, we will explore the *Bluesky* Project for experimental orchestration, which encompasses an *a la carte* scientific Python ecosystem for data acquisition, management, analysis, and advanced feedback loops including machine and human dependencies. This enabling technology has empowered users since first light to drive their experiments and has grown to support cutting-edge AI-driven experimentation that keeps the experts as an important collaborator.

8.1   INTRODUCTION TO NSLS-II AND *BLUESKY*

A scientific user facility is a research center or laboratory that provides access to specialized equipment, resources, and expertise to scientists, engineers, and researchers from academia, industry, and government. The goal of a scientific user facility is to advance scientific knowledge and innovation by providing researchers with access to cutting-edge technologies and resources that they might not have access to otherwise. The DOE Office of Science currently maintains and operates 28 such facilities, including 5 light sources, 2 neutron sources, and 6 facilities dedicated to materials science. Although the DOE user facilities service researchers globally, there are many other centers around the world outside of this scope, including more than 50 light sources worldwide [1]. Beyond these facilities for the public good, there is a growing number of service science organizations in the private sector including experiments on demand and contract research organizations [27]. We turn our focus here to light sources, as they are exceptionally well-positioned facilities to develop, provide and integrate the technologies necessary for accelerated discovery through analysis MAPs [294].

DOE User Facilities

- Light Sources

    - National Synchrotron Light Source II (NSLS-II)
    - Advanced Photon Source (APS)
    - Advanced Light Source (ALS)
    - Stanford Synchrotron Radiation Light source (SSRL)
    - Linac Coherent Light Source (LCLS)

- Neutron Sources

    - Spallation Neutron Source (SNS)
    - High Flux Isotope Reactor (HFIR)

- Materials Science

    - Center for Functional Nanomaterials (CFN)
    - Center for Nanoscale Materials (CNM)
    - Center for Integrated Nanotechnologies (CINT)
    - Center for Nanoscale Materials (CNM)
    - The Molecular Foundry (TMF)
    - Environmental Molecular Sciences Laboratory (EMSL)

Light sources, such as synchrotron radiation facilities, are among the most prominent types of user facilities, producing high-intensity beams of light for a wide range of scientific applications. NSLS-II is one of the newest, brightest, and most advanced synchrotron light sources in the world [28] (Figure 8.1). NSLS-II enables its growing

---

[1]https://lightsources.org/lightsources-of-the-world/

Figure 8.1 Aerial photograph of the National Synchrotron Light Source II situated at Brookhaven National Laboratory. (Courtesy of Brookhaven National Laboratory.)

research community to study materials with nanoscale resolution and exquisite sensitivity by providing cutting-edge capabilities. A schematic of the facility can be seen in Figure 8.2, showing both existing and available space for future beamlines. The source brightness coupled with advanced multidimensional detector technology leads to experiments being performed i) much faster and with higher throughput than ever before, ii) with higher resolution for both imaging and spectroscopy, iii) with an unprecedented signal-to-noise ratio thereby enabling studies of previously unobservable signals. This has led to an increased rate of data production, amounting to many petabytes of data per year produced at NSLS-II [239].

Together with visiting researchers from all around the world, interdisciplinary teams at NSLS-II uncover the atomic structure, elemental makeup, and electronic behavior of materials. By creating this new, deeper understanding of materials, these research teams advance our knowledge in a wide range of scientific disciplines such as life sciences, quantum materials, energy storage, advanced materials science, physics, chemistry, and biology. NSLS-II enables a collaborative, holistic approach to advance scientific endeavors by offering free access to highly advanced instruments and unique expertise. The instruments are used to reveal the electronic, chemical, and atomic structure as well as function of materials using a broad spectrum of light beams,

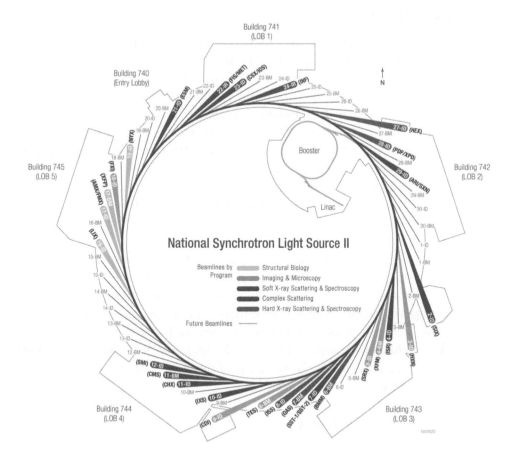

Figure 8.2 Layout of the individual beamlines at the NSLS-II, indicating the different scientific areas covered. (Courtesy of Brookhaven National Laboratory.)

ranging from infrared to hard X-rays. This all is done under incredibly stable and reliable operating conditions.

Despite the huge data generation rates, approaches to experimental control and data analysis have not kept pace. Consequently, data collected in seconds to minutes at a user facility may take weeks to months of analysis to understand. These limitations often divorce knowledge extraction from the measurement process. The lack of real-time feedback forces users into "flying blind" at the beamline, leading to missed opportunities, mistakes, and inefficient use of beamtime as a precious resource—as all beamlines are oversubscribed. This is underscored by the increasing transition to partially or fully autonomous operation for safe and effective experiments. As such, there has been a surge of interest in the optimal use of experimental resources, particularly in applying artificial intelligence and machine learning (AI/ML) [608, 446, 297]. While AI/ML enable experiments at the light source to be performed more efficiently, intelligently, and safely [82], there is a mismatch in expertise and accessibility: the predominant users of beamlines are experts in their scientific domain and not necessarily in AI, computer science, or controls engineering.

8.1.1 *Bluesky's* Ahead

Coincident with the construction of NSLS-II, a common software framework was developed for driving the beamlines: the *Bluesky* Project for data acquisition, management, and analysis [15]. *Bluesky* is a toolbox of individually useful software components that enable experimentalists to build specialized capabilities and developers to collaborate on a shared core. Modeled on the success of the scientific Python ecosystem—which has had an enormous impact in many areas of science—*Bluesky* enables experimental science at the lab-bench or facility scale. The project empowers scientists to meet their own needs and provide unique capabilities while sharing software infrastructure developed by themselves and others. Instruments and facilities typically adopt *Bluesky* piecemeal, extending, customizing, or developing new components as needed. In turn, the community of users continues to benefit from these external developments.

New developments are open source from the very first line of code, which enables facilities to discover potential collaboration points early and avoid wasteful duplication of effort. This open development mode (together with a widely used, industry-standard, OSI-approved open-source license) has contributed greatly to the project's success and has fueled its worldwide adoption. *Bluesky* has demonstrated success in overcoming the technological and sociological challenges of distributed collaboration. As a result, there are now effectively hundreds of developers of *Bluesky* that NSLS-II and other user facilities can benefit from.

Institutions known to be using Bluesky

- NSLS-II

- LCLS (widespread use)

- SSRL (few instruments)

- APS (scaling from a few beamlines to dozens)

- ALS (scaling from a few beamlines to dozens)

- Diamond Light Source (leading new developments, adoption in progress)

- Australian Synchrotron (several beamlines)

- Canadian Light Source (one beamline)

- PSI (evaluating)

- Pohang Light Source II

- BESSY II

- Various academic labs

We will explore the components of this project in depth throughout this chapter, and describe how in tandem with facility infrastructure they can be used to create MAPs or SDLs. The components permeate all aspects of the experimental governance at NSLS-II. As previously noted, experimental orchestration is significantly more challenging at a user facility that provides many different types of experimental end stations, many of which are first-in-class or first-in-world. The *Bluesky* project addresses this diversity through generic interfaces that can be shaped to meet an experiment's need. These interfaces include a toolkit for the facile integration of AI/ML. Since all experiments at NSLS-II already possess some degree of automation, the same toolkit that is driving an automated experiment can be used to upgrade it into an autonomous experiment.

8.2 INFRASTRUCTURE UNDERPINNING ADVANCED EXPERIMENTS

One aspect of a facility that is often overlooked in its importance is that of the underlying IT infrastructure. Many state-of-the-art laboratories are still plagued by isolation and outdated IT infrastructure. Many SDLs rigidly vertically integrate—often confined to a single room or all-in-one instrument—with data transfer accomplished by email or copying files to a portable hard drive. Industry-standard IT enables communication between different and distant systems (e.g., experiments in different places); however, it is costly. There isn't an overabundance of resources for scientists that reduce the barrier to entry to deploying IT solutions such as networking, security, remote access, data storage, and synchronization.

Nonetheless, a common, reliable, and flexible infrastructure is essential for deploying the solutions as described in this chapter, ensuring data that is findable, accessible, interoperable, and reusable (FAIR), and efficient interoperability between instrumentation. This ethos has seen increased adoption in industry [27] and academic labs. Tools like ESCALATE [401] and ARChemist [143] have been built with remote access and distributed data in mind, and others such as ChemOS [447] have been embedded into MAPs with increasingly robust IT. In order to provide this foundation, the NSLS-II has embarked on a multi-million dollar project to modernize its infrastructure and provide a platform upon which complex systems can be built. The main goals that were accomplished are described in the sections below.

Throughout this section and the remaining chapter, we will highlight some of the technological products currently used at NSLS-II. The use of any tool is the result of an evaluation and engineering decision that was suited to NSLS-II at the time of its implementation and is subject to change. The mention of any commercial product does not represent an endorsement by the authors, NSLS-II, the U.S. Department of Energy, or the National Institute of Standards and Technology (NIST). We seek to highlight the design considerations faced when implementing certain tools, as opposed to the prescribing of a particular product.

8.2.1   Networking

The NSLS-II was initially constructed with a non-routable network. This meant that all communication had to be proxied via a 4×1 Gbps connection to the BNL campus network, which itself had an 80 Gbps connection to the outside world. Protocols and programs that couldn't use the proxy were unable to connect outside of NSLS-II. This resulted in multi-hop connections and indirect methods to copy data out of the facility, creating a bottleneck for distributed access and interaction both during and after an experiment.

The networking upgrades have seen the NSLS-II move to now having a routable network that enables connectivity both within BNL and to external sites. The connection is now 4×100 Gbps to the BNL High Throughput Science Network (HTSN) which has 400 Gbps connectivity to the outside world. By moving to a routable network we are now able to connect to resources outside of the local beamline network which opens up the opportunity to make use of central computational clusters, advanced data storage, and the ability to access richer metadata from external information sources (e.g., proposal systems and sample databases). Furthermore, this enables the communication between beamlines within NSLS-II and beyond, opening up opportunities for geographically distributed MAPs [294].

This networking infrastructure also facilitates the exploitation of the Energy Sciences Network (ESNet), which powers high-speed connections (100 gigabits per second) and enables fast communication between different DOE laboratories [2]. ESNet is focused on coupling simulation, AI, and experiments between central facilities. This requires labs to maintain high-speed connections within their own facilities. Clearly, this demonstrates that networking infrastructure is not just an important priority for an experiment or user facility, but for the broad collaboration of advanced instrumentation globally.

8.2.2   Automation and Provisioning

Cyber security is paramount for a scientific user facility to maintain confidentiality, integrity, availability, reputation, and compliance [3]. These facilities often handle sensitive and confidential information, such as research data and proprietary information, which must be protected from unauthorized access. Furthermore, any cyber attack that alters or corrupts data can compromise the validity of scientific findings and harm the reputation of the facility. Since facilities rely on digital systems to operate and provide access to researchers, security must ensure that these are available without undue interruption. While reputation is not the primary motivator for security, erosion of public and community trust is a real danger of a cyber attack. To facilitate cyber security across the DOE, all Office of Science user facilities are subject to regulations and standards that require them to implement robust cyber security measures to protect sensitive information and critical infrastructure [380].

---

[2]https://www.es.net/
[3]https://www.directives.doe.gov/terms_definitions/cybersecurity

In order to provide a consistent and secure computing environment, all of the computers at NSLS-II were reinstalled with an up-to-date and supported operating system, namely Red Hat Enterprise Linux (RHEL).[4] Given the number of computers at the NSLS-II (850 and rising), it is not feasible to reinstall all the machines without implementing some degree of configuration control and automation. For the initial provisioning and life cycle management of machines, we make use of Red Hat Satellite Server [5]. The configuration control is defined using Ansible and automated using Red Hat Ansible Automation Platform. This gives the ability to control the configuration and behavior of all the machines by pushing changes to a git repository that holds the Ansible source files, while still maintaining a safe and secure system.

8.2.3 Deployment Through Virtualization

Not only are there hundreds of computers to manage at NSLS-II, but there are also thousands of applications and environments to manage. Currently, most applications at NSLS-II are managed through virtual machines on centrally managed hardware infrastructure.

The virtualization of computer systems has been a huge leap forward for software of all types, optimizing resource utilization, scalability, disaster recovery, application testing, security, and development agility. This ranges from the developer wishing to verify capabilities on different operating system versions, patch levels, etc., to the deployment of thousands of near-identical systems using automation. VMWare 1.0 was released in 1999 [250], with FreeBSD offering the first version of chroot jails in 2000. VMWare Server was first released in 2006 to go beyond simple desktop virtualization, along with Virtualbox releasing its first open-source edition in 2007. As is clear from the release dates virtualization is a mature technology at this point in time with a proven track record.

NSLS-II makes extensive use of VMWare clusters to offer virtualized instances of predominantly RHEL systems, with Windows and some other images provided by vendors. VMWare at this juncture features advanced networking which has been extensively used to offer access to appropriate VLANs depending on the machine's purpose. VMWare clusters also offer much better hardware utilization; previously, machines would often be largely idle performing specialized tasks. On top of that, the clusters offer automatic failover should any member of the node suffer a hardware failure enabling restoration of services within seconds (less if required and the appropriate hardware/licensing is put in place).

VMWare offers a level of familiarity that helps teams transition away from a historically bare-metal environment, with much better reliability than could ever be offered with Virtualbox which was used for some virtualization in the past. VMWare has continued to grow and innovate into an enterprise product whereas Virtualbox has

---

[4]Certain commercial equipment, instruments, products, or materials are identified in this paper to foster understanding. Such identification does not imply recommendation or endorsement by the National Institute of Standards and Technology, nor does it imply that the materials, products, or equipment identified are necessarily the best available for the purpose.

[5]https://www.redhat.com/en/technologies/management

largely remained a desktop-focused product with few features aimed at production deployment. All of these solutions suffer one particularly high computational cost that makes them better suited to legacy applications: the virtualization of an entire operating system. This leads to all of the costs of booting, installing, and offering virtual hardware. Nonetheless, this cost is often seen as a great strength, as software rarely needs modification and scientists can treat the virtual images in much the same way as they treated the bare-metal systems they are accustomed to.

8.2.4   Outlook on Deployment: Virtualization vs. Containerization

Virtualization and containerization are two different approaches to running multiple applications on the same physical hardware. Virtualization involves creating a virtual machine (VM) that runs its own operating system (OS) and applications, isolated from the underlying physical hardware. Each VM operates as a standalone computer and has its own dedicated resources such as CPU, memory, and storage. Containerization, on the other hand, involves packaging an application and its dependencies into a single container that runs on a host operating system. Containers share the same underlying host OS and hardware resources but are isolated from each other at the application level. This makes containers more lightweight and efficient than VMs, as they do not require a separate OS and do not have the overhead associated with virtualizing hardware resources.

Containers stem from the early use of chroot jails, commonly used to isolate software processes since the early 2000s, along with offering more reproducible software build/deployment environments. Containers offer a full software environment, but they do not virtualize everything and so offer a computationally cheaper way to isolate software services. They are tightly coupled to the Linux software stack and so cannot be used for other operating systems, and software often needs some adaptation to run within a container.

Instead of installing a virtual image into a simulated operating system, containers are built using recipes that can easily be placed under version control. Containers also have a base image they start from, and can be built up in layers. Though it is possible to get a running shell within a container, that is not how they are intended to be used. Another excellent but often unfamiliar detail is that running containers are ephemeral—that is, any changes made in a running image will be lost upon restarting the container and it will revert back to the state it was built in. This offers a huge amount of reproducibility, and for things that should be retained, it is possible to mount storage inside the container, save to databases, or use other approaches. This means that starting a container usually involves passing in some configuration if any given image should run differently, or building in all configurations to a given image. This is where software often needs some adaptation, and there is usually some orchestration that sits above the images. All of these details make using containers excellent opportunities for highly collaborative scientific environments.

One of the most powerful capabilities offered by containers is the ability to build a container image and share that image with the world. That image contains a full binary software stack from the operating system's libraries and up. So when a developer tells you "it works on my system" you now have the ability to ship said system. For science, this is very powerful as it is possible to offer a recipe to build the image along with a binary copy of the built image that anyone can download and run. That image is lighter than a VMWare image, usually without all the licensing issues associated with using a full operating system, and it contains tested, validated software environments that might be newer or older than the host operating system.

Instead of one big VMWare image, containers often favor more focused images to offer different services. Kubernetes is a commonly used open-source platform for automating the deployment, scaling, and management of containerized applications. Many people also use Ansible to configure or interact with containers. Other orchestration technologies include Docker Compose, along with capabilities offered within the Podman project that provides an image-compatible service on Linux systems. Singularity, Shifter, and several other technologies have built upon the beginnings in Docker to offer more specialized containers for use in high-performance computing (HPC) where network isolation is not as important, for example.

For many software deployments, containers are the future. At many of the industry's largest technology companies, they have been the favored deployment strategy for many years. Any RHEL system can run containers using Podman, software services can be deployed using Kubernetes within OpenShift clusters or one of a number of other products. Critically, the images themselves are not so tied to an individual product or project—the image format is an open standard and they can be hosted, converted, and reused.

VMWare and virtualization will likely be used for many years at NSLS-II, but containers can and should be used with more modern software deployments to go beyond what virtualization can offer. There are even standard base container images to use CUDA from within containers that can effectively share valuable resources such as CUDA accelerators to offer faster data processing speeds where experiments strive for faster feedback loops with humans and/or AI. While both virtualization and containerization provide routes to serve the software stack needed for autonomous experiments at scale, containerization is more likely to effectively manage a facility with a rapidly growing diversity of tools and shared resources.

8.2.5 Data Storage and Compute Access

The NSLS-II maintains a dedication to findable, accessible, interoperable, and reusable (FAIR) data. While NSLS-II does not compel users to maintain FAIR data practices, the facility enables and encourages such behavior throughout the data life cycle. All measurements at the facility contain rich metadata and are stored immutably in redundant MongoDB databases for at least one year following all experiments[6].

Databroker [83] is currently used for data access and management, with *Tiled* [44] being deployed as a next-generation tool for rapid access irrespective of data storage approach. *Tiled* also enables the customized storage of analyzed data with rich

---

[6]This period of time for data retention may change in the future, and is governed by the data management policy of the NSLS-II which is available on its website: https://www.bnl.gov/nsls2/

metadata (e.g., a processed spectrum that corrects for background and experimental aberrations, or even the decision-making of an AI/ML agent). Though these tools serve as the foundation for making data findable and accessible, more work is needed to ensure that the data is related to samples robustly to ensure it is inter-operable and reusable, especially if it is to be used by the community for the training of new data-driven models or for scientific reproducibility purposes. Together, these capacities open the door for concerted efforts that engage multiple disciplines in human-machine collaboration, produce FAIR data, and—most importantly—rapidly develop materials for urgent problems in energy and climate.

8.2.5.1 Current Scale of Resources

At the risk that this section will almost certainly be out of date even before the volume that this chapter is published, in this section we will give a brief outline of the scale of resources that are available currently at the NSLS-II. All of the available resources are currently (2023) physically distributed across the facility, with the bulk being situated in the NSLS-II data center or the BNL data center. The central storage capacity for data is approximately 7PB, with additional smaller central storage appliances for standard use cases such as user home directories (for over 4,500 users) and software shares. There are 85 HPC cluster nodes that are split over four different clusters, together with a small number (currently 7) dedicated GPU machines that contain dual A100 (80GB) for workloads that require it. These workloads have the ability to be deployed as containers using either *podman* [7] or *kubernetes* [8]. There are also more than 850 Linux machines across the facility which are used for a wide variety of workloads, from hardware interfacing and data acquisition through to workflow execution and data interpretation.

8.2.5.2 Outlook for Large Scale Computational Resources

The landscape for compute resources has changed over the past decade, where traditional HPC systems were the only feasible choice for facilities (and therefore its users) to have access to large-scale computational and data storage resources. While HPC and traditional batch clusters still have an important role to play, the rise of cloud computing and related technologies has seen a huge shift in the flexibility that can be leveraged by a facility. Large HPC and High Throughput Computing (HTC) deployments are extremely effective for scientific use cases and user communities where there is a fixed number of relatively static workflows, this can be seen for example in the effectiveness of the Large Hadron Collider (LHC) computational and data infrastructure [117]. However, in the case of experimental user facilities, the number of workflows that need to be run changes constantly with the unique demands of each user's experiment. Factoring in the nearly 30 beamlines running simultaneously, the number of potential workflows quickly grows into a number too unwieldy to build individual static workflows on large systems. This is where the flexibility of cloud

---

[7]https://podman.io/
[8]https://kubernetes.io/

technologies comes into play, where we have the ability to spin up and tear down resources and capabilities quickly and easily. By using Software as a Service (SaaS) and Infrastructure as a Service (IaaS) solutions, the definition and specification of the infrastructure can be stored as "code" in a source code repository and therefore has all the benefits this entails, such as revision control and history, and Continuous Integration and Deployment (CI/CD). Different institutions have employed various solutions ranging from using a commercial cloud vendor to a purely on-premise private cloud, but most seem to have a solution that has a mixture of the two. This shift in the industry has not been ignored by the US Department of Energy's HPC user facilities, where they all have substantial efforts to meet this new expectation in users' needs.

The NSLS-II is actively working with colleagues across the other light sources in the DOE complex on a proposed project, DISCUS (Distributed Infrastructure for Scientific Computing for User Science). This would be a joint BES (Basic Energy Sciences) Light Source, ASCR (Advanced Scientific Computing Research), and Laboratory computing effort to develop a computational fabric for scientific computing that spans the data life-cycle are required. We envision a data analysis pipeline capable of analyzing data at its natural production rate. This analysis may be user-driven and interactive or presented to the user as push-button analysis. While not all facilities will require linkages to ASCR facilities to interact with and analyze data, most will need hardware and software infrastructure to stream data from its source to the computing required to analyze it, whether local or remote.

A user facility is unlikely to have the level of funding necessary to provide computational resources that meet the combined peak performance requirements for all end stations. This is the main driver for exploring novel solutions, leveraging collaborations, and sharing resources across the DOE complex. Our philosophy at NSLS-II is to provide a minimum baseline of resources within the facility in order to meet standard operational requirements, and then to "burst" out and make use of additional resources—from an HPC facility or the commercial cloud—during periods of high demand.

8.3 EXPERIMENTAL ORCHESTRATION AND DATA LIFECYCLE

Experimental orchestration, in conjunction with data acquisition, management, access, and analysis is the critical digital component of a scientific user facility. These software segments enable efficient and effective use of the facility's resources, as well as accurate and comprehensive data collection and analysis. Here, experimental orchestration refers to the coordination of all aspects of executing an experiment that collects data, including potential feedback loops. A "self-driving beamline" can be considered an experimental orchestration task that is autonomous. The data lifecycle, on the other hand, refers to the entire process of data creation, storage, processing, analysis, and dissemination. This includes the collection of raw data, its processing and reduction, and its analysis and interpretation. It also includes the long-term storage and archiving of data, as well as the sharing of data with the scientific community.

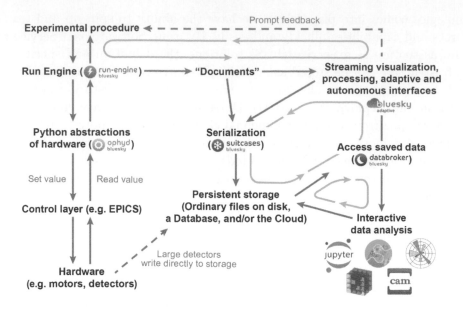

Figure 8.3 Diagram of Bluesky components. The leftmost column is the data acquisition software and hardware, the center column is the data management and storage, and the rightmost column is analysis and visualization, both prompt "live" and posthoc. Because the Run Engine provides a live stream of the data as it is taken the experiment can be visualized or reduced in real time. The system was designed from the ground up to support adaptive feedback to experiments.

As shown in Figure 8.3, the *Bluesky* project approaches these tasks through a multi-facility collaboration that develops and supports a collection of Python libraries. These are co-developed but may be used *a la carte* to leverage existing open-source scientific software (in general) and the scientific Python software ecosystem (in particular) to improvise cutting-edge experiments and data analysis at the beamline. The Bluesky Project is an end-to-end solution, encompassing hardware integration (*Ophyd*), experiment specification and orchestration (*Bluesky Run Engine*), online visualization and analysis, data export (*Suitcase*), data storage (*Databroker*), and data access (*Tiled*). *Tiled* is not explicitly pictured in Figure 8.3, as it is an alpha project that will inevitably replace *Databroker* for fast data access (Sec. 8.3.5). Also not pictured is the *Run Manager*, which is a distributed service to manage the run engine through the *Queue Server* (Sec. 8.3.4).

The following sections will introduce each of these software packages from a high level. For further details and tutorials in using each package, we refer the reader to the rich documentation and tutorials maintained for each package. These will be up to date as the packages grow and change to meet the needs of the scientific community.

8.3.1 Device Abstraction with Ophyd

There is a great diversity of hardware that exists on a beamline—e.g., motors, temperature controllers, pin diodes, 2D detectors, and many more—all of which must seamlessly work together for the beamline to operate and generate scientific data. Within each of those broad classes of devices, there is a further range of vendor and implementation variation. Uniform management of this diversity of instrumentation is extremely challenging. To enable scientists to write generic data collection plans, we have developed the *Ophyd* library to provide a unified interface to the hardware.

Here are some highlights of the Ophyd's features:

- Abstract the details specific to a device, control system, or software system behind a high-level interface with methods like `trigger`, `read`, and `set`.

- Group individual control channels (such as EPICS Channel Access Process Variables (PVs)) into logical "Devices" to be configured and used as units with internal coordination.

- Assign readings with names meaningful for data analysis that will propagate into metadata.

- Categorize readings by "kind" (primary reading, configuration, engineering/debugging) which can be read selectively.

In addition to providing an abstraction to the details of the hardware *Ophyd* is agnostic to the underlying control system. Although many of the facilities that have adopted *Ophyd*—including NSLS-II—use EPICS[9], we have demonstrated implementing `ophyd` objects on top of real hardware in a range of underlying control systems. In addition to EPICS, we have Tango[10], direct communication, and simulations, from naive in-memory signals to sophisticated beamline simulations using Sirepo[11]. Because all of these objects implement the same API, it is possible to use them interchangeably with `plans` and the Run Engine (Sec. 8.3.3), and even to mix them in the same process, without any other code being aware of the differences.

The *Ophyd* documentation details abstraction over the EPICS control system. The EPICS-specific signals are built on top of generic Signal classes (`ophyd.SignalRO` and `ophyd.Signal`). Devices then bundle multiple signals and add specific logic via the `set`, `trigger`, `stage`, etc. methods, extensively being customized in the `ophyd.areadetector.*` modules.

In Table 8.1, we provide a brief—and by no means comprehensive—list of the kinds of device interfaces that have been integrated using *Ophyd*. These resources can serve as a starting motivation to the reader looking to integrate their devices with *Bluesky*. In this section, we will cover in more detail *Ophyd* abstractions over the following hardware/software control systems:

8.3.1.1 *The Sirepo-Bluesky Library*

Sirepo is a browser-based framework that allows for various types of simulations (mainly accelerator physics and physical/geometrical optics codes) [420]. The framework provides a way to programmatically communicate with the backend server using

---

[9]http://www.aps.anl.gov/epics

[10]https://github.com/bluesky/ophyd-tango

[11]https://www.sirepo.com

TABLE 8.1   Ophyd abstraction layers over various communication protocols.

Communication protocol	General category	Example implementation
EPICS	Channel Access / pvAccess	github.com/bluesky/ophyd-epics-devices
Tango	Channel Access / pvAccess	github.com/bluesky/ophyd-tango
RS-232	Serial	[237]
USB	Serial	github.com/BNL-ATF/ophyd-basler
HTTP	TCP/IP	github.com/NSLS-II/sirepo-bluesky
Socket	TCP	github.com/BNL-ATF/atfdb
Labview	Varied	github.com/als-computing/bcs-api

the HTTP(s) protocol and the JSON exchange format between the server and the client parts. The Sirepo-Bluesky library [421] wraps a given Sirepo simulation into a set of conveniently abstracted *Ophyd* objects enabling control and manipulation of various components of the simulation. For example, this includes representing an optical element (such as an aperture or a mirror) as an *Ophyd* Device with multiple components for the corresponding element's properties and parameters.

Those *Ophyd* Devices and their components are dynamically constructed using the JSON model for the corresponding counterparts in Sirepo. For that reason, special Sirepo-aware classes for *Ophyd* Signals and Devices were derived, which allow for coordination of the changes on both the *Ophyd* and Sirepo sides of the integrated system. For example, when a user changes an *Ophyd* value of an optical element's component via the `set` method, the same value is automatically updated in the JSON model responsible for a given simulation.

Currently, three simulation codes are integrated with the Sirepo-Bluesky library: SRW for physical optics simulations [88], Shadow3 for geometrical optics simulations [465], and MAD-X for the accelerator beamlines calculations [89]. The first two codes share a very similar model that represents a beamline as a sequence of optical elements and the corresponding information about the source of X-rays. This feature enables an easy conversion of a beamline simulation from one type to another on the Sirepo level (with some limitations) and a convenient way to compare results. Besides optical elements, a "Watchpoint report" exists in the SRW and Shadow3 applications of Sirepo. This is a simulated equivalent of a physical 2-D area-detector and is implemented in the Sirepo-Bluesky library as a detector object allowing to stage, trigger, and unstage the simulated detector in the same fashion as a physical detector on a beamline (Figure 8.4).

There are other report types that represent other flavors of output data from the simulation codes (such as a 1-D spectrum report). The Sirepo-Bluesky library enables users to switch between the physical and digital twin beamlines smoothly and empowers them to prototype data acquisition plans and analysis tools before their beam time. The simulated backends can be helpful for prototyping offline beamline optimization/alignment tools and applying them later to a physical beamline, saving valuable time for experiments. The MAD-X simulation code's model is significantly different, but it also fits into the *Ophyd* abstraction using the Sirepo-Bluesky

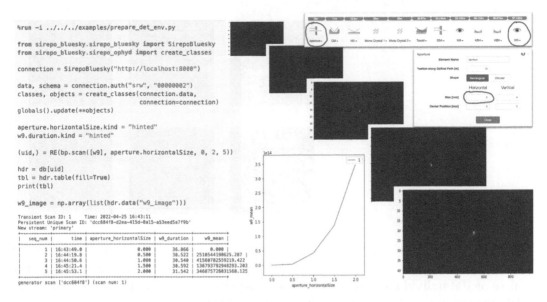

```
%run -i ../../../examples/prepare_det_env.py

from sirepo_bluesky.sirepo_bluesky import SirepoBluesky
from sirepo_bluesky.sirepo_ophyd import create_classes

connection = SirepoBluesky("http://localhost:8000")

data, schema = connection.auth("srw", "00000002")
classes, objects = create_classes(connection.data,
                                  connection=connection)
globals().update(**objects)

aperture.horizontalSize.kind = "hinted"
w9.duration.kind = "hinted"

(uid,) = RE(bp.scan([w9], aperture.horizontalSize, 0, 2, 5))

hdr = db[uid]
tbl = hdr.table(fill=True)
print(tbl)

w9_image = np.array(list(hdr.data("w9_image")))
```

Figure 8.4   A snippet of the code to execute a *Bluesky* scan to capture the intensity distribution on the Sirepo-Bluesky abstracted detector w9 (Watchpoint report) while scanning the horizontal aperture size of the virtual beamline in Sirepo between 0 and 2 mm (on the left), the live table and live plot representing the progress of the scan (bottom), the resulting intensity distribution (the diagonal images), and the beamline layout in the Sirepo interface, where the scanned objects are circled (top-right).

library smoothly, allowing to extracting of the valuable electron beam propagation information programmatically and manipulating it using a familiar *Ophyd/Bluesky* approach.

8.3.1.2   The Ophyd-Basler Library

The Basler industrial cameras[12] have a native graphical application and Python API for programmatic control of the cameras. The advantage of this software ecosystem is that it provides a convenient environment to integrate with Ophyd, and multiple emulated camera objects without physical cameras. The approach contained here serves as a model for how to integrate an existing Python API for device communication into an *Ophyd* object that works with the Run Engine.

In the ophyd-basler Python package, the communication with a camera is done using the API, with which users can obtain the camera's properties (model, serial number, image dimensions, etc.) and set the parameters (such as the exposure time, pixel format, etc.). In the emulated mode the camera allows users to select between the existing test pattern images that change on every single shot, or to even feed the custom predefined images to the camera. This feature can be used to "replay" the experimental results so that they can be emitted using the native camera interface.

---

[12]https://www.baslerweb.com

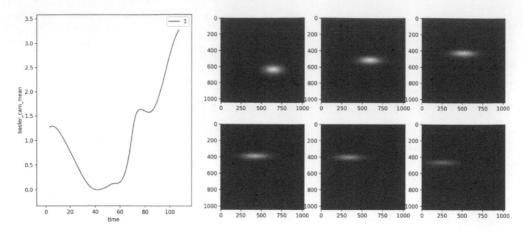

Figure 8.5 Left: Live plot produced during a *Bluesky* `count` plan displaying the averaged signal from the emulated Basler camera with predefined images over the course of 100 individual counts. Right: Selected images extracted using a *Databroker* API (the second row's sequence is from left to right).

An example of such an approach is depicted in Figure 8.5, where a Gaussian beam with varying shape is randomly "wandering" on the camera images. In all cases, the images from the camera were saved to an HDF5 file with a simple schema.

8.3.1.3 Socket-Based Control System

Another example of diverse *Ophyd* integration is with a socket-based data acquisition system, implemented at the Accelerator Test Facility of the Brookhaven National Laboratory, which uses the socket server-client communication protocol. The system exposes the parameters of the steering magnets, controlling the trajectory of the electron beam as well as statistical parameters of the observed signal on the "frame grabber" detectors, which can be read over the network by connecting to the server using the server's socket. The server update rate is 1.5 Hz. In the `atfdb` Python package, we implemented custom *Ophyd* `Signal` classes which allow communication using the socket client API. Like in EPICS, the system has PV-like strings with special commands to the ATF control system to set and read the values, which are leveraged in these custom classes. For convenience, we developed a test socket server which enabled us to perform continuous integration testing without access to the actual control system at the facility.

Figure 8.6 displays the results of the scan performed at ATF where the *HeNe* laser intensity was captured by a Basler camera (using the *Ophyd* Basler API) as the laser intensity increased (via the socket-backed *Ophyd* objects) in the same scan, thanks to the flexibility provided by *Ophyd* to abstract two different control backends to have the same high-level API.

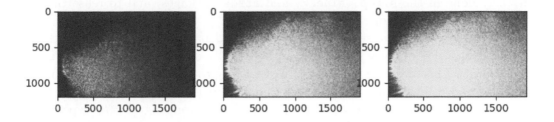

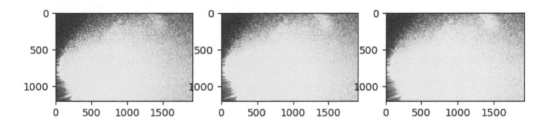

Figure 8.6 Results of the Bluesky scan performed at ATF where the *HeNe* laser intensity was captured by a Basler camera as the laser intensity increased.

8.3.2 OpenCV-Based Integrations

The EPICS AreaDetector framework has limited support for streaming cameras. However, in the case of the URL-based streaming Axis camera[13] used at the Tender Energy X-ray Absorption Spectroscopy (TES) beamline of NSLS-II, we observed the IOC lock-ups which prevented the use of AreaDetector IOC-based Ophyd-EPICS integration. We decided to write the integration purely in *Ophyd*, which resulted in the `vstream` *Ophyd* object implementation for the camera. This has a very similar implementation to the Ophyd-Basler integration discussed above. The specific camera was sending 14-15 frames per second to the network, which were captured using the OpenCV library, averaged over the user-controlled "exposure time", and eventually saved to an HDF5 file. USB and embedded laptop cameras can be integrated similarly using the OpenCV library[14].

8.3.2.1 Ophyd's Future

All the above examples utilize the *Ophyd* v1 API. An "Ophyd v2" redesign to better separate the control layer from the *Ophyd* objects' logic, is under active development to refactor the *Ophyd* library. In the new version, it will be more convenient to define the specific communication protocol via a separate library (in a similar fashion it is

---

[13] https://www.axis.com
[14] https://github.com/mrakitin/Berlin-2019-tutorials-workshop/blob/master/demo/LaptopCam.ipynb

currently done in the `sirepo-bluesky` and `atfdb` packages) and to reuse the core *Ophyd* components without much modification.

To date, the most substantive utility of *Ophyd* has been in the abstraction of beamline-related equipment and what can be broadly considered as materials analysis platforms. This direction is continually expanding to include the scope of synthesis platforms and reactors that are often relevant to experiments at the beamline [294]. This is motivated in part by the capabilities of synchronizing *in situ* experiments directly with the analysis, and increasingly by the capability of autonomous experiments at beamlines [291, 294, 296]. Most "closed-loop" beamline experiments have the complete sample library synthesized [297], or make post-synthetic modifications at the beamline [44]. Treating the beamline as an analysis MAP, there is a pressing need to integrate synthesis platforms or MAPs into the beamline or into a control layer that includes the beamline [294]. As such, the vision of *Ophyd* development includes increased integration of reactors and synthesis platforms that can be run using *Bluesky*.

8.3.3 The Run Engine

Bluesky is a library within the *Bluesky Project* that holds its original namesake and primary mechanisms for experiment control and collection of scientific data and metadata. It emphasizes the following virtues:

- Live, streaming data: available for inline visualization and processing.

- Rich metadata: captured and organized to facilitate reproducibility and searchability.

- Experiment generality: seamlessly reuse a procedure on completely different hardware.

- Interruption recovery: experiments are "rewindable", recovering cleanly from interruptions.

- Automated suspend/resume: experiments can be run unattended, automatically suspending and resuming if needed.

- Pluggable I/O: export data (live) into any desired format or database.

- Customizability: integrate custom experimental procedures and commands.

- Integration with scientific Python: Interface naturally with NumPy and Python scientific stack.

The Run Engine is responsible for orchestrating the actual execution of the experimental plan and emitting the collated measurements—as "Documents"—to the downstream consumers as they are available. The experimental intent of the user is expressed as a sequence of commands to the Run Engine, colloquially called `plans`, that capture the scientifically interesting details. The `plans` are typically implemented as

Python co-routine generators[15] which allow for the sequences to be dynamically generated and for two-way communication between the plan and the Run Engine. Leveraging this Python language feature, we are able to achieve composable and adaptive plans (which will be discussed at length below).

The Run Engine relies on the hardware abstractions provided by *Ophyd*, as discussed in the previous section, to remain agnostic to the details of the underlying hardware. Thus, a `plan` can be parameterized by the devices that provide both the dependent and independent variables which provides significant re-use between instruments and facilities.

In the common operating mode, *Bluesky* is used interactively from an IPython environment. A session would start by loading startup scripts that define beamline-specific devices and `plans`, and then interactively running `plans` one-by-one from IPython prompt or executing a script that runs a sequence of `plans`. This has satisfied the needs for most interactive experimentation that have close human engagement in the decision-making and control loop; however, will not be sufficient for more advanced Autonomous Experimentation.

8.3.4 The Run Manager

In *Bluesky*, `plans` provide a way to express each individual element of an experiment, such as a temperature ramp or an energy scan at a specific point on a sample. While `plans` can be trivially composed to express an entire experimental campaign, this can be inconvenient if the scientist changes their mind about future experiments they would like to do based on the results as they come in. This is exceptionally common in adaptive experiments, or Autonomous Experimentation which abstracts the independent variables and observable query from the data acquisition, then chooses a new set of independent variables to query. To support this we have developed the *Queue Server* which provides a way to schedule `plans` for future execution. Until a `plan` is taken off the top of the queue to be executed, the parameters of the `plan` and its position in the queue can be modified by the user.

The *Queue Server* allows *Bluesky* to run in a dedicated Python process (Run Engine worker environment). The worker environment is created and managed by a Run Engine (RE) Manager, which can be run as an application or a service. As in the IPython workflow, the startup code is loaded into the environment and the beamline devices and `plans` are available in the environment namespace. Bluesky `plans` are executed by populating and starting the plan queue maintained by RE Manager. The queue is editable and can be modified by users at any time: queue items may be added to the queue, replaced, edited, moved to different positions, and removed from the queue.

The *Queue Server* includes the core `bluesky-queueserver` package and `bluesky-queueserver-api`, `bluesky-httpserver` and `bluesky-widgets` packages that implement additional functionality (Figure 8.7). The first package implements the RE Manager and control over a message bus. This distributed control allows decoupling the `plan` executions from any process—such as a graphical user

---

[15]https://tacaswell.github.io/coroutines-i.html

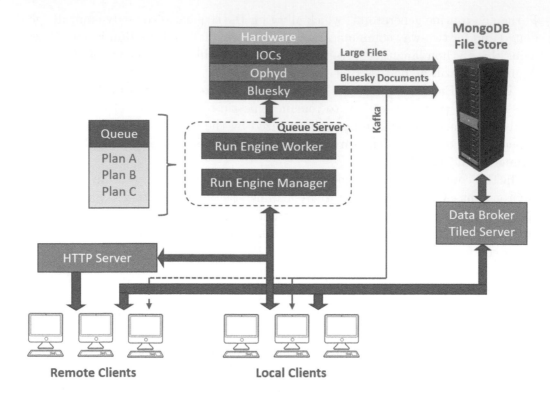

Figure 8.7 Diagram of Queue Server components. The Queue Server is running two processes: RE Manager process and RE Worker process. The RE Manager process is responsible for maintaining and controlling the tplan queue and 0MQ communication with clients. The queue is stored outside RE Manager (in Redis) and persists between restarts. The local clients (with access to the local network) communicate with the Queue Server using 0MQ API. Remote clients connect to HTTP server and use REST API to control the Queue Server. As the `plans` are executed, the *Ophyd* code communicates with EPICS IOCs (to control hardware) over the network and *Bluesky* code generates documents that are saved into MongoDB and/or published to Kafka depending on the Run Engine subscriptions.

interface—that should fail independently. The *Bluesky HTTP Server* is designed to control the RE Manager from outside the local network and provides a matching set of REST API and basic authentication, authorization, and access control. The API for the *Queue Server* supports a more convenient and universal interface for communication with RE Manager. The package contains an API for synchronous and asynchronous (asyncio) communication either directly to RE Manager, or over the HTTP interface remotely. From a security perspective, the RE Manager provides permissions (who can add which `plans` to the queue), whereas the HTTP interface can provide authentication (validating you are who you say you are). Front-end development is ongoing, with some early efforts contained in `bluesky-widgets`, albeit should be left to the customization for individual end stations or applications.

8.3.5 Data Model and Access with *Tiled*

Any given beamline instrument or experiment can present unique requirements, but we can generalize around two common archetypes: during the experiment and after the experiment.

During the experiment, we often want streaming access to incremental updates, supporting real-time decision-making about how to steer and when to stop an experiment. These decisions may involve a human "in the loop", or they may be partly or fully automated. Humans will need live analysis and visualizations to support their decision-making. Scientists or systems that are driving experiments often have privileged access (i.e., they are inside the firewall); granular access control is not a concern. In the pursuit of low latency, these systems can typically take on some amount of specialization, adapting themselves to work with the data in something close to its raw form, using specialized formats and protocols if necessary. This cluster of requirements is well served by Bluesky's event-based data model[16]. Metadata and small or low-rate data are published as JSON over a message bus, like Kafka. Large arrays or high-rate data are sent over whatever systems match the detectors' capabilities, and the locations of these artifacts are references within the JSON documents.

After the experiment, we tend to want random access to the data set with a premium on convenience. The data should be available in formats and via protocols convenient to the analysis tools. We need granular access controls on the data. This family of requirements is well served by a service, starting with web (HTTP) and extending to more specialized and performant services. This cluster of requirements drove the development of *Tiled*, a web service for structured data access.

Tiled enables efficient search and structured, chunk-wise access to data in an extensible variety of appropriate formats, providing data in a consistent structure regardless of the format the data happens to be stored in at rest. The natively supported formats span slow but widespread interchange formats (e.g., CSV, JSON) and fast, efficient ones (e.g., C buffers, Apache Arrow, and Parquet). Thus, the same underlying data can be accessed in the format most convenient for the user. This may vary based on the data analysis application being used, the user's comfort level with various formats, and speed considerations.

Notably, *Tiled* is a data access service that is agnostic to the data storage approach. It puts an emphasis on structures, rather than formats. Driven by the growing needs of users of *Databroker*, this tool was developed to be fully separate from the *Bluesky* ecosystem as a generic access tool for scientific data, with *Databroker* being rebuilt with a *Tiled* back-end to reduce friction for current users of *Databroker*.

Tiled implements extensible access control enforcement based on web security standards, similar to JuptyerHub[17]. Like Jupyter, *Tiled* can be used by a single user or deployed as a shared public or private resource. It has been deployed at the facility scale and on researchers' laptops for personal use.

---

[16]https://blueskyproject.io/event-model
[17]https://jupyter.org/hub

8.3.6   Workflow Management

The is an enormous number of workflow management tools, designed to represent logical chunks of computation as a chain of tasks. In our experience, many that are developed specifically for scientific use cases work well for the domain of origin but struggle to generalize. After surveying dozens of workflow managers, the team at NSLS-II piloted a project to use *Prefect*[18], a general-purpose workflow manager. *Prefect* is an open-source project backed by a for-profit company, with managed deployments and support available. We employed it for data engineering tasks (e.g., data movement) as well as data analysis tasks, including:

- Validating at the end of a scan that all data is readable

- Exporting data in formats convenient to the user in a shared directory

- Performing zero-parameter automated data reduction and processing

- Performing tunable, parameterized data processing

Prefect emphasizes managing task failure well, enabling intelligent retires. It includes a sleek web interface, as well as Python APIs.

8.3.7   Adaptive Experiments in Bluesky

When imbuing experiments with any range of autonomy, it is worth considering the diversity of computational agents that can process an experiment's results and/or execute control. Every useful agent may not be considered AI, nor equipped with the grammar common to that community (e.g., PID loops, behavior trees, or finite state machines). Agents will operate on distinct timescales, degrees of responsiveness, quality of service, and production scale. Importantly, agents will vary in production quality, ranging from routine facility scale deployment to experimental and provided *ad hoc* by scientists or users. It is with these considerations in mind, that an adaptive harness and grammar has been designed for integration with the *Bluesky Run Engine* and *Run Manager*: bluesky-adaptive[19].

It is useful to have a generic interface to expect for all agents, so that the same useful agents can be easily deployed in different experiments, and diverse agents can be deployed alongside of each other in a single experiment. We designed a **tell–report–ask** interface between experimental orchestration and any arbitrary agents. In a Python-wrapped agent, each agent was part of some object that had a **tell** method to tell the model about new data, a **report** method to generate a report or visualization, and an **ask** method to ask the model what to do next. While the latter method is required with adaptive learning in mind, it enables simple adaptations such as a model detecting an anomaly and wishing to pause the experiment. The methods are prescribed to generate documents following the *Bluesky* event model, for facile reload and replay of decision-making after an experiment. This generic

---

[18]https://www.prefect.io/
[19]https://github.com/bluesky/bluesky-adaptive

interface suits most needs for AI at a beamline, and allows users to "plug-and-play" models they have developed without considering how the data is being streamed or other communication protocols.

Before considering examples of agents driving *Bluesky* orchestration, it is worth exploring the degrees of required "adaptiveness" in plans. Fixed plans (e.g., `count` or `scan`), are sufficient for many scientific use-cases, but in cases requiring agent feedback between the data and the orchestration, the feedback can happen at many levels of fidelity. Broadly these can be broken down to intra-plan for agents that provide feedback inside a plan, and inter-plan for agents that provide feedback at the completion of one (or many) plans. The former approach requires lockstep behavior between datum and directive, whereas the latter approach can be asynchronous.

TABLE 8.2 Adaptiveness can be inserted into the data collection process at several levels and is arranged below in order of rate of reaction.

Application	Implementation	Integration
In or below the control system	FPGA, control system	Intra
Within a plan	*Run Engine*	Intra
Per-event	*Run Engine*	Intra
Per-run	*Run Engine*	Intra, Inter
Asynchronous and decoupled feedback	*Run Manager*	Inter
Across many runs or end stations	*Run Manager*	Inter

As shown in Table 8.2, adaptive behavior can be exploited by the data collection process at several levels. Here we use the language of the *Bluesky* event model, where a "run" is composed of many "events" between a "start" and a "stop" document. The start and stop documents carry the metadata of the experiment and the run, and each event document carries the measured data. Each level of fidelity and interaction has a use, and which ones to pick will depend on the requirements and constraints on a per-facility, per-beamline, and per-experiment basis. A given experiment may even make use of adaptiveness from multiple levels!

In or below the control system. If decisions are needed on very short time scales (and have a computation that can fit in the time budget), then building the adaptive computation into or below the control system is a good choice. One example of this is in the scaler devices that are used as the backend electronics for integrating point detectors on many beamlines. Typically they are configured to take a fixed length exposure; however, they can be configured to gate on any of the channels. Thus by gating on the I/0 (incoming photon flux) channel, your otherwise fixed plan would "adapt" the exposure time to account for upstream fluctuations in photon intensity.

Within a plan. At the most granular level, *Bluesky* gives the plan author access to the data extracted from the control system before it is processed through the event model documents. This is the level that we use in the `adaptive_scan` which is bundled with *Bluesky*. This level has also been used at LCLS, implementing the frame-dropping logic described above at the plan level.[20] This level gives the author a tremendous amount of flexibility and can be used to prevent "bad" data from entering the document stream, but quickly becomes very plan-specific and difficult to generalize and re-use.

Per-event. In cases where the computational cost to recommend the next step is fast compared to the time it takes to collect a single data point (aka an event), then it makes sense to run the recommendation engine on every event. At the end of the plan, we will have 1 run whose path through phase space was driven by the data. Examples of this are a 1D scan that samples more finely around the center of a peak or a 2D scan across a compositional gradient that samples more finely at phase boundaries. In these cases, there is a 1:1 mapping between an event collected and a recommendation for the next point to collect.

Per-run. When the data we need to make a decision about what to do next maps more closely to a run, we do the same as the per-event case but only expect a recommendation once per-run. An example of this could be a 2D map where at each point we take an X-ray-Absorption Near-Edge-Structure (XANES) scan and then focus on points of interest within the map. Much of this depends on how an experiment is atomized between the start and stop documents (some scientists prefer to design plans that encapsulate every event in run). At this level of fidelity, we can— but do not require—a 1:1 mapping between a completed run and a recommendation for the next run.

Asynchronous, decoupled, and distributed feedback. An increasingly desirable operating paradigm for Autonomous Experiments considers the directives of many agents, or even networks of agents, including multiple human agents. The previously described lock-step approaches to experiment and analysis, leave no room for human experts to engage in the loop, incorporation of information from complementary techniques, or the integration of multiple computational agents. In this more complex paradigm, various agents must be able to process the captured data stream, suggest plans to be executed, and create reports for human consumption. This is exemplified in the case where multiple passive agents are performing dataset factorization or AI-based compression algorithms that provide visualization tools for users, multiple active learning agents are providing suggestions for the next experiments as they complete their computation, and human agents are also guiding the experiment [291]. Here, the same `tell–report–ask` grammar can be used in conjunction with *Kafka*, *Tiled*, and the *RunManager*. Furthermore using *Kafka* for distributed communication, one can construct meta-agents, or adjudicators, which coordinate between a collection of agents in more sophisticated ways than the *RunManager* priority queue and provide an additional avenue for human intervention. Lastly, new developments include running these agents and adjudicators as services with a REST API, so web-based user interfaces can be developed to manage agents at run time.

Across many runs or end stations. A natural extension of decoupled feedback is to link multiple experimental end stations or unit operations in multimodal feedback loops. If all end stations are already controlled by a *RunManager*, then

---

[20]https://github.com/pcdshub/nabs/blob/master/nabs/streams.py

this amounts to a networking task and agent design task. The network must allow communication between the agents and multiple *RunManagers*. And the agents must be designed to take advantage of both end station data streams. In a multi-fidelity arrangement—where one end station provides a slower, yet higher resolution characterization—a monarchsubject relationship can be used that lets an agent subscribing to one beamline dictate the plans of the opposite beamline [291]. This design opens up opportunities for autonomous multimodal or multi-fidelity experiments that make use of correlated data from many experiments to elucidate a more holistic understanding of a scientific system.

8.4   AI INTEGRATIONS AT NSLS-II

By leveraging the *Bluesky* software ecosystem inherent to the beamlines at the NSLS-II, rapid prototyping and integration of Python-based data analysis and control schemes are often straightforward. The facility uses AI and ML to automate tasks and streamline operations, allowing greater flexibility, extensibility, and reliability. Examples of AI/ML methods being developed and deployed include such as automating beamline operations, performing data analysis, or even fully autonomous beamline control.

NSLS-II's Beamline for Materials Measurement (BMM) is an example of where AI/ML is used to automate tedious and repetitive tasks [239]. A supervised learning model is used to evaluate every X-ray Absorption Fine Structure (XAFS) spectrum as it is measured, to distinguish measurements that look like XAFS spectra from failed measurements. A dataset of over 800 spectra has been tagged by beamline staff as either successful or failed measurements (Figure 8.8), and using this corpus, a classifier was trained to evaluate newly measured data. This evaluation tool is now incorporated into all XAFS measurements at the beamline. Measurements recognized as successful are subjected to further data reduction before being presented to the user, while a negative evaluation triggers an alert to staff for potential operational problems at the beamline.

At the Submicron Resolution X-ray Spectroscopy (SRX) beamline, pre-trained 3D convolutional neural networks (CNN) are being used to provide autonomous data acquisition for X-ray fluorescence (XRF) imaging and mapping [348]. To build up an image, the sample is rastered through a focused X-ray beam. This technique is sufficient for 2D imaging but is slow for 3D imaging and tomography. Advanced algorithms use AI/ML methods to improve the reconstruction quality and data collection efficiency. In this project, called HyperCT, NSLS-II is employing super voxel model-based tomographic reconstruction (svMBIR) algorithms that require fewer projections to resolve the sample reconstruction quality [44]. Combining this with artificial intelligence and adaptive scans that can identify future projections that will have the greatest impact on improving reconstruction quality, will decrease the time necessary to collect hyperspectral, multi-element volumes of new and exciting materials and samples.

NSLS-II also uses ML methods to automate the analysis of streaming spectral data. Non-negative matrix factorization (NMF) is an appealing class of methods

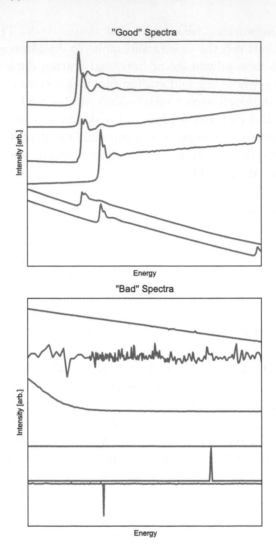

Figure 8.8 Examples of good and bad spectra used to train the classifier at the BMM beamline (shown in photograph). Note that data measured on any element are transformed onto a common, unit-less abscissa as required for classification by the machine learning model. The examples show the transformed data [239]. Reproduced with permission from the Royal Society of Chemistry.

for performing unsupervised learning on streaming spectral data, and Constrained Matrix Factorization (CMF) has been developed to improve the reconstruction of true underlying phenomena [295] (Figure 8.9). Another example of AI for analysis developed at NSLS-II is the crystallography companion agent (XCA) [293, 43]. This is an open-source package initially developed for the feed-forward classification of diffraction experiments. The classification approach is pseudo-unsupervised and does not require labeled data to be trained. Instead, XCA simulates a realistic dataset that encompasses the perturbations and physics of the measurement. Starting only from

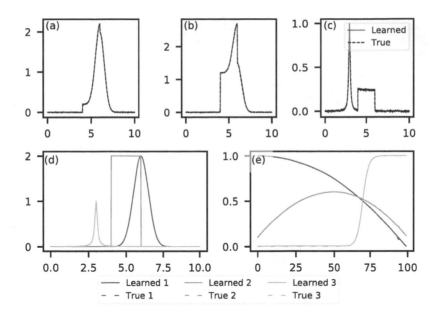

Figure 8.9 Users commonly need to decompose their datasets in real-time but posses knowledge of some underlying details of the data. The reconstruction of the (a) first, (b) median, and (c) last pattern of a dataset of mixed Gaussian, Lorentzian, and box functions using constrained matrix factorization. (d) Prior knowledge of the functions fixes the underlying components. (e) The learned and true weights are plotted against the dataset index. The learned weights match the underlying mixing function and create an accurate reconstruction of the dataset [295]. Reproduced with permission from AIP Publishing.

proposed crystalline phases, XCA overcomes the challenge of degenerate solutions by training an ensemble of agents that can accurately predict phase existence and uncertainty.

X-ray Photon Correlation Spectroscopy (XPCS) is a technique used to study sample dynamics in a quantitative manner at the Coherent Soft X-ray scattering (CSX) beamline. However, XPCS algorithms cannot distinguish sample dynamics from intensity changes caused by instrumental effects, affecting the accuracy of results. Researchers at NSLS-II have developed AI/ML models that detect artifacts, remove noise, and extract quantitative results with an improved resolution for non-equilibrium dynamics (Figure 8.10) [240, 242]. These models are sample and instrument agnostic and are integrated into a variety of results-reporting schemes with outlooks for data collection approaches. Unsupervised clustering of frames using 1D time series leads to more efficient computation but increases the number of datasets to investigate. Unsupervised grouping methods remain part of the essential toolbox since data masking may be applied to the full downstream result.

The facility has also deployed algorithms for autonomous X-ray scattering experiments and high-throughput measurements. The autonomous X-ray scattering

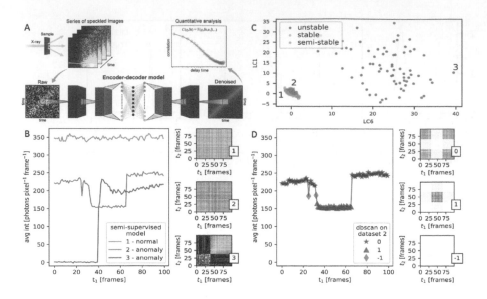

Figure 8.10 The CSX beamline with examples of AIML for XPCS analysis. A. XPCS workflow with deep learning B. Application of semi-supervised outlier models (IFT, LOD, EE) from derived 1D time-series (average intensity) to streamline the review of automated C2 results [239] C. Identifying truly anomalous data with unsupervised clustering on latent space coordinates from CNN-ED for C2 denoising [240] D. Unsupervised clustering of derived 1D time-series data to dynamically mask/unmask pre-computed C2 results. Reproduced with permission from Taylor and Francis.

experiments utilize automated sample handling and data acquisition, real-time data processing, and decision-making algorithms based on Gaussian process regression in a Bayesian optimization framework to select the next experimental point to be measured. Among others, this adaptive learning approach has been deployed at the Complex Materials Scattering (CMS), Soft Matter Interfaces (SMI), and Pair Distribution Function (PDF) beamlines [127, 34, 291] (see Chapter 10). Other high-throughput measurements at the PDF beamline use reinforcement learning to automate and optimize data collection, with an RL agent trained via a dynamic interaction loop and a data quality metric employed to ensure the decision-making logic learned by the RL agent is broadly generalized to any scenario that can output a similar formatted data quality metric [297]. These AI/ML approaches have been shown to significantly increase efficiency in measuring critical structural information.

Scientists at NSLS-II are using AI/ML methods to improve operations, expedite analysis, and facilitate discovery. NSLS-II has been working with other DOE light source facilities to establish a computational fabric that covers the full lifecycle of data generated, connecting instruments with a multi-tiered computing landscape. The vision is to best serve the thousands of DOE light source users per year. Much of this success is dependent on the developments of robust infrastructure at the facility and software tools in the *Bluesky Project*.

8.5   ACKNOWLEDGMENTS

This research used resources of the TES beamline (8-BM) and the National Synchrotron Light Source II, a U.S. Department of Energy (DOE) Office of Science User Facility operated for the DOE Office of Science by Brookhaven National Laboratory under Contract No. DE-SC0012704.

This research was also supported by a BNL Laboratory Directed Research and Development (LDRD) projects 22-031 "Simulation-aided Instrument Optimization using Artificial Intelligence and Machine Learning Methods", 20-032 "Accelerating materials discovery with total scattering via machine learning", 20-038 "Machine Learning for Real-Time Data Fidelity, Healing, and Analysis for Coherent X-ray Synchrotron Data", and 23-039 "Extensible robotic beamline scientist for self-driving total scattering studies".

We acknowledge support from and the Small Business Innovation Research grant (DE-SC0020593) from the DOE Office of Science in Basic Energy Sciences.

This research also used resources from the DOE projects "Intelligent Acquisition and Reconstruction for Hyper-Spectral Tomography Systems (HyperCT)" and "Integrated Platform for Multimodal Data Capture, Exploration and Discovery Driven by AI Tools (AIMM)".

GLOSSARY

BNL: Brookhaven National Laboratory

NSLS-II: The National Synchrotron Light Source II

CHAPTER 9

Introduction to Reinforcement Learning

Yixuan Sun

Argonne National Laboratory, Lemont, Illinois, USA

Krishnan Raghavan

Argonne National Laboratory, Lemont, Illinois, USA

Prasanna Balaprakash

Oak Ridge National Laboratory, Oak Ridge, Tennessee, USA

CONTENTS

DOI: 10.1201/9781003359593-9

OVERVIEW

Reinforcement learning (RL) is concerned with teaching an agent how to achieve specific goals in an environment where the only feedback is due to the reinforcement/reward signal. Typically, in an RL task, an agent seeks to learn "what actions to choose" such that the reward is maximized. In recent years, RL has achieved many breakthroughs in artificial intelligence, as evidenced by the superhuman performance achieved by RL on AlphaGo and AlphaZero. Such success illustrates that RL may be the most natural way of imitating human-level learning. In this chapter, we aim to briefly introduce RL and take a Markov decision process (MDP) viewpoint of RL. We will begin this chapter with the basic MDP-driven problem formulation of RL and describe the fundamental elements of RL, such as environment, reward, actions, policy, and value, followed by a description of foundational algorithms for solving RL problems. Next, we will illustrate different learning methods in RL, such as Monte Carlo and temporal difference learning methods. Subsequently, we will illustrate different RL algorithms and discuss recent advancements in deep RL. We will then describe the different applications of RL and provide perspectives on some of the key challenges in RL. We will end this chapter by providing perspectives for future work.

9.1 INTRODUCTION TO REINFORCEMENT LEARNING

Machine learning (ML) has become increasingly ubiquitous in recent years, permeating the scientific community with its transformative potential, as highlighted in the detailed discussion on challenges and opportunities in [494]. Data plays a pivotal role in the development of machine learning (ML), where the availability of informative data in large quantities often results in impressive outcomes. A central aspect of ML involves training models by evaluating a loss function based on training data. The loss function typically quantifies the model's correctness, which is facilitated by labels. These labels can be viewed as feedback provided by the data on the model's performance. ML methods are generally classified based on the type of feedback they use. For example, supervised learning methods are employed when exact feedback is available, while unsupervised learning is used when no feedback is present. Reinforcement learning (RL) constitutes a unique class of ML techniques that rely on partial feedback in the form of reinforcement or rewards.

RL involves an ML model referred to as an "agent", operating within an "environment" that generates data or "states", which the agent consumes and uses to generate "actions". In traditional RL, the states were generated as sensor information or numerical data. However, in model-based RL, states are typically snapshots of the environment The agent's objective is to generate actions based on states, and depending on the actions taken, the environment provides feedback in the form of reward signals. The agent learns to take actions that maximize the reward, making RL unique among ML techniques.

Typically, rewards in RL are constructed as integer values, with 1 indicating a positive outcome, -1 indicating a negative outcome, and 0 indicating no effect.

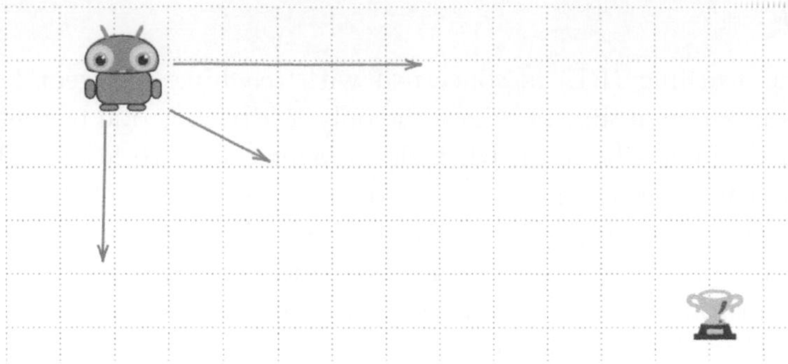

Figure 9.1   Illustration of the reinforcement learning problem.

However, recent research in control literature has described RL signals as continuous values [344]. To illustrate the RL problem, let us consider an example involving a robot (i.e., an agent) learning to travel a distance as quickly as possible to reach a prize, as depicted in Figure 9.1. The state of the environment is communicated to the agent via the position of the robot on a grid, where the robot can take one of three actions (i.e., move right, move down, move diagonally) with the goal of reaching the prize. Learning in the RL problem is divided into episodes, where each episode involves taking a sequence of steps to reach the prize. The sequence of actions taken by an agent is called a policy. In an episodic setting, the RL agent completes an episode upon reaching a terminal state associated with an episode length T. Reaching the terminal state with the prize earns a positive reward, while the terminal state without the prize results in negative rewards. The agent learns to reach the prize through successive episodes, as illustrated in this example of episodic RL. Alternatively, in continuous RL, there is no terminal state (i.e., $T = \infty$), and the robot can continuously collect different prizes. This setup is commonly studied in robotics and control [344]. The problem of RL, as described above, is one of the most important paradigms in science applications. In science applications such as autonomous discovery, data can be noisy, generated in huge quantities, and may lack precise feedback. Despite these challenges, the unpredictable nature of data must be studied to advance science and make informed decisions that account for this nature.

This chapter aims to provide an overview of the components of RL through a Markov decision process formulation in Section 9.2. We will present a taxonomy of various RL algorithms in Section 9.3 and explore different applications of RL in Section 9.4. Section 9.5 will focus on describing challenges associated with utilizing RL in science applications, while Section 9.6 will offer insights into future perspectives.

9.2   MARKOV DECISION PROCESSES FORMULATION OF RL

This section aims to provide a mathematical formalization of Reinforcement Learning (RL) by introducing the fundamental concepts of Markov Decision Processes (MDP). To achieve this goal, we will first delve into the basics of MDP and then proceed to describe how it relates to RL.

9.2.1 MDP

The fundamental components of a MDP consist of a state-space $\mathcal{S}$, an action space $\mathcal{A}$, and a state-transition function $T : \mathcal{S} \times \mathcal{A} \rightarrow \mathcal{S}$. In the specific example depicted in Figure 9.1, the state space represents the total number of positions in the grid, while the action space comprises the three actions that the robot can take. The transition functions describe how the robot's position changes as it takes an action. At each time step k, the current state of the robot is denoted as $s(k) \in \mathcal{S}$, and the robot takes an action $u(k) \in \mathcal{A}$, causing the environment to transition from the current state to the next state $s(k + 1)$ based on the transition function T. The transition from one state to another is typically a stochastic process, governed by T, where the agent's position can vary across episodes. When this stochastic process satisfies the Markov property, the system becomes an MDP. The Markov property implies that the future states of the agent only depend on the present state and not on the past history of states. In other words, the transition function depends solely on the current state, making it possible to derive future states solely from the present. This property is crucial to the reinforcement learning problem, where optimizing the current action can affect the choice of future actions and corresponding rewards. This concept underpins the Bellman principle [50], which we will explore later in this chapter. Before that, however, we will introduce the notion of rewards in RL, as an MDP is incomplete without them.

9.2.2 MDP Formulation of RL

An MDP problem becomes an RL problem through the introduction of the reward function R [503]. During the learning process, an RL agent perceives environmental information described by the state $s \in \mathcal{S}$ and takes an action a, typically based on a policy $\pi : \mathcal{S} \rightarrow \mathcal{A}$, which defines how the agent will choose actions. Once the agent selects an action, it interacts with the environment and receives an external scalar reward/reinforcement signal $r \in \mathcal{R}$ from the environment. The agent-environment interaction is described by the tuple s, a, s', r. A collection of these tuples forms the experience of the RL agent. We previously noted that MDP models a stochastic process due to the agent's choice of action and the corresponding state that results. This process is typically governed by a probability space defined over all possible future states given the current state and a set of rewards, $\mathcal{R}$. The probability distribution for this space is $p(s', r|s, a)$, which encodes the stochasticity of the transition function. This probability distribution represents the probability of transitioning to the next state s' from state s given action a and reward r. The central idea behind RL is that the agent seeks to maximize rewards, which translates to the agent's decision-making process. The probability of transitioning from one state to another, known as the state-transition probability, depends directly on the rewards through the expression:

$$p(s'|s, a) = \sum_{r \in \mathcal{R}} p(s', r|s, a). \tag{9.1}$$

Summing over all possible rewards yields the probability of transitioning from one state to another. Therefore, the probability of moving to a state with a higher reward is higher as the number of experiences grows. This is an important concept in RL, where an agent exploring actions with higher rewards initially in the learning process is more likely to transition to states that provide more rewards because the state-transition probability is higher. This concept is central to RL, where better exploration leads to more significant rewards in the long run. The formalism of this notion provides the goal of obtaining maximum accumulated reward, a concept formalized through the return function. In RL, the return function $(R(k))$ is defined as:

$$R(k) = r(k) + r(k+1) + \cdots + r(K), \tag{9.2}$$

where K is the terminal time. An important aspect of the return function is that it does not account for past rewards; it starts from the current time k and sums over all future instances. Typically, two types of problems are considered in this context: finite horizon and infinite horizon. In finite horizon problems, K is a natural number, while in infinite horizon problems, K approaches infinity. In infinite horizon scenarios, the sum described in (9.2) is unbounded. To address these cases, a discount factor $0 \leq \gamma \leq 1$ is introduced. The discount factor provides a weight to rewards at each instant k, with a value closer to one implying that the agent gives more weight to immediate rewards, while a value closer to zero implies that the agent provides more weight to future rewards. The return function, in the presence of a discount factor, becomes:

$$R(k) = r(k) + \gamma r(k+1) + \gamma^2 r(k+2) + \cdots = \sum_{i=0}^{\infty} \gamma^i r_{k+i+1}. \tag{9.3}$$

A discount factor introduces a weight to rewards at each time step k. A value closer to one indicates that the agent places more emphasis on immediate rewards and is more likely to select actions that improve immediate rewards. Conversely, a discount factor closer to zero implies that the agent places greater importance on future rewards and tends to select actions that improve the likelihood of obtaining higher rewards over longer time horizons. This trade-off between immediate and future rewards is a crucial aspect of RL, as it allows the agent to balance short-term gains with long-term objectives, making it a powerful tool for decision-making in complex and dynamic environments.

The discounted return is computed over all possible state-action pairs, and based on the evaluation, the agent must decide which actions to take. To facilitate this decision-making process, the agent is equipped with a set of rules that form a policy in RL terminology. The policy, denoted by π, is essentially defined by the state-transition probability, and to determine the optimal policy, the agent must learn the likelihood function of the state-transition probability. Mathematically, the goal then translates to finding the optimal policy π^* that will maximize the discounted return.

$$\pi^* = argmax \, E_\pi[R(k)] \tag{9.4}$$

Solving 9.4 is not feasible directly because of two reasons. First, the discounted return requires complete knowledge of how each action will influence each state at every instant in the future. While this might be feasible for systems with small state-action space, it is certainly not possible in real-world applications where the total number of states and actions is large. With a large state action space comes another issue. The policy has to determine how each action would interact with each state of the system and the corresponding reward to calculate the return function and make a decision. However, this requirement increases the number of choices and exponentially increases the search space size with the number of states and actions. This phenomenon is known as the curse of dimensionality in the RL literature, where the compute and memory requirement for completely traversing the search space becomes insurmountable for most real-world applications. To address this curse of dimensionality, certain constructs are required that simplify the search problem. One of the main pillars of this construct is Bellman's principle of optimality, as mentioned earlier [50]. Bellman's principle of optimality states that

"An optimal policy has the property that whatever the initial state and initial decision are, the remaining decisions must constitute an optimal policy with regard to the state resulting from the first decision."

This landmark intuition by Bellman allows us to break the curse of dimensionality by simplifying the search problem. Within this construct, an optimal policy over the long horizon can be reached by just taking an optimal step at the current k without considering the effects of future rewards and actions on the current step. Therefore, it is enough to consider the effect of the current policy on the discounted return given the present state. This effect is summarized by a construct known as the value function. A value function of the given state is the expected return of the given state under the policy, as follows

$$v_\pi(s) = \mathbb{E}_\pi[R(k)|s] = \mathbb{E}_\pi[\sum_{i=0}^{\infty} \gamma^k r_{k+i+1}|s], \quad \forall s \in \mathcal{S}. \tag{9.5}$$

Note here that the expected values is obtained over the complete policy. For this reason, this function is also called the state value function given a policy π. Similarly, if the effect of an action is explicitly considered by conditioning the expected value on the current action, we obtain the state-action value function ($Q-$function) given a policy π as

$$Q_\pi(s, a) = \mathbb{E}_\pi[R(k)|s, a] = \mathbb{E}_\pi[\sum_{i=0}^{\infty} \gamma^k r_{k+i+1}|s, a], \quad \forall s \in \mathcal{S}, \forall a \in \mathcal{A}. \tag{9.6}$$

It can be observed that the value function can be obtained from the q-function by integrating all the possible actions and vice versa by explicitly considering the effect of the present action on the value function. Note that even though the value/Q function summarizes the effect we need, we still have terms $> k,$, which is the future. We can further expand the state-value function by leveraging Bellman's principle here. By basically splitting the terms for the present time step and all the future time steps, we obtain the following (detailed derivation can be found in [503])

$$v_\pi(s) = E_\pi\left[r(k) + \gamma v_\pi(s')\right]. \tag{9.7}$$

Equation 9.7 is called the *Bellman equation for v_π*, which describes a recursive relationship between the value function of the current state and the next. A similar relationship can be found on the $Q-$function such that

$$Q_\pi(s, a) = E_\pi \left[r(k) + \gamma Q_\pi(s', a') \mid s, a \right]. \tag{9.8}$$

Now, the problem of RL is translated to solving the recursive relationship between the value function and the $Q-$function and maximizing these functions to obtain a policy π. In other words

$$v_\pi^*(s) = \max_\pi E_\pi \left[r(k) + \gamma v_\pi^*(s') \right].$$

$$Q_\pi^*(s, a) = \max_\pi E_\pi \left[r(k) + \gamma Q_\pi^*(s', a') \mid s, a \right]$$

The two equations above summarize the most basic constructs in the RL literature used to build all the algorithms in RL described in the next two sections.

9.3 RL ALGORITHMS

Prior to describing the algorithms, we will state the simplest example in the RL literature—a cart pole balancing problem. Most algorithms discussed in this section can be implemented on the cart-pole balancing problem, and code can be found online easily. We recommend starting with OpenAI Gym [72] and [503].

9.3.1 Cartpole

A cart-pole problem consists of a cart with a small mass attached to the cart using a rod. The problem is maintaining the pole at a specific angle or ensuring that the pole stays upright (refer Figure 9.2). The state of this system is described by the angle θ and the (x, y) coordinate of the cart. The action in this system is to choose the force F. The RL goal in this problem is to ensure that the pole on the top does not tip over. The problem is episodic, where we determine the success or failure of the system at the end of each episode. When the pole tips over a certain angle, we view it as a failure, and a reward of -1 is ordained. On the other hand, when the pole stays upright, a reward of $+1$ is ordained. We seek to control the force F on the cart by generating an action. The state-transition probability of the system determines how the force affects the pole on the cart. For the rest of the chapter, we will use the cart-pole balancing problem to demonstrate different approaches and algorithms for solving RL problems.

9.3.2 Learning for Cartpole

In a standard RL sense, there are two usual ways to generate the actions. These are the Monte Carlo method and temporal difference learning methods.

9.3.2.1 *Monte-Carlo Methods*

Monte Carlo methods derive their basic structure from sampling, where repeated sampling is utilized to evaluate the value functions and policies. In an episodic setting

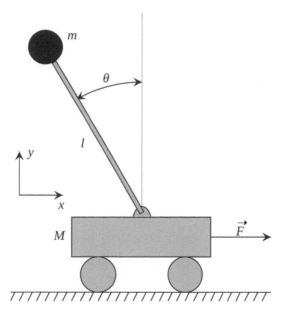

Figure 9.2 Cart-pole balancing. (https://danielpiedrahita.wordpress.com/portfolio/cart-pole-control/)

of RL, each episode has a starting and terminal point. Throughout the episode, the agent is provided with a reward after each action. Based on the accumulated reward, the agent evaluates the policy by sampling from the experiences (that state, action, and reward from each episode of the cart-pole) and estimating the expected value of the action (the force on the cart). Based on an update to the policy/value, we improve the policy, and the next episode is played using an updated policy. This manner of evaluating the policy (using the policy to play—policy evaluation) and updating the policy (improving the policy—policy improvement) repeats in a cycle until convergence is reached. This method is also known as generalized policy iteration in the RL literature, with the key point that the improvement and evaluation step is performed using samples. The same process can also be repeated with the value function where the agent seeks to update itself with sample (the most popular sampling strategy is importance sampling), the method known as value iteration. Value and policy iterations form the central tenet of learning with reinforcement signals. These methods can be applied to both state-value functions and the value function.

9.3.2.2 Temporal Difference Learning (TD)

The key difference between the Monte Carlo method and the TD learning is that, while Monte Carlo updates are only performed at the end of each episode, TD updates [530] are done at arbitrary steps. This algorithm is called the $TD(\lambda)$ algorithm. The most basic case is $\lambda = 0$, also known as the $TD(0)$. This method requires an update after each step. To make these updates, one usually defines a construct known as temporal difference error TD where the equations in (9.2.2) are modified to write sort of an error such that

$$\delta(s, s') = \left[\underbrace{r(k) + \gamma v_\pi(s')}_{\text{TD target}} - v_\pi(s)\right],$$

(9.9)

A similar construct of TD error can be written with the $Q-$ function as well

$$\delta(s, s', a) = \left[\underbrace{r(k) + \gamma Q_\pi(s', a)}_{\text{TD target}} - Q_\pi(s, a)\right],$$

(9.10)

These constructs can then be utilized as an update to the value and the $Q-$ function to obtain

$$Q_\pi(s', a) = Q_\pi(s, a) + \alpha\delta(s, a, s')$$ (9.11)

$$v_\pi(s') = v_\pi(s) + \alpha\delta(s)$$ (9.12)

The usual method of learning involves starting with an estimate of the value/Q function and utilizing the update at each iteration for recursively obtaining policies that must be utilized. Again, both value and policy iteration algorithms can be adapted to TD-based learning.

9.3.2.3 Trade-Offs Between These Two Methods

The key advantage of the TD over Monte Carlo is that they do not require a model of the environment or a model of the reward and/or the probability distributions corresponding to the next state. All these constructs are required in the Monte Carlo setting to ensure that the samples are informative. The second advantage of TD is that they are implemented fully online, and we do not have to wait for the episode to end such that rewards are obtained. Therefore, TD methods are more nimble in real-world applications in continuous and rapidly changing environments. Furthermore, the learning in Monte Carlo methods is exceptionally slow as they have to ignore many experimental actions. On the other hand, every time step update strategy is able to learn effectively from transitions at each step. Furthermore, the $TD(0)$ method has been shown to converge given α in (9.12) is sufficiently small and has been found to be faster than Monte Carlo methods in stochastic environments.

9.3.2.4 Exploration-Exploitation Trade-Off

A significant need for any RL system is to be able to take actions that result in useful rewards. An agent that keeps on taking actions that led to good rewards in the past will result in an agent that will perform very well in states that it has observed before—this is known as exploitation. However, such an agent cannot adapt to changes in the environment. To adapt to a change, an agent must take actions that it did not take before. This is known as exploration. However, if an agent explores more, its cumulative reward will never improve. As evident, an agent must trade-off exploration and exploitation to retain nimbleness to changing environments and increased performance. A detailed analysis of the exploration-exploitation trade-off can be obtained from [77].

9.3.3 RL Methods

All RL methods can be categorized into actor-critic, policy-based, and value-based. We start by describing the value-based methods.

9.3.3.1 Value-Based Methods

The goal of these methods is to improve value. However, in most cases, the value of a problem cannot be analytically evaluated. Therefore, we must estimate or approximate the value through an iterative update procedure. We consider the case when we do not have the complete model that describes the environment, and the state transition probability is unknown. In other words, the agent has to experience its environment by trial and error. These methods are known as model-free learning methods, and the optimal policies are discovered using value function-based update procedures.

Q-learning: Q-learning is one of the most popular RL methods where we directly approximate the optimal action-value function. This is done by converting the $Q-$ function update rule in (9.12) into a recursive model that allows us to learn the $Q-$function. Typically, it is an off-policy temporal difference learning method defined as the following

$$Q(s,a) \leftarrow Q(s,a) + \alpha \underbrace{[r_{k+1} + \gamma \max_a Q(s',a) - Q(s,a))]}_{\delta(s,a)}, \qquad (9.13)$$

where the Q-function estimate is updated by the current estimate, $Q(s,a)$, plus the difference between the Q-function for the next state and the current estimate. Note that the estimate for the next state Q-function value is not necessarily obtained by following the current policy derived from the current Q values but by taking the action that maximizes the value. Therefore, Q-learning is an off-policy method. This update will guarantee convergence to the optimal Q-value, $q*$ [561].

In the cart-pole balancing problem, the state space contains finite combinations of position and angles, and the actions are the forces toward left or right. With Q-learning, we initialize the Q values for each state-action pair and then select the action, a, for the current state, s, greedily based on the Q values. Now the system returns a reward, $r(t+1)$, and next state, s'. To obtain the Q value for the next state, we select the action, a, that maximizes the value instead of following the current policy. The process will repeat until reaching the terminal state. The implementation of Q-learning for the cart-pole balancing problem can be found in [503].

SARSA: Like Q-learning, SARSA estimates the optimal Q-function for each state-action pair and then derives the policy using the updated Q values. The difference is that, in SARSA, the next state Q value is associated with the action following the current policy instead of the action that produces the maximum value. It has the following update rule,

$$Q(s,a) \leftarrow Q(s,a) + \alpha[r(k+1) + \gamma Q(s',a') - Q(s,a))], \qquad (9.14)$$

Thus, compared to Q-learning, to update the current Q-values, SARSA selects the next action, a', based on the current policy instead of the Q-values. In the cart-pole balancing problem with SARSA, we first obtain the action (horizontal forces) for the given state (angles) from the current policy and again select the action for the next state based on the current policy to update the current Q-values. If the next state, s', is terminal, then $Q(s',a') = 0$.

Q-learning and SARSA are model-free value-based methods, where Q-learning is off-policy, and SARSA is on-policy. Being an off-policy method, Q-learning can learn the optimal policy regardless of the current policy, but converging takes longer. On the other hand, SARSA can converge faster and handle noise in the environment but may converge to a sub-optimal policy. Therefore, directly updating the policy is preferred in such contexts, which leads to our next category of methods, which is the policy-based methods.

9.3.3.2 Policy-Based Methods

The value-based methods aim to estimate the optimal state-action values and then derive the optimal policy based on these values. The policy would not exist without the value estimates. Policy-based methods, on the other hand, directly estimate the policy, bypassing the need to obtain the value functions. The policy-based methods have gained significant attention due to their inherent flexibility, ability to handle continuous action spaces, and convergence properties [403]. In this section, we introduce the policy gradient method, one of the most basic methods in this domain. The key difference between policy-based and other methods is that we parameterize the policy function with a differentiable function and some parameters w and find the optimal parameters via gradient-based optimization methods.

Policy Gradient: In the policy gradient approach [403], we directly approximate the policy function, $\pi_w(s|a)$, with a set of parameters, w. The objective is to find the optimal policy by searching in a compact space $\mathcal{W}$ but with gradient-based optimization methods. A stochastic policy is typically utilized instead of a deterministic to ensure that the policy explores enough of the environment. Therefore, the policy function, $\pi_w(s|a)$, is a probability distribution. When the action space is discrete, it is straightforward to parameterize the policy function that is endowed with a soft-max distribution $\pi_w(a|s) = \frac{e^{h(s,a,w)}}{\sum_b e^{h(s,b,w)}}$, where $h(\cdot)$ is the action preference function and can be parameterized arbitrarily. On the other hand, if the action space is continuous, the common way is to assume a Gaussian distribution and then parameterize

9.3.4 Recent Advances in Reinforcement Learning

Here is an overview of some common approaches to deep learning in reinforcement learning: The first RL method that demonstrated effectiveness in the presence of images was a deep-Q network in [322], where a neural network approximates the Q-function directly from the raw sensory input, allowing for more effective handling of high-dimensional state spaces. DQNs were one of the first successful examples of using deep learning in reinforcement learning and have been applied to a wide range of tasks, including game playing and robotics. This work has been extended to use double neural networks in [542] to address some sampling challenges. These deep learning methods belong to the class of value-based methods where two separate neural networks are utilized to estimate the action and the target in the temporal difference error. Essentially, the Q-function in the temporal difference error is approximated through a neural network. However, these works can only be effective in small action spaces.

Alternatively, policy gradient methods such as deep distributed policy gradient (DDPG) can use a neural network to optimize the policy function directly [47, 388]. These methods addressed the problem of DQN when the action space is continuous. However, similar to the actor-critic methods, these methods utilized an actor with a separate critic to achieve performance on several robotic tasks. While DDPG also borrows the ideas of experience replay and separates the target network from DQN, DDPG seldom performs exploration for actions. To fix this issue, noise is added to the parameter or action space.

The key backbone of all these deep learning-based methods is the policy gradient theorem introduced in [504] and the natural policy gradients in [221]. However, the policy gradient method suffers from several challenges. The first challenge is sample inefficiency, where the samples are utilized only once, and a new trajectory is sampled each time a policy is updated. Moreover, because sampling is expensive, this becomes computationally prohibitive in the long run. In addition, after a large policy update, the old samples no longer represent the problem at hand. The second challenge is due to the inconsistency of policy updates, where the updates tend to miss the reward peak, and the algorithm does not recover. Finally, high variance is present in the reward due to the fact that policy gradients s a Monte Carlo learning approach, and the convergence gets hampered.

While these challenges can be addressed by introducing a critic, [471] introduced "trust region-based policy optimization (TRPO)", which improves upon natural policy gradients by introducing a KL divergence-based penalty into the update rule. TRPO introduces and improves many of the challenges of policy gradients. However, TRPO involves second-order optimization and is ineffective for large parameter spaces that are obviously very important in deep RL scenarios. This issue was later corrected in the [475], which introduced a proximal policy gradient (PPO) optimization approach that introduces an adaptive penalty to the KL-divergence term and introduces heuristics such as clipped objective to make the optimization cheaper to implement and fast to converge. Since its inception in 2017, PPO has been the standard go-to algorithm for continuous control problems in RL.

As we mentioned earlier, introducing a critic can address the high variance of the policy gradient methods. These lead to the set of actor-critic methods where the policy in policy gradient methods is rewritten using an advantage function to obtain an advantage actor-critic (A2C) and its asynchronous version, which is the asynchronous advantage actor-critic (A3C) [321]. The difference between the two methods is the presence of asynchronous policies in A3C which makes the search effective.

9.4 REINFORCEMENT LEARNING APPLICATIONS IN SCIENCE

Due to the use of deep neural networks, RL has become very popular and has led to numerous breakthroughs in many applications. We discuss some of these applications below.

9.4.1 Robotics and Control Systems

Learning to control complex systems with feedback data is an active field of systems and control theory research. Various RL approaches have been utilized in control applications, ranging from RL in adaptive control [345] to more modern reinforcement learning techniques such as actor-critic methods [57, 265].

Traditionally, robots have been controlled with RL through sensor measurements generated over time [503]. However, in many emerging robotic systems, the feedback information is only available in the form of snapshot images [322]. Therefore, modern RL algorithms aim to learn dexterous robotic manipulation directly from raw pixels. This is crucial in cases where no explicit state measurements or desired trajectory are available, with the camera and neural network extracting the necessary information from the images. In this context, recent advancements in deep Q learning, spurred by [322], have demonstrated that RL models using deep neural networks can perform complex decision-making tasks, and they have been extended to robotic systems (see [349] for a good review). RL plays a significant role in two major classes of robots: rigid link and soft continuum robots.

Rigid-link robots   Reinforcement learning has a rich history of research in the control of rigid link robots [505]. Several RL algorithms have been developed for robotic manipulators, including traditional algorithms such as [402, 403] and more modern ones that depend on image feedback as in [341, 340, 264]. These algorithms aim to control each link in the robot embedded with motors. The control is generated in such a way that the combination of these links performs the optimal action. However, since rigid link robots have limited degrees of freedom, the problem is severely underactuated, making it difficult to learn to grasp with RL [408]. This remains one of the most significant open problems in robotics, as the flexibility of the biological hand is much superior to that of rigid link robots, and even grasping is a difficult task for rigid link robots.

Soft-link robots: Conventional rigid-link robots can be controlled directly by motor commands on each joint. The pose of all points, including the end-effector, is fully determined once joint angles and link lengths are available. In contrast, "continuum" mechanisms leverage continuous arcs to produce a bending motion without a skeletal structure, providing additional degrees of freedom and flexibility to robots, particularly in surgical devices and healthcare [445, 556]. However, controlling such robots using RL is challenging due to the need for tracking continuous arcs, resulting in a significantly larger state space compared to rigid link robots and the curse of dimensionality problem becomes severe [408]. To address this challenge, model-based and model-free approaches have been explored in this field [408]. For example, [520] leveraged a model-based policy learning algorithm to simulate a tendon-driven soft manipulator capable of touching dynamic targets, while [579] accomplished the position control of a cable-driven soft arm using Deep Q-learning. Other examples include actor-critic methods [408, 343], value-based methods such as SARSA [23], Q-learning [213], and DQN and its extensions such as DDQN and Double DQN [470, 579] [594].

Perspectives: There are numerous challenges for effective RL in robotics, such as sample efficiency, overfitting, bias, and safety and reliability. Safety and reliability are of prime importance to robotics, where there is no scope for errors as they can lead to the loss of human life. Therefore, safe and reliable learning is the most important RL challenge for robotics applications. While every control algorithm comes with some mathematical analysis that ensures that the control algorithm can keep the system in bounds while the RL algorithm is learning, recent trends in deep reinforcement learning do not provide such guarantees [74]. Therefore, providing empirical and mathematical safety guarantees is essential for the RL algorithm in applications such as robotics.

9.4.2 Natural Language Processing

Natural language processing (NLP) is concerned with the interaction between computers and human languages, including text and dialogue generation, machine translation, and text summarization. With the advent of machine learning techniques, various supervised learning models such as recurrent neural network-based language models [500, 288, 275], and transformer-based language models [210, 12, 419] have shown significant progress in these tasks. However, the need for high-quality labeled data and handling context coherence remain challenging. Reinforcement learning algorithms, which can optimize complex decision-making processes based on rewards received by the agent, are flexible and adaptive and have been applied to various NLP tasks. RL has the potential to address some of the challenges in NLP, including generating coherent and informative text with limited labeled data and improving the performance of machine translation and text summarization.

Dialogue generation   is an essential component of natural language processing, which seeks to generate natural language responses that are relevant and coherent to a given context or user prompt. However, achieving response relevance and coherence remains one of the main challenges in the field. To address this, reinforcement learning has emerged as a promising approach, enabling the learning of policies that optimize the relevance and coherence of generated responses based on pre-defined reward criteria. These criteria can include informativity, coherence, and grammatical measures, among others [267, 268, 589]. Handling long-term dependencies is another challenge in dialogue generation. With the utilization of RL methods, the agent can learn a policy by taking into account the entire conversation history instead of only the most recent context. In [461], it has shown the efficacy of using RL to capture long-term dependencies using a hierarchical structure to model the conversation history. In addition, RL can handle user preferences and feedback. This capability can fine-tune large language models (LLMs) [101]. In this approach, the agent receives feedback from humans to improve its understanding of the goal and the corresponding reward function. The RL algorithm then trains the agent to achieve its goal. As the agent improves its performance, it continues asking for human feedback for actions with higher uncertainty, allowing it to refine its understanding of the goal further. The GPT family of LLMs, including GPT-4, have benefited from this technique and have achieved outstanding performance [101, 387, 73].

Machine translation. Machine translation aims to translate natural languages accurately and coherently. Deep learning, particularly recurrent neural network-based sequence-to-sequence (seq2seq) models, has significantly improved this task [501]. However, seq2seq models face challenges like exposure bias and inconsistency between training and test objectives [51]. Reinforcement learning (RL) methods, with additional reward formulation, can address these issues [571, 232]. RL agents maximize reward functions, such as BLEU [392], for better translation quality. Exposure bias arises when models train on ground truth but generate tokens based on predictions at test time [423], causing errors and reduced translation quality. The REINFORCE algorithm can mitigate exposure bias by exposing models to their predictions during training [503]. Inconsistency between objectives occurs when training loss functions, like BLEU, don't optimize evaluation metrics. Incorporating RL helps optimize reward functions aligned with evaluation metrics [577, 573].

Language understanding Language understanding is a challenging problem in NLP, and RL has emerged as a promising technique to solve it. RL algorithms can handle large amounts of data and learn from them to make decisions in the form of an optimal policy. In a natural language understanding (NLU) problem, grammar can be used to determine the meaning of a sentence. RL can then be applied to select the optimal sequence of substitutions during the parsing process, producing a vector-type representation of relevant variables. This representation can determine the next appropriate action. One of the early successes of RL in NLU was demonstrated in [186], where a deep reinforcement relevance network (DRRN) was proposed. The model showed the ability to extract meaning from natural language rather than simply memorizing it. As a result, DRRN could summarize news articles by learning to generate summaries that are both relevant to the article and concise. DRRN is trained to select the most important sentences from an article by considering the rewards of selecting a sentence and the next state. It is then optimized to maximize the expected reward of selecting all the sentences in summary. Another work in NLU that leverages RL is the Deep Reinforcement Learning for Sequence to Sequence Models (DRESS) [354]. DRESS is an RL-based sequence-to-sequence model that can generate polite responses in a dialogue system. The model learns to optimize a reward function that encourages the generation of polite responses by assigning positive rewards for polite responses and negative rewards for impolite ones. In addition, in [75], the authors proposed a model that uses RL to generate natural language explanations for the output of a neural network. The model generates a sequence of sentences by selecting the most relevant sentence given the current state and then predicts the next sentence using an RL-based policy.

9.4.3   Image and Video Processing

RL methods have successfully been applied to tasks such as object detection, image segmentation, image enhancement, visual navigation, and video summarization in image and video processing.

Object detection   In object detection, an RL agent will interact with the image of the environment. At every time step, it will decide which region to focus attention on with the objective of locating the object in the minimum possible number of time steps. For example, an RL agent can learn to deform a bounding box using transformation actions to determine the most specific target object location [81]. In another work, an RL agent was trained to decide where to pay attention among predefined region candidates in a given image window, following the idea that the focused regions contain richer information [76].

Image enhancement   RL agents can be trained to denoise the images in image enhancement tasks. With the formulation of pixel-wise reward, the agent can effectively take action to change the pixel values to achieve certain goals. PixelRL [154] adopted deep RL to perform image denoising, image restoration, local color enhancement, and saliency-driven image editing. RL models can utilize learned latent subspaces to recursively enhance the quality of camera images through patch-based spatially adaptive artifact filtering and image enhancement for camera image and signal processing pipelines. In Recursive Self Enhancement Reinforcement Learning (RSE-RL) [37], the authors utilized a soft actor-critic approach to govern the adaptive learned transformation for filtering and enhancing the noisy patches. This resulted in effective performance in removing heterogeneous noise and artifacts from the image.

Object segmentation   Segmenting objects in videos is an important but challenging task in computer vision. In [460], the authors proposed a deep RL-based framework for video object segmentation, which divides the image into sub-images and applies the algorithm to each sub-image. A set of actions are proposed to modify local values within each sub-image, and the Q-values are approximated using a deep belief network. The agent receives a reward based on the change in the quality of the segmented object within each sub-image across the action. Furthermore, a deep RL model trained with hierarchical features achieved a high edit score and frame-wise accuracy in recognition of surgical gesture [277]. The method treated the classification and segmentation as an MDP where the state while the actions are optimal step size and choosing the gesture class. The reward function facilitated a larger step size and the correct gesture class.

Visual navigation   RL has been successfully used in visual navigation to tackle problems such as target-driven visual navigation, where an agent learns to navigate toward a target object in an environment. In one study, an RL-based method called the Neural Map was proposed, which combined a spatial memory system with an attention mechanism to learn target-driven navigation policies [394] efficiently. The Neural Map allowed the agent to store and access spatial information about the environment, improving its navigation performance. In another work, the authors used deep RL to train an agent capable of navigating a 3D environment to find target objects, demonstrating the potential of RL for solving complex visual navigation tasks [609].

Video summarization   is another area where RL has demonstrated success. In the work of Zhang et al., a deep RL-based method called DR-DSN was proposed for unsupervised video summarization [606]. This method formulated video summarization as a sequential decision-making problem and used a deep Q-network (DQN) to learn an effective summarization policy. The authors showed that the proposed method could generate video summaries with higher quality and diversity than several state-of-the-art methods.

RL has been applied to predict binding affinity and optimize the binding process in protein-ligand binding. One such method, RL-Dock, was proposed to address molecular docking problems by training an RL agent to navigate the protein-ligand binding landscape, optimizing both the pose and the scoring function [215]. RL has been used in drug discovery to optimize drug design parameters, predict drug-target interactions, and identify new drug targets. For example, the MolDQN model used deep RL to design new molecules with specific properties, such as low toxicity and high efficacy [607].

Overall, the application of RL in bioinformatics and drug discovery holds great promise for accelerating the development of new drugs and therapies and improving our understanding of biological systems at the molecular level. However, many challenges still need to be addressed, such as developing more efficient RL algorithms and integrating RL with other computational and experimental approaches. With continued research and development, RL has the potential to transform the field of bioinformatics and drug discovery in the coming years.

9.4.4 Climate Modeling and Prediction

Climate modeling has become a critical area of research in recent years due to the urgent need to address climate change. These models typically involve spatiotemporal problems that require careful consideration of changes in the spatial environment and the temporal spectrum. In the literature, these problems are commonly modeled as point processes through time series data or field processes through raster data. Point processes involve a set of time-ordered events with an underlying partial differential equation, which can be modeled for one or multiple spatial locations using reinforcement learning (RL)([269, 532]). Raster data, on the other hand, is collected using two-dimensional partial differential equations that impose boundary conditions on a 2D grid over time([559, 578]).

While standard machine learning models have been widely applied in climate modeling (see [272] for a review), there is not enough literature on the application of RL methods for temporal point processes or spatiotemporal problems. As far as we know, RL has been primarily applied in social climate modeling, which seeks to understand the impact of social policies on the climate through RL. One of the first works to explore the marriage of RL and social climate modeling was presented in [495], where the authors used various deep Q-learning variants to model the interactions between decisions on a holistic level using an earth systems model ([353, 236]). The goal was to generate undiscovered holistic policies that could influence decisions. [574] extended this work by modeling climate as a control problem and demonstrating the effect of decisions on climate. This work has significant implications for the earth as a whole.

While RL is one of the most important paradigms for science, the application of RL in science has not really taken off due to many challenges that exist. We discuss some of these challenges below.

9.5 CHALLENGES AND LIMITATIONS OF REINFORCEMENT LEARNING IN SCIENCE

Reinforcement learning for complex systems, such as observed in science applications, is of absolute importance. There are many applications in science where labeled data is limited. Due to this, learning in a supervised fashion is infeasible, and learning from abstract rewards, such as what is done within the reinforcement learning paradigm, is more suitable for such cases. Motivating this line of thought, we presented several algorithms and methods that were presented in the previous section. Although impressive performance has been observed, several challenges exist and must be discussed to achieve the fullest potential of reinforcement learning in science.

9.5.1 Sample Inefficiency and Exploration Challenges

A typical RL task is composed of three steps: (1) an agent selects an action and utilizes it to actuate or effect change in the environment, (2) the environment, as a consequence of the action, transitions from its current state to a new state, and (3) the environment then provides a scalar reinforcement/reward signal as feedback. By performing these three steps repeatedly, the agent explores the environment and accumulates experiences or knowledge regarding the environment.

Based on these experiences collected over time, the agent learns the best sequence of actions from any situation/state [206] to achieve a desired objective/goal, usually described as a function of the temporal difference (TD) error. A typical DRL learning scheme such as the one introduced in [322], or its variants in [542, 471] aims to nudge toward the optimal value function successively. Consequentially, efficient learning requires an accurate approximation of the optimal Q-function. The learning then requires storing experiences and exploring the search space using these experiences. In the end, efficient learning is pivotal on two key conditions: efficient use of experiences and sufficient exploration of the parameter or weight space.

To enhance the process of collecting and efficiently utilizing experiences within the high-dimensional state (such as images and audio inputs) and action spaces [261, 189, 188, 306], an experience replay strategy [274, 225, 453, 405] is usually employed in RL schemes. However, the memory space to store the experiences may be limited or expensive in many applications, and therefore, a mechanism to omit irrelevant experiences are desirable and has been considered [558, 405, 201]. Despite appropriate sample selection strategies, useful experiences may be discarded. For instance, experiences collected initially near the initial states may be useful, but when some experiences are omitted based on time or reward, these experiences may be discarded. Even when these initial experiences are not discarded, they have the tendency to bias the learning procedure, [519, 322] as the agent may spend more time exploring states dictated by the initial experiences, but these experiences do not contribute positively to the overall objective of the agent. Furthermore, this inappropriate exploration of the sample space leads to the need for a significant number of samples for learning. In summary, despite several successful applications of DRL, these strategies of discarding or utilizing experiences, in general, are typically sample-inefficient [471, 214].

9.5.2 Overfitting and Bias

Bias-variance tradeoff is an important phenomenon in ML where this phenomenon can be understood in the context of a model: a model that underfits the data has high bias, whereas a model that overfits the data has high variance. A model with high bias fails to find all the patterns in the data, so it does not fit the training set well and will not fit the test set well either. A model with high variance fits the training set very well, but it fails to generalize to the test set because it also learns the noise in the data as patterns. In most Reinforcement Learning methods, the goal of the agent is to estimate the state value function or the action value function. Value functions describe how desirable the action is with respect to the state and the environment. In most common scenarios, the state-value function/action-value function is iteratively estimated to accurately predict the impact of the agents' actions on an environment.

The idea of Bias-variance tradeoff in this context appears when analyzing the accuracy of the state-value function/action-value function. A bias and variance are defined through the difference between the expected value of the estimated state value function/action value function with their true values considering which regions of the space were explored during training. Initially, the agent takes random actions, and the learning of the agent is completely biased by the returns the agent receives based on these initial experiences. If the agents do not experience new regions of the space, the final performance of the agent will be biased toward these initial experiences. Therefore, the agent will not do very well in unexplored regions of the sample space, leading to poor generalization. While reducing the bias can help in dealing with unexplored regions of the sample space. The associated drawback is increased variance in the training data and, subsequently, poor accuracy in the environment.

The variance of an estimator denotes how "noisy" the estimator is. Because this estimator is a (discounted) sum of all rewards, it is affected by all actions. If two trajectories have different returns, the associated value function will provide different values leading to high variance. Large variance for the agent leads to underfitting, while large bias leads to overfitting. Another way to understand the table above is to use the idea of underfitting and overfitting. Suppose that our agent in a "real life" environment died. To improve, we need to consider what actions led to our death so we can perform better. Its death was likely caused by actions close to its death: crossing the sidewalk at a red light or falling down a cliff. However, it is also possible that a disease it contracted several months ago is the cause of our death, and every action after had minimal impact on its death.

9.5.3 The Perils of Backpropagation in the Context of RL

On the other hand, a tractable and efficient training strategy to use the experiences can improve the exploration of the environment and subsequently better accuracy [602, 110]. The most common strategy for training DNNs is the backpropagation/stochastic gradient descent (SGD) algorithm [322]. Despite promising results, SGD suffers from issues such as the vanishing gradient problem and slow learning [257, 244, 183] when dealing with RL problems.

Many alternative approaches have been introduced in the literature to target these issues [244, 258, 376, 131, 433]. One prominent method is a direct error-driven learning (EDL) approach that was proposed to train DNNs [244] in the context of supervised learning. However, the EDL approach was not evaluated in DRL applications with image pixels as the NN inputs, and a theoretical analysis of the EDL algorithm is a missing component in [244].

Similar to the SGD, the EDL approach focused on developing explicit targets for training each layer of the DNN by introducing a linear transformation of the error and replacing the gradients in the updated rules. In summary, sample efficiency and reduction in the learning time in a large, uncertain environment could potentially be resolved if an algorithm can: a) efficiently reconstruct a virtual model that can store more information, especially in a stochastic state-space; b) reduce the vanishing of learning signal that can stagnate the learning; and c) improve the exploration rate in a large stochastic state space.

9.6 FUTURE DIRECTIONS AND RESEARCH OPPORTUNITIES

Reinforcement Learning (RL) has shown great promise in solving complex problems across various domains. As research progresses, several future directions are likely to shape the development of RL:

- *Multi-Agent Reinforcement Learning.* Multi-agent RL involves learning policies for multiple agents that interact with each other and the environment. Multi-agent RL is important for game-playing, robotics, and social network analysis applications.

- *Transfer Learning and Meta-Learning.* Transfer learning and meta-learning involve using knowledge acquired from one task to improve performance on another task. These approaches are important for enabling RL algorithms to learn more efficiently from limited data and to adapt to new tasks and environments.

- *Safe and Robust Reinforcement Learning.* Safe and robust RL involves developing algorithms that can operate in uncertain and unpredictable environments while minimizing the risk of catastrophic failure. Safe and robust RL is important for autonomous driving, healthcare, and finance applications.

- *Efficiency and scalability of RL algorithms.* Developing new evaluation metrics and benchmarks and integrating RL with other machine learning and optimization techniques.

- *Ethical and societal considerations.* Addressing the ethical implications of RL applications, ensuring fairness, accountability, and transparency in AI decision-making, and understanding the broader societal impact of RL systems.

III

Applications

CHAPTER 10

Autonomous Synchrotron X-Ray Scattering and Diffraction

Masafumi Fukuto

National Synchrotron Light Source II, Brookhaven National Laboratory, Upton, New York, USA

Yu-chen Karen Chen-Wiegart

Department of Materials Science and Chemical Engineering, Stony Brook University, Stony Brook, New York, USA
National Synchrotron Light Source II, Brookhaven National Laboratory, Upton, New York, USA

Marcus M. Noack

Applied Mathematics and Computational Research Division, Lawrence Berkeley National Laboratory, Berkeley, California, USA

Kevin G. Yager

Center for Functional Nanomaterials, Brookhaven National Laboratory, Upton, New York, USA

CONTENTS

DOI: 10.1201/9781003359593-10

I N THIS CHAPTER we discuss applications of Autonomous Experimentation methods in synchrotron X-ray scattering and diffraction experiments. Synchrotron X-ray scattering and diffraction provide information about material structure at length scales ranging from Ångstroms to hundreds of nanometers, which is critical to determining material properties. Synchrotron facilities hold the unique advantages of providing bright X-ray sources for rapid measurements and offering highly automated instrumentation, both of which are well suited to enabling Autonomous Experiments.

The chapter is divided into four sections. In the first section, we provide an introduction to the motivation, highlighting the potential benefits of enabling Autonomous Experiments at large-scale user facilities. The anticipated impact specific to the synchrotron facilities will be discussed, followed by a brief introduction to the synchrotron X-ray scattering and diffraction techniques. In the second section, we describe the implementation of an Autonomous Experimentation framework at the National Synchrotron Light Source II (NSLS-II), a U.S. Department of Energy (DOE) Office of Science user facility located at DOE's Brookhaven National Laboratory (BNL). This framework is based on an automated measure-analyze-decide feedback loop in which the *gpCAM* software package is used to enable on-the-fly machine-guided decision-making. In the third section, we discuss the types of Autonomous Experiments that have been conducted at NSLS-II and illustrate them with examples. The list of examples is not comprehensive and captures only a small subset of all the efforts being made at NSLS-II and other facilities to leverage artificial intelligence (AI) and machine learning (ML) methods. We have chosen these examples with the intention of illustrating certain advantages of the Autonomous Experimentation approaches. The chapter concludes with remarks on key ongoing and future directions to be pursued, including extending the Autonomous Experimentation methods to *in-situ/operando* experiments, incorporating ML-enhanced data analysis and domain-knowledge-aware decision-making into the autonomous loop, and developing multimodal workflow designs.

10.1 INTRODUCTION

Integrating advanced AI/ML algorithms provides great opportunities to significantly accelerate materials research, which has an impact in the fields of chemistry, physics, and materials science. The broader impacts include various engineering disciplines that rely on the advancement in materials design, such as energy technology, alloy design, microelectronics, bioengineering, mechanical engineering, and civil engineering, as well as education [456, 184, 282, 182]. A materials research workflow typically encompasses various tasks including materials synthesis and processing, device construction, characterization at various stages, and then providing feedback to improve the materials preparation or device design. Throughout this workflow there are opportunities to gain fundamental scientific knowledge as well as to develop engineering strategies based on such knowledge. Autonomous Experimentation (AE) methods have the potential to accelerate not only various stages of this workflow, but more importantly fundamentally shift the paradigm by implementing an autonomous closed-loop workflow, where the characterization feedback is closely integrated with making research decisions, including those that guide theory and simulations [493]. Below we focus our discussion on Autonomous Experiments at synchrotron facilities, large-scale user facilities that provide bright X-ray sources, and advanced instrumentation for cutting-edge material characterizations.

10.1.1 Potential of Autonomous Experimentation for Enhancing Experiments at Large-Scale User Facilities

The development of new materials is of great importance to industry and society, as they play a critical role in modern technology. However, designing new materials is a complex and time-consuming process that often relies on traditional trial-and-error approaches. The reason for this is that the design parameter space for modern materials is simply too vast to be fully explored even with the help of theoretical computation and simulation. To address this challenge, government-funded initiatives like the Materials Genome Initiative (MGI) [516] and the Designing Materials to Revolutionize and Engineer our Future (DMREF) program funded by National Science Foundation, along with partnerships between academic and industrial research institutions, are working toward accelerating the discovery, design, and deployment of materials. Autonomous Experimentation has emerged as a promising paradigm to provide a faster and more efficient approach to synthesizing and processing materials, as well as characterizing their structure and properties. The recent advances in Autonomous Experimentation and data-driven approaches have revolutionized materials design by providing researchers with an unprecedented amount of information on material behavior. By integrating AI/ML-based data analytics with advanced characterization tools, scientists can now efficiently design materials with specific properties targeted for particular applications and gain a deeper understanding of the fundamental mechanisms that govern their behavior. These advances point to the potential of Autonomous Experiments to transform the research enabled at large user facilities. In particular, synchrotron and Autonomous Experiments together can address each aspect of the *materials paradigm* (Figure 10.1), which consists of elucidating materials' processing, structure, property, and performance relationships:

- **Processing:** Autonomous Experiments can enable researchers to perform synthesis and processing of new materials under *in-situ* and *operando* conditions more effectively. By automating the process intelligently, scientists can explore a wider range of processing conditions in a shorter time, leading to more efficient and effective discovery of new materials.

- **Structure:** Synchrotron techniques provide high-resolution imaging and characterization of material structure across scales, facilitating a better understanding of the underlying mechanisms that drive material behavior. Coupled with Autonomous Experiments, researchers can effectively analyze a vast library of materials more intelligently, allowing more comprehensive investigations into the relationship between the structure and other characteristics of materials.

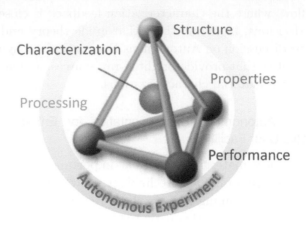

Figure 10.1 The integration of autonomous experiments in materials science, supported by initiatives such as the Materials Genome Initiative (MGI) [516, 102], offers a pathway toward efficient and accelerated materials discovery and design. The approach provides an avenue to studying materials across the paradigm of processing, structure, and property relationships, especially enabling a closed-loop approach, ultimately leading to the rational design of materials with tailored properties for specific applications [567].

- **Properties and Performance:** Autonomous Experiments can facilitate efficient exploration of materials properties, allowing scientists to identify promising candidates for further analysis. Moreover, the key to accelerating materials discovery and design lies in creating a closed-loop cycle between synthesis, characterization, and property testing, which can be achieved by integrating synchrotron and Autonomous Experiments.

Furthermore, the combination of experimental data and simulations can further enhance our understanding of materials behavior and pave the way for the rational design of materials with desired properties. By leveraging these new approaches, we can accelerate the development of innovative materials and move toward a more efficient and sustainable future with new materials design.

Automated vs. Autonomous. Before we discuss the advantages of Autonomous Experiments, it is important to differentiate the concept of Autonomous Experiments from what may be traditionally considered automation. Automation means having the ability to repeat pre-defined tasks precisely and reliably. The nature of automation is simply repetitive. While conducting an Autonomous Experiment requires the measurements, data processing, and analysis to be automated, it also needs to involve an intelligent decision-making process that is based on the ability to automatically learn and adapt to the experimental results. Figure 10.2 depicts and contrasts these two types of experiments.

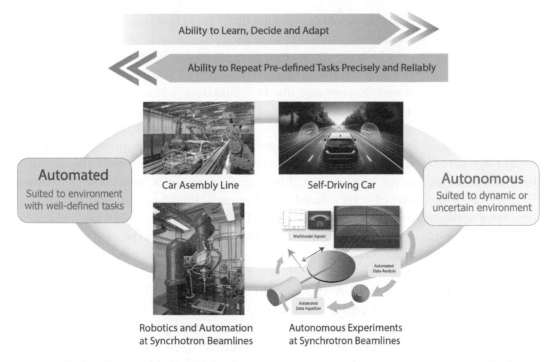

Figure 10.2 Comparison of autonomous experiments and automation, highlighting the difference in decision-making processes involved.

Having automated data collection and analysis infrastructures is a prerequisite for Autonomous Experimentation. At modern synchrotron facilities, instrumentation controls and data collection, management, and analysis capabilities are integrated and fully programmable. As such, synchrotron facilities are inherently well suited to enabling Autonomous Experiments.

10.1.2 Why Autonomous Experimentation? A Synchrotron User Facility Perspective

This chapter describes our recent experience with the application of Autonomous Experimentation (AE) methods to X-ray scattering and diffraction experiments at NSLS-II. Since it started operation in 2015, NSLS-II has steadily expanded its portfolio of specialized experimental stations, called the "beamlines", each of which offers unique, cutting-edge research tools to users. NSLS-II has 28 operating beamlines currently accepting users and a capacity to accommodate ~30 more beamlines in the future. The facility is in user operation for over 4,500 hours every year, and different beamlines operate simultaneously and independently during operation. During the operation period, the beamlines are available for experiments 24 hours a day. For the past year, NSLS-II received over 2,500 beam time proposals and hosted over 1,300 unique users.

Introduction to Synchrotrons. Before discussing the potential of AE methods to transform the experiments at synchrotron facilities, we first summarize salient features of synchrotrons. As scientific user facilities, synchrotrons around the world are dedicated to advancing scientific and technological research by enabling state-of-the-art experiments in which the interaction of synchrotron light with matter is used to probe the structure and property of materials [18]. Synchrotron light typically comes in the form of very intense and highly collimated X-ray beams, whose intensities are orders of magnitude greater than achievable with laboratory X-ray sources and whose properties (e.g., photon energy/wavelength, beam size, divergence, coherence) can be tuned by using appropriate X-ray source devices and optics. Synchrotron X-rays are generated by streams of electrons that are circulating near the speed of light inside a synchrotron storage ring. The ring circumference, which varies between facilities, is of the order of several hundreds of meters. Positioned around the storage ring are a series of X-ray source devices, each of which consists of a lattice of magnets and causes the trajectory of electrons to oscillate side to side; the associated acceleration of the electronic charge results in the emission of X-ray photons. The X-ray beam emitted from a source device is directed toward a beamline, which is located just outside the storage ring. A typical beamline is equipped with a set of X-ray optics—such as a monochromator and mirrors/lenses, to select the X-ray energy and focus the beam—and an endstation where a host of instruments are in place to allow users to perform X-ray experiments on their samples.

Now the question is: Why do we need machine-guided AE capabilities for synchrotron experiments? In our view, the answer arises from two key challenges that synchrotron facilities are starting to face, namely, the high rates and complexity of data generated by modern synchrotron experiments.

First of all, continual advances in synchrotron X-ray sources, detectors, and automation have enabled tremendous increases in the data collection rates at beamlines. For example, the 11-ID Coherent Hard X-ray Scattering (CHX) beamline at NSLS-II can collect 20,000 X-ray scattering patterns, or several GB of raw data, in just a few seconds, which in turn are automatically processed to generate over 100 GB worth of data sets of so-called two-time correlation functions to characterize time-resolved dynamic phenomena. As a result of increasing data collection and processing rates, experimental decision-making by human experimenters is starting to become the time-limiting step in many experiments at beamlines.

Secondly, the data spaces probed in modern synchrotron experiments are highly heterogeneous, and this gives rise to the challenge of maximizing the *value* of each measurement or datum while exploring these complex data spaces. Data may span real space (e.g., scanning-probe images of multi-component materials), or experimental parameter spaces (e.g., nanoscale structure measured as a function of temperature and composition). As the increasing data throughput propels the accessible data space of interest to expand in size, dimensionality, and complexity, traditional experimental strategies based on intuition or exhaustive exploration become increasingly inefficient and wasteful. In order to probe complex data spaces efficiently, intelligent experimental guidance methods—wherein the information content of a measured data point is maximized—are just as important as throughput. Such methods are also critical to enabling experiments with limited budget for time (e.g., finite beam time) or X-ray dosage (e.g., radiation-sensitive materials).

Recent advances in data science and high-performance computing offer a great opportunity to address these challenges by enabling machine-guided decision-making as an integral part of synchrotron experiments. What we envision is a future in which AI/ML-integrated experimental workflows are routinely used to automatically analyze and learn the data and autonomously decide how to steer the experiments as the data is being collected.

There are aspects of synchrotron facilities and their operations which make beamline experiments particularly well positioned to adopt the AE paradigm. First of all, computer control is already a crucial part of enabling experiments at beamlines. Since synchrotron X-rays are sources of radiation hazards, most X-ray endstations are enclosed inside steel- or lead-walled hutches, and experimenters are not allowed inside the hutch while the X-ray beam is on. Because of this, nearly all the experiments at beamlines are controlled from outside the endstation hutch. Second, many beamlines are already leveraging automation of experimental and data workflows in order to promote efficient use of limited beam time. Users gain access to a facility through a competitive beam-time proposal process, and the proposals submitted to given beamlines often outnumber those that are allocated time by a factor of two or more. Driven by the limited access due to high demand and the finite amount of beam time available for each experiment, computer-controlled automation of experimental hardware and software processes has become commonplace and already played a key role in enhancing the efficiency and throughput of synchrotron X-ray experiments over the years.

10.1.3 Introduction to Synchrotron X-ray Scattering and Diffraction

Our recent work on implementing the AE methods at synchrotron beamlines has focused on X-ray scattering, which includes **small- and wide-angle X-ray scattering (SAXS/WAXS)** for studying materials with varying degrees of structural order and **X-ray diffraction (XRD)** for crystalline materials. X-ray scattering represents one of the most mature and commonly used techniques to characterize the atomic to nanoscale structure of materials. As such, most synchrotron facilities have beamlines dedicated to X-ray scattering experiments. In a typical synchrotron X-ray scattering experiment, a highly collimated, monochromatic X-ray beam strikes the sample of interest and the X-rays scattered by the sample are recorded on area detectors as scattering/diffraction patterns, i.e., intensity as a function of scattering angles. Since X-rays interact with electrons within the sample material, the scattering pattern encodes the spatial distribution of electrons inside the material [18]. To a first approximation, the scattering intensity is directly proportional to the Fourier transform squared of the position-dependent electron density $\rho(\mathbf{r})$ of the material under X-ray illumination:

$$I(\mathbf{q}) \propto \left| \int d^3r \rho(\mathbf{r}) \exp(i\mathbf{q} \cdot \mathbf{r}) \right|^2 \tag{10.1}$$

where the scattering wave vector $\mathbf{q} = \mathbf{k}_{out} - \mathbf{k}_{in}$ corresponds to the momentum transfer between the incident and elastically scattered X-ray photons with wave vectors

$\mathbf{k}_{in}$ and $\mathbf{k}_{out}$, respectively, and $k = |\mathbf{k}_{out}| = |\mathbf{k}_{in}| = 2\pi/\lambda$, where λ is the X-ray wavelength (~ 1 Å in typical scattering experiments). Thus, by analyzing the scattering pattern, the researcher can learn about the structural details that are captured in the electron density $\rho(\mathbf{r})$. Due to the reciprocal nature of the Fourier transform, scattering at small scattering angles 2θ or small $q = 2k\sin(\theta)$ signifies the presence of structural features with large characteristic lengths, and scattering at large q corresponds to small scales. To give a simple example, if the material contained a periodic structure with period d, the scattering pattern would exhibit a diffraction peak at an angle 2θ corresponding to the so-called Bragg condition, $q = 2k\sin(\theta) = 2\pi/d$. The intensity of the peak provides a measure of how much of the illuminated material contains the corresponding periodic structure. The width of the peak (in q or in angle) is inversely proportional to the length scale ξ over which the periodic structure is correlated or extended (e.g., crystalline grain/domain size), such that the greater the grain size, the sharper the diffraction peak.

As will be discussed further below, one of the key prerequisites to implementing the AE approach is the availability of real-time data processing and analysis tools as part of the autonomous feedback loop. Given that synchrotron X-ray scattering measurements at beamlines are fast, many X-ray scattering beamlines are already equipped with automated real-time data processing and analysis workflows in order to aid users with understanding the results of the latest data being collected and leveraging them to plan, optimize, or redirect the next set of experiments during their beam time. This makes X-ray scattering beamlines well poised to become early adopters of the autonomous paradigm as well as testbeds for its applications at synchrotron beamlines.

10.2 IMPLEMENTATION OF GAUSSIAN PROCESS-BASED AUTONOMOUS EXPERIMENTATION FRAMEWORK AT NSLS-II

We describe here an AE approach that has been implemented at scattering beamlines at NSLS-II. This work is based on a collaboration between the Center for Advanced Mathematics for Energy Research Applications (CAMERA) at Lawrence Berkeley National Laboratory, BNL's Center for Functional Nanomaterials (CFN), and NSLS-II. The aim of this effort has been to establish AE methods that are targeted for broad use at synchrotron facilities where a multitude of diverse experiments are hosted. The effort recently led to successful demonstrations of autonomous X-ray scattering and diffraction experiments at NSLS-II [367, 373, 374, 370, 603, 129].

In general, possible AE methods can be considered to span a spectrum between two extreme cases, one aiming to leverage as much of existing domain-specific knowledge (DK) about the materials or phenomena under study as possible in the AI/ML-guided experimental decision-making, and the other being completely agnostic about such knowledge. For the purpose of the present discussion, these extremes are denoted as "DK-informed" and "DK-agnostic" approaches, respectively. Prior to our AE work, successful implementations of AI/ML-assisted experimental guidance (e.g., [443, 582, 91, 350]) tended to rely on DK-informed approaches that closely integrated existing knowledge, such as structural databases and a physical theory, into the experimental control algorithm. Despite the success of these approaches, we recognized that such close integration could be disadvantageous from the perspective of broad deployment and applications at large user facilities. First, by design, the applicability of a DK-informed approach tends to be limited to those material systems or physical phenomena for which substantial knowledge already exists. Second, when the decision-making algorithm is tightly coupled to a specific class of materials, the corresponding AE workflow would not be generally applicable. Even when the workflow could be repurposed for studying new or different materials, it would involve considerable work to reconfigure the decision algorithms to integrate a new set of knowledge.

In order to explore alternative AE approaches that have the potential for general applicability at large user facilities, we chose to tackle this problem from the DK-agnostic end of the spectrum. Specifically, our goal has been to develop a modular AE workflow that is based on DK-agnostic baseline decision-making algorithms but is also sufficiently versatile to provide ways to incorporate domain knowledge as needed.

The schematic in Figure 10.3 captures the basic features of the AE workflow that we implemented. The workflow consists of three fully automated steps that are

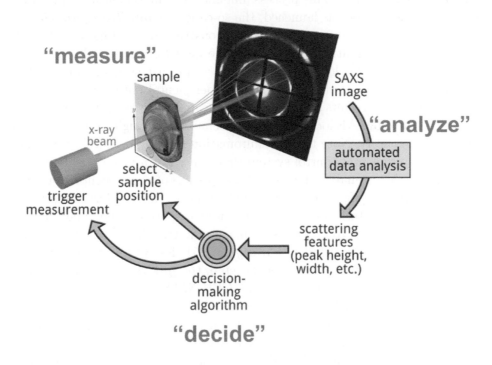

Figure 10.3 A schematic illustration of an autonomous experimentation loop orchestrated by automated sample handling and data collection, real-time data analysis, and AI/ML-guided decision-making. Each step in the "measure-analyze-decide" feedback loop is automated. The specific sample and SAXS pattern shown correspond to Experimental Example 1, which is described in Section 10.3.1 and Figure 10.4. Adapted from [367].

clearly separated out but interact with each other to orchestrate an autonomous "measure-analyze-decide" feedback loop. The **"measure"** step executes automated sample handling and data collection. The next **"analyze"** step carries out real-time data processing and analysis. A key role of this step is to perform *dimensionality reduction* to convert high-dimensional raw data (e.g., scattering patterns on pixelated area-detector images) to a limited set of physically meaningful quantities (e.g., intermolecular spacing, crystalline grain size) that can serve as the "signals" to guide experimental decisions. These analysis-derived signals are then passed on to the **"decide"** step, which is executed by decision-making algorithms. The algorithms that we developed are based on the use of **Gaussian process (GP) regression** [431] in a Bayesian-like optimization framework to facilitate active learning-driven, real-time steering of experiments [370]. These algorithms take the "measured" signals based on the data collected so far, use GP to generate a surrogate model of the signal as a function of the experimentally controlled parameters (e.g., sample position, composition, temperature), and at the same time estimate the uncertainty distribution associated with the model. Then the algorithms use the knowledge gained from the GP modeling and uncertainty quantification to select a set of experimental parameters at which to do the next measurement. This process amounts to a mathematical optimization problem: before the AE loop is launched, the so-called *acquisition function* that is appropriate for the experimental objective is constructed as a function of the GP-derived uncertainty distribution and/or surrogate model of the signal(s) of interest; at each iteration of the AE loop, the algorithms locate a new maximum in the acquisition function and select this point in the experimental parameter space for the next measurement. The results are then sent to the "measure" step to close the loop.

In practice, the above AE loop has been implemented through three software processes working in concert. Note that the automation of the "measure" step depends on the specific experimental control system that is used at a given facility or beamline. At NSLS-II, this is enabled by the *Bluesky* suite of experimental control and data collection software [62], as described in Chapter 8. The "analyze" step depends on the particular characterization techniques used to perform the experiment. For the experimental examples to be discussed below, the real-time data processing and analysis were performed by using *SciAnalysis* [251] for SAXS/WAXS and *XCA* [299] for XRD, respectively. For the "decide" step, the *gpCAM* software [370, 178], which was developed through the CAMERA-CFN-NSLS-II collaboration, was used to execute the automated GP-based decision-making. These pieces of software are all Python-based, and the interfacing between them is accomplished through message queue or object transfer services. In particular, NumPy arrays or lists of Python dictionaries are used to pass the analysis-derived signals and metadata to the "decide" step and then the decision-derived experimental parameters for the next iteration of measurements to the "measure" step.

The benefits and effectiveness of the implemented AE framework are highlighted below through the discussion of experimental examples in the next section. Here, it is worth emphasizing two key features of the GP-based decision algorithms. First, in their basic form, these algorithms enable purely data-driven, DK-agnostic decision-making. The key here is the decoupling of the decision algorithms from the analysis of experimental data. Since the inputs to the decision algorithms are not raw experimental data but a limited set of signals extracted from the analysis, the algorithms are independent of the specific characterization technique used in the experiment or particular material systems being studied. Thus, these algorithms are applicable to a wide range of techniques or material systems. This general applicability is very important for use at large user facilities.

Second, the acquisition function provides a facile and versatile means to tailor the decision algorithms according to the particular objective of a given experiment [367, 373, 374, 370]. Simply by constructing different acquisition functions and allowing the experimenter to select the appropriate function, the AE workflow can adapt a suitable decision mode and autonomously drive a series of iterative measurements toward achieving the experimental objective. Simple examples of commonly used decision modes include:

- **Survey** mode aims to efficiently explore and quickly grasp the landscape of an experimental parameter space. The simplest way to implement this is to set the acquisition function to the GP-derived uncertainty distribution. At each iteration of the AE loop, the algorithms select the unmeasured point in the parameter space with highest uncertainty in the surrogate model as the next experimental point to be measured, thus maximizing the information gain per measurement.

- **Feature optimization** mode aims to search for a set of experimental parameters that maximizes a certain physically meaningful quantity (e.g., grain size), which is reflected in the selected signal. One way to implement this is to set the acquisition function to the GP-derived uncertainty distribution weighted by the surrogate model. This allows the algorithms to achieve a balance between the parameter space exploration and the search for maxima in the signal.

- **Boundary detection** mode aims to locate and characterize internal boundaries (e.g., between phases) in the experimental parameter space. One way to implement this is to set the acquisition function to the GP-derived uncertainty distribution weighted by the gradient of the surrogate model. This allows the algorithms to achieve a balance between the parameter space exploration and the search for maxima in the signal gradient.

- **Cost optimization** mode aims to take into account experimental cost, e.g., as measured in *time* associated with preparing follow-up measurements. For example, next possible measurements could involve a variety of sample-handling processes in the "measure" step, ranging from simply translating the sample (to measure another spot on it), to changing the sample temperature, robotically exchanging the sample, or even making a new sample. The time costs associated with these processes can be very different. When available experimental time is limited (e.g., synchrotron beam time), it may be reasonable to require that a measurement with time-consuming sample processes be selected only when there is a great need for it. One way to impose such a requirement is to incorporate a weight based on the experimental time cost into the acquisition function, such that time-consuming measurements tend to be disfavored during the parameter space exploration.

Note that the use of the GP-derived surrogate model in defining the acquisition function (e.g., feature optimization and boundary detection modes) provides a simple way to enable decision-making that is sensitive to local variations in the data. Some of the above decision modes are used in the experimental examples to be discussed in the next section.

We wish to close this section with a few remarks. First, although the DK-agnostic nature of the implemented AE framework is emphasized above, this framework does provide avenues to incorporate domain-specific knowledge, at the expense of reduced agnosticism. This is a topic of ongoing research and outside the scope of this article. Nevertheless, it is worth noting that a few different ways to impart domain knowledge into the GP-based AE framework have been identified [370]. Second, the concept of building the AE framework from three separate but interacting components—with each fully automated and serving a distinct function (i.e., the "measure-analyze-decide" steps in the AE loop)—is well suited to deployment at large user facilities since it allows each of these components to be modular. Even the "decide" step can be modular. We focus here on a GP-based approach as it provides a simple way to enable an active learning-based AE framework. Nevertheless, it should be recognized that the "decide" step can and should utilize decision algorithms based on other AI/ML approaches, or even pre-programmed non-AE routines (e.g., grid scans), if the latter turns out to be more effective or efficient in achieving the objective of a given experiment.

10.3   EXPERIMENTAL EXAMPLES

In this section, we illustrate the benefits of the implemented GP-based AE framework and the potential of the AE paradigm to transform synchrotron beamline experiments by providing a few recent examples of autonomous X-ray scattering and diffraction experiments conducted at NSLS-II. These experiments were performed at the 11-BM Complex Materials Scattering (CMS), 12-ID Soft Matter Interfaces (SMI), and 28-ID-2 X-ray Powder Diffraction (XPD) beamlines of NSLS-II.

10.3.1   Real-Space Imaging of Material Heterogeneity by Scattering

Characterization of material heterogeneity in real space represents one of the simplest examples of parameter space exploration. In this case, the parameter space consists of sample coordinates, say, (x, y) for two-dimensional (2D) cases. If the length scale of the heterogeneity is greater than the X-ray beam size, scattering measurements as a function of the sample position provide an effective method to probe the spatial variations of nanoscale structures. X-ray scattering techniques are inherently multimodal, with each scattering pattern providing access to multiple features (e.g., scattering peak intensities, positions, and widths) and corresponding physical quantities (e.g., phase fractions, intermolecular spacing, and grain sizes). When combined with X-ray beam scanning, any of these quantities or "signals" can serve as a contrast mechanism to generate real-space images of materials. Thus, scattering-based 2D mapping of material heterogeneity in films is well suited to the application of our GP-based AE approach.

Example 1: Advantages over pre-programmed scans. Figures 10.3 and 10.4 show one of our first proof-of-principle demonstrations of autonomous synchrotron X-ray scattering experiments. In this example, the AE loop deployed at the CMS beamline was used to drive autonomous SAXS experiments in order to probe the nanostructure in a solution-cast film of nanoparticles [367, 373]. The sample consisted of nanoparticle coating on a microscope cover glass created by evaporating a droplet of a nanoparticle solution. The drying front created a characteristic "coffee-ring" pattern, with the darker stains in the optical image (Figure 10.4a) indicating

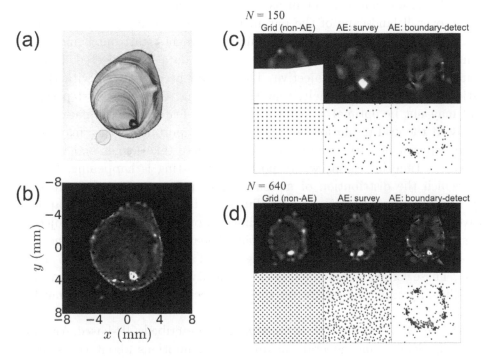

Figure 10.4 Results of an autonomous SAXS imaging experiment, based on work described in [367]. (a) Optical image of the sample used, consisting of a "coffee-ring" pattern that arises from evaporation of a nanoparticle-containing solution. The dark stripes are regions of significant nanoparticle deposition. In these regions, nanoparticles are locally (at the nanoscale) packed into arrays. A typical transmission SAXS pattern from such a region is shown in Figure 10.3. (b) 2D map of the SAXS peak intensity distribution based on all the data acquired during the experiment, taken at more than 4,000 distinct (x, y) positions. The X-ray beam size, corresponding to the area illuminated for each individual measurement, was 0.2 mm × 0.2 mm in size. (c, d) Comparison between three different data collection strategies, with each after (c) 150 and (d) 640 measured data points: a conventional grid scan (left panels) and AE methods based on "survey" (middle) and "boundary detection" modes (right).

regions of high nanoparticle concentration. Drying-induced self-assembly of ordered nanoparticle superlattices gave rise to strong diffraction peaks in the SAXS pattern, often with the six-fold symmetry consistent with 2D hexagonal packing of nanoparticles. As a measure to quantify the presence of these ordered nanostructures, the circularly averaged intensity of the first-order SAXS peak was obtained by real-time fitting. The extracted peak intensity as a function of sample position (x, y) was used as the "signal" to steer the autonomous 2D mapping experiments.

In Figure 10.4c,d, we compare three different types of experiments at two different stages. The left panels show the results of the control experiment based on a grid scan, which is pre-programmed and hence non-autonomous. The middle and right panels show the results of Autonomous Experiments performed using the survey and boundary-detection modes, respectively.

The advantages of Autonomous Experiments are immediately evident. First, for the two AE modes shown, fewer measurements are needed for the experimenter to grasp the overall landscape of the parameter space, as compared to the grid scan. Second, the resolutions of the Autonomous Experiments start out being coarse and then improve naturally and continuously as more data are collected, in a manner that suits the experimental objective. This should be contrasted with the general limitations of grid scans, namely, that the scanning resolution or step size is pre-set and may not be optimal for capturing the key features of the parameter space efficiently. If the grid resolution is too coarse, an important feature may be missed, while an experiment with too fine a resolution may take longer and wastes much measurement time in regions where nothing interesting is happening. Finally, the way in which the distribution of measured data points evolves differently for the survey and boundary-detection modes illustrates the power of the AE approach in adoptively steering the experiment, according to both the experimental objective and the existing data collected up to a given point. The results also highlight the ease with which the AE loop can be tailored by designing an appropriate GP-based acquisition function. These comparisons demonstrate the utility of the AE approach in efficiently and intelligently exploring parameter spaces, especially those for which much is still unknown.

Example 2: Multimodal imaging. The described GP-based AE approach is extremely flexible and general, making few assumptions about the structure or dimensionality of the parameter space to be searched or the signals being measured. As such, it can readily be extended into multimodality, where many measurement signals are considered simultaneously. The most straightforward way to incorporate multimodality is to compute a set of signals-of-interest, and essentially treat them as independent signals for the purpose of AE decision-making. Conceptually, n signals can be analyzed by n parallel GP learning processes, which means that on each iteration of the AE loop, there are n predictions for new points to measure in order to improve the underlying models. All the n newly measured points then serve as inputs for the next iteration of modeling each signal. While these models are computed independently, which allows for the signals to be probing distinct and unrelated aspects of the given material problem, in practice there will often be interesting and useful correlations between signals. This means that a point selected to improve a given

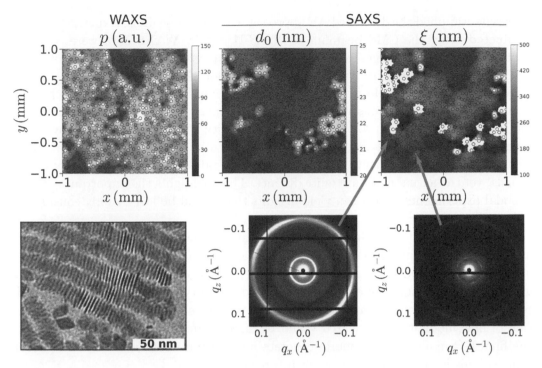

Figure 10.5 Autonomous imaging of a heterogeneous sample using multiple signals. Small-angle (SAXS) and wide-angle (WAXS) X-ray scattering was measured as a function of position across a thin film formed from nanoscale platelets that self-assemble into a superlattice (electron microscopy image in lower-left). The SAXS and WAXS images are analyzed to extract several signals of interest. Three signals were selected for autonomous control: the WAXS scattering intensity associated with the platelet atomic unit cell (p), the repeat-spacing of the platelet packing in the nanoscale superlattice (d_0), and the grain size of the superlattice grains (ξ). The combination of signals allows efficient AE mapping of multiple aspects of material heterogeneity. Reproduced from supporting information of [370].

signal will tend to yield useful information for the other models being constructed. This thus yields, in a very straightforward manner, a surprisingly robust exploration strategy.

As a concrete example of this method, we consider the imaging of heterogeneous films formed by the assembly of nano-platelets into superlattices transferred onto solid substrates (Figure 10.5) [370, 138]. Self-assembly processes are stochastic and assembled films correspondingly exhibit heterogeneity at many different length scales. Ensemble measurements average over the distribution of ordering motifs, providing an incomplete picture of these rich materials. Spatially resolving the ordering (packing, defectivity, etc.) provides a much more meaningful understanding of ordering; however, a priori the relevant scale of heterogeneity may not be known. A simple grid scan would require an arbitrary selection of measurement spacing, leading to data collection that is sub-optimal or that might even miss features of interest. Instead, researchers applied the described AE approach to study these films. Experiments were conducted using the CMS beamline at NSLS-II. SAXS/WAXS data were collected simultaneously, and images were automatically analyzed to extract a set of physical signals. Three separate signals were selected for simultaneous modeling, which allowed the AE loop to select three new meaningful measurement points on each iteration. This AE rapidly reconstructed a high-quality map of the sample across multiple modalities (Figure 10.5), allowing the heterogeneity to be assessed. Moreover, the maps of different signals can be compared to identify similarities and differences. In the described experiments, there are common scales for the heterogeneity across signals, yet the actual maps are quite distinct. This highlights the importance of multimodal experiments, which can yield insights that would be lost by only considering a single modality.

In the presented example, signals were considered separately, and compared only after AE completion. More advanced AE strategies are possible, where the correlations between modalities are explicitly modeled. Such an approach can yield even greater AE performance since insights from one modality can be leveraged by the GP to better reconstruct another modality. Moreover, the scientifically interesting question of how modalities are related would be directly answered by such an AE, since the final GP model would contain concrete information about correlations between signals. This is an exciting avenue for future studies.

10.3.2 Combinatorial Parameter-Space Exploration and Material Discovery by Scattering

We now turn to applications of the GP-based AE approach to exploring and elucidating abstract material parameter spaces. Examples of our initial success with these applications came from two types of experiments. The first type corresponded to measuring a large number of individual samples that had been prepared to populate a parameter space of interest, with each sample representing a discrete point in this space. A typical experiment consisted of loading many such samples in a queue and then using the AE loop coupled with a robotic sample exchanger to perform autonomous SAXS measurements; in particular, the order in which these samples were measured, was prioritized by "snapping", i.e., by selecting on each iteration the available sample that is closest to the point suggested by the GP-based decision algorithms [367]. This type of approach is useful when the available beam time is limited and can accommodate measurements of only a subset of available samples. These experiments also demonstrated that by leveraging the "snapping" concept, the GP-based AE approach can work effectively even when the available samples are constrained to a discrete set of points in the parameter space.

The second type of experiment utilized combinatorial sample libraries. With the use of combinatorial samples, the exploration of abstract material parameter spaces is turned into a real-space mapping problem, akin to the imaging of material heterogeneity discussed in the preceding section. In particular, a combinatorial thin film on a substrate, which is fabricated to exhibit gradients in certain specific material parameters of interest over the surface, provides a convenient platform to elucidate the effects of these parameters on the film structure. Here, we highlight two recent examples of autonomous SAXS experiments on combinatorial nanostructured films.

Example 3: Transition from exploration to exploitation. Large-scale, ordered assembly of polymer-grafted nanorods (PGNR) has attracted much interest due to their potential for sensor and photonic-coating applications. In order to explore scalable manufacturing of PGNR-based materials, researchers studied blade coating (also known as "flow coating"; Figure 10.6a) of PGNR inks on a substrate,

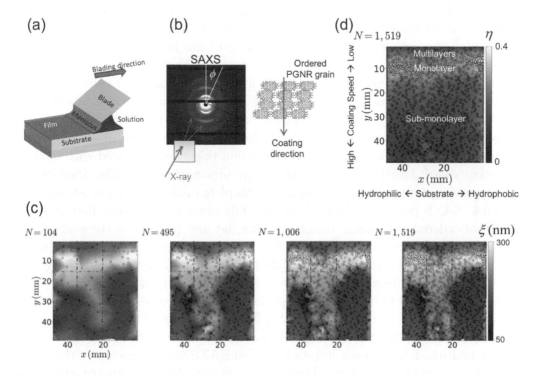

Figure 10.6 Results of an autonomous SAXS experiment on a combinatorial nanocomposite sample, as a function of coordinates (x, y) representing a 2D sample-processing parameter space; based on work described in [374]. The sample consisted of a flow-coated film of polymer-grafted nanorods (PGNR) on a surface-treated substrate (a), where the substrate surface energy increased linearly from 30.5 mN/m (hydrophobic) at $x = 0$ to 70.2 mN/m (hydrophilic) at $x \approx 50$ mm, and the coating speed increased at constant acceleration (0.002 mm/s²) from 0 mm/s (thicker film) at $y = 0$ to 0.45 mm/s (thinner film) at $y \approx 50$ mm. (a) Schematic illustration of the flow coating process. (b) Representative transmission SAXS pattern and a schematic illustrating an ordered PGNR domain with in-plane orientation consistent with the SAXS pattern. (c, d) 2D heat maps of GP-based surrogate models for (c) grain size ξ and (d) grain orientation anisotropy η, where N denotes the number of measurements taken and the points show the locations of measured data points. The Autonomous Experiment successfully identified a well-ordered region (between red lines) that corresponded to uniform monolayer domains. Blue lines mark the region of solution-meniscus instability (see text). Panel (c) was adapted from [374].

which yields thin films of surface-parallel PGNRs upon solvent evaporation [374]. A primary interest was to fabricate highly ordered films in which nanorods were well aligned with each other, with a well-defined inter-rod spacing and a high degree of in-plane orientation anisotropy. Understanding the effects of fabrication parameters on the resulting nanostructures was a prerequisite to controlling the nanoscale ordering within the PGNR film. To tackle this problem, researchers conducted autonomous SAXS experiments on a combinatorial PGNR film at the CMS beamline [374], with two specific aims. The first aim was to explore how the nanoscale ordering in the PGNR films depended on two parameters, namely, the degree of hydrophobicity/hydrophilicity of the underlying substrate and the film coating speed. The second aim was to exploit the knowledge gained to identify and focus in on the regions in the 2D parameter space that yielded the most highly ordered nanostructures.

To facilitate this study, a 2D combinatorial sample library (Figure 10.6) was created in which the substrate surface energy increased along the x-direction (from being hydrophobic at $x = 0$, to hydrophilic at $x = 50$ mm) and the coating speed increased along the y-direction (from being low at $y = 0$, to high at $y = 50$ mm). The film casting conditions were such that the film thickness decreased with increasing coating speed; based on optical microscopy observations, the film transitioned with increasing y from multilayers through a complete monolayer to a sub-monolayer with holes. SAXS patterns exhibited arc-like diffraction features that indicated the presence of ordered domains of nanorods lying flat and parallel to the surface and aligning with their neighbors (Figure 10.6b). For the autonomous SAXS experiment, three analysis-derived signals were used to drive the AE loop: the grain size ξ of ordered domains; the degree of anisotropy $\eta \in [0, 1]$ in the in-plane orientation of the domains, where $\eta = 0$ for random orientations and $\eta = 1$ for perfect alignment; and the azimuthal angle ϕ or the factor $\cos(2\phi)$ for the direction of the average in-plane domain orientation, respectively. As in the case of Example 2, the implemented AE loop was multimodal, and each iteration yielded SAXS measurements at three new sets of sample coordinates (x, y). The experiment was divided into two stages. In the first stage ($N < 464$, first 4 hours), the survey mode was used for each of the three signals. In the second stage ($464 \leq N \leq 1520$, additional 11 hours), the feature optimization mode was used for the anisotropy signal η to determine regions with the highest degree of domain orientation order, while continuing to use the survey mode for ξ and $\cos(2\phi)$.

The results of the autonomous SAXS experiment are summarized in Figure 10.6. The series of ξ maps in Figure 10.6c, indicating the distribution of measured data points at different stages of the experiment, demonstrate that the AE loop focused initially on the parameter-space exploration but transitioned to concentrating more points at specific locations as the experiment progressed. The multimodal AE search revealed important aspects of the nanoscale ordering in the blade-coated PGNR film. For example, the results indicate that the ordered domains tend to orient such that the nanorods' long axes are aligned perpendicular to the coating direction [i.e., $\cos(2\phi) \approx 1$], and that there is a strong correlation between large grain sizes (i.e., large ξ) and high domain orientation anisotropy (i.e., large η; Figure 10.6d). From the perspective of optimizing the fabrication process, the experiment was useful in: (i) identifying the optimal coating-speed range for large domain size and high in-plane orientation anisotropy, which coincided with the regions of uniform monolayer coating ($5 < y < 15$ mm); (ii) revealing a region of poor ordering at intermediate x, which likely resulted from solution meniscus-induced instability in the coating process; and (iii) providing evidence that aside from this region of instability, the nanoscale ordering appears to be qualitatively independent of whether the underlying substrate is hydrophobic or hydrophilic. It is worth emphasizing the efficiency of the GP-based AE approach; it made the above observations possible without painstakingly taking high-resolution scans, placing more measurements only where needed. If high-resolution grid scans had been taken, say, at the resolution of the beam size used (0.2 mm × 0.2 mm) and starting at the high coating-speed side ($y = 50$ mm), it would have taken more than a week to find the highest ordered regions, instead of a fraction of a day using the AE approach.

Example 4: Machine-guided nanomaterial discovery. The AE method can very naturally be adapted to material discovery problems by combining it with combinatorial material synthesis. The creation of gradient thin films enables the formation of a single sample that is essentially a library of all possible materials across the selected gradient. Methods have been demonstrated for gradients in film thickness, composition, substrate surface energy, and processing temperature. As a concrete example, we consider the use of autonomous X-ray scattering to study the ordering of block copolymer (BCP) blends ordering across a range of substrate chemical patterns. Block copolymers are polymer chains with chemically distinct regions. The natural tendency for the different parts of the BCP chain to chemically demix is frustrated by the linkage between these blocks, and the compromise is instead to form nanoscale morphologies with locally separated chain-ends. BCPs naturally form phases such as nanoscale cylinders or lamellae. It has recently been discovered that mixing different BCPs (that form different morphologies when pure) can give rise to novel structures, such as coexistence phases [583, 35]. It was moreover found that when these BCP blends order on top of a substrate with chemical stripe patterns, the morphology that forms depends on the underlying chemical grating pitch and stripe width [492]. This responsive behavior suggests that novel morphologies could be accessed if one selected the correct underlying substrate conditions [378]. The associated exploration space for this problem is large and complex, since one can vary blend composition (materials and ratios), the underlying surface (average chemistry, grating pitch and duty cycle), and processing conditions (annealing temperature and time). This is thus an ideal problem for AE exploration.

To study this problem, researchers fabricated a combinatorial sample with a chemical grating, where the pitch was systematically varied along one axis and the stripe width varied along the orthogonal axis. This single sample thus enables all possible directing gratings to be studied. Researchers cast a uniform BCP blend film on this substrate, annealed the sample, and studied it using autonomous microbeam SAXS at the SMI beamline of NSLS-II (Figure 10.7a) [370, 129]. The AE signal was the scattering intensity of the BCP morphology, with an objective to both minimize error and focus on regions of strong intensity (which normally correlate to good nanoscale ordering). An initial exploration identified significant heterogeneity across the sample,

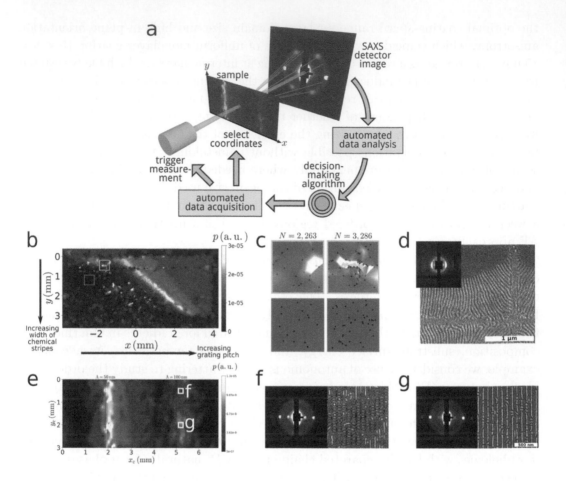

Figure 10.7 Autonomous nanomaterial discovery accomplished by combining X-ray scattering with combinatorial sample preparation; based on the work described in [370] and [129]. (a) The AE loop iteratively selected positions on a sample, and used SAXS to characterize ordering. The AE method considered the scattering intensity (p) associated with the repeat-spacing of the polymer nanostructure that forms. The sample consisted of a block copolymer blend thin films cast on a chemical grating, where grating pitch and stripe width varied systematically across the sample x and y directions, respectively. (b) An initial mapping experiment demonstrated curious heterogeneity. (c) The AE identified and focused attention on several regions of high scattering intensity. (d) Follow-up electron microscopy studies in these regions demonstrated spurious ordering, which could be attributed to errors in the lithographic preparation of the underlying chemical grating. The AE thus rapidly identified sample synthesis errors, allowing them to be corrected. (e) A subsequent AE using an optimized synthesis protocol yielded meaningful self-assembled order throughout the combinatorial space. Researchers identified several unexpected regions of good ordering, especially when the underlying chemical grating pitch was roughly double the intrinsic ordering length-scale of the block copolymer. (f, g) This enabled the identification of several novel self-assembled morphologies that had never been seen before—indeed that had not been predicted by any prior work.

including small regions of high scattering intensity and regions of unexpected four-fold symmetry. This enabled targeted follow-up studies using electron microscopy (Figure 10.7d) where it was rapidly discovered that the underlying lithographic exposure had generated errors at the boundaries between different patterning regions, which in turn gave rise to spurious BCP ordering. Although this experiment was a failure in terms of material discovery, it demonstrates the power of AE, which was able to rapidly guide experimenters and identify the source of synthesis problems [370]. Armed with this information, the researchers refined their sample synthesis protocol and conducted a second AE [129]. In this case, the AE identified good BCP ordering throughout the exploration space (Figure 10.7e). In addition to identifying the previously known BCP blend morphologies, this autonomous search discovered several new nanostructures that had never been seen before (nor, indeed, predicted); including a skew morphology, an alternating structure, and a nanoscale "ladder" formed by combining long "rails" with cross-connecting "rungs." This discovery of new morphologies emphasizes the power of AE, which can uncover ordering that might have been missed in a conventional study where, e.g., researchers arbitrarily selected a small set of conditions to fabricate and image. In fact, the entire AE was concluded using just ∼6 hours of beam time, which can be compared to a more conventional manual and non-combinatorial approach that might have involved months of sample preparation and imaging.

10.3.3 Toward a Closed-Loop Strategy: Machine Learning Prediction Guiding Autonomous Experimentation

The integration of machine learning (ML) prediction with AE-based validation is a key concept in accelerating the development of new materials. ML models can be trained on existing data sets to make predictions on new material systems, thus enabling the rapid exploration of large parameter spaces that would be impossible to cover using traditional trial-and-error approaches. These predictions can then be incorporated into the materials design model, which can guide the synthesis or processing of new materials with specific properties tailored for targeted applications. However, to be confident in the accuracy of these predictions, it is crucial to validate them experimentally. This is where Autonomous Experiments come into play. Autonomous Experiments involve the use of automated and self-directed experimental tools to carry out complex experiments with minimal human intervention. These experiments can be designed to test the ML predictions.

To elaborate on the concept, we now discuss a use case based on recent work by Zhao et al. [603] in which an ML approach was introduced for designing nanoarchitectured materials by dealloying (Figure 10.8). The use of machine learning in materials design has gained much attention in recent years. For nanoarchitectured materials with a large and complex design parameter space, ML methods can offer great potential, with prior work done with data mining and automated image analysis [309]. Metal-agent dealloying is a promising materials design method for fabricating nanoporous or nanocomposite materials, but progress has been slow due to reliance on the trial-and-error approach. The research team established a workflow for designing

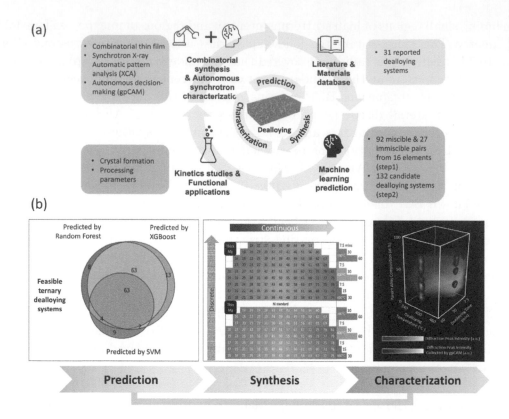

Figure 10.8 Scientific case of designing nanoarchitectured materials via solid-state metal dealloying [603]: (a) The design of solid-state metallic dealloying is facilitated by an ML-augmented framework that involves several steps forming a closed-loop cycle. (b) Illustration of the concept of working toward a closed-loop of prediction, synthesis, and characterization framework enabled by Autonomous Experiment, involving ML-enabled prediction, combinatorial synthesis, and Autonomous Experimentation.

nanopores/nanocomposites by using ML, combinatorial sample preparation, and autonomous synchrotron characterization [603]. The workflow involved applying several ML methods to predict new ternary dealloying systems and analyzing the underlying materials design principles in the metal-agent dealloying method. The ML models predicted 132 ternary dealloying systems from 16 selected metal elements.

With the materials predicted by the ML models, a proof-of-concept for AE-based validation was tested by dealloying a Ti-Cu alloy with Mg. A combinatorial Ti-Cu thin film was prepared to cover a wide range of parent alloy compositions, and autonomous synchrotron X-ray diffraction characterization of the film was conducted at the XPD beamline to explore the kinetics in Ti-Cu/Mg dealloying systems. The crystallographic companion agent *XCA* was used to automate the XRD analysis, and Autonomous Experimental control was used to drive XRD characterization for faster materials discovery. Overall, the ML-augmented workflow developed in this study shows promise in improving the data acquisition accuracy and rate for validating ML predictions and developing nanoporous/nanocomposite materials. In the future, Autonomous Experiments can contribute further to this workflow by facilitating efficient synthesis/processing and characterization of the structure-property relationship. By achieving a closed-loop between these experimental aspects and adding simulation, the potential for designing a wider range of solid-state materials with limited published results can be realized.

10.4 FUTURE PROSPECT AND NEXT STEPS

The future prospect of Autonomous Experimentation in advancing science and accelerating discovery at large user facilities is promising, as researchers continue to explore new methods to achieve more complex, efficient, and intelligent experimental and data workflows that are automated. With increasing deployment and utilization of AE methods at synchrotron facilities, the roles of beamline scientists and users will likely evolve. However, it should be emphasized that Autonomous Experiments are meant to not replace but *liberate and empower* beamline scientists and users. Just as automation has significantly enhanced the efficiency and scope of beamline experiments, Autonomous Experiments are expected to allow beamline scientists and users to be more deeply involved in the scientific aspects of research, design more ambitious experiments, and manage them at a higher level.

Nevertheless, the application of the AE paradigm to synchrotron beamline experiments is still in its infancy and requires much further development to realize its potential to substantially expand the scope of user science. We close this chapter by noting several key development directions that we believe are important.

Autonomous in-situ/operando experiments under material processing conditions. The ability to perform *in-situ/operando* experiments is one of the major strengths of synchrotron experiments. This ability is essential to tackling common challenges of modern materials science, which include: design and discovery of material synthesis and processing methods for new functional materials; understanding the non-equilibrium spatiotemporal behavior of complex materials during their formation process and in response to environmental changes or external stimuli, which often depends on the processing history or kinetic pathway; precise, adaptive control and optimization of material architectures and processes toward desired functionality and enhanced performance. So far, most of our successful autonomous X-ray experiments at beamlines are based on measurements on pre-made static samples, as illustrated by the examples given in Sec. 10.3. In order to extend the AE methods to *in-situ/operando* experiments, it is critical to integrate material synthesis and processing platforms into the beamline control system and make them automatable, such that autonomous decision-making algorithms guide not only the X-ray characterization but also the on-the-fly generation and processing of high-value samples to be studied at the beamline (e.g., [565]). Also important is the integration of non-synchrotron ancillary characterization methods into the "measure" step of the AE loop. Besides providing another possible set of "signals" to guide autonomous *in-situ/operando* experiments at the beamline, they can serve as proxy probes which can be correlated with the physical quantities derived from synchrotron X-ray experiments and be leveraged subsequently to inform non-synchrotron experiments or even optimize industrially relevant synthesis and fabrication processes being carried out at researchers' home institutions.

Incorporating advanced ML-enabled analysis. Reliable automated data analysis is crucial for Autonomous Experiments. It is therefore an important current and future development to continue advancing automated data analysis. Recent work done at NSLS-II include ML-enabled automated data analysis. Examples involve tackling challenges with noise in the experimental data and deciding which analysis techniques to use for a given dataset. A new smart analysis workflow (SAW) has been developed to analyze out-of-equilibrium X-ray photon correlation spectroscopy (XPCS) data, which includes a denoising autoencoder (DAE) to reduce noise, a classifier to distinguish between different dynamics, and trust regions to assess fit quality and domain expertise [241, 242]. The SAW eliminates manual binning and preserves the time resolution set by the data acquisition rate. This workflow's model-agnostic extraction of quantitative dynamics could be useful in future Autonomous Experiments, and similar machine-learning algorithms may be applied to automate analysis for other scattering techniques. Beyond X-ray scattering and diffraction, for X-ray imaging and microscopy, a machine learning-based image processing method has been developed for sub-10 second fly-scan nano-tomography, achieving sub-50 nm pixel resolution with reduced data acquisition time and X-ray dose [600]. The algorithm was successfully applied to study dynamic morphology changes in a lithium-ion battery cathode during thermal annealing. These developments point to the potential of incorporating smart, ML-driven automated data analysis into future Autonomous Experiments, working in tandem with theory and computational inputs to enable more intelligent steering of Autonomous Experiments in the future. Various ML models for on-the-fly analysis have been demonstrated at NSLS-II, with a focus on integrating the models into existing experimental workflows for easy implementation by the user community, which could further promote the growth of Autonomous Experiments [242].

Incorporating domain-specific knowledge into decision-making. A key direction for the "decide" step in the AE loop is to develop methods to incorporate varying degrees of existing domain-specific knowledge into the autonomous decision-making process, to further enhance its ability to guide experiments, while preserving the general applicability of the decision-making framework as much as possible. The data-driven, DK-agnostic approach described above ignores existing knowledge about the material system or phenomenon under study. Depending on the given study, the nature of domain knowledge that could enhance the experimental decision can range from being qualitative or intuitive to highly quantitative, based firmly on a large database, a well-developed theory, and/or extensive simulation results. A key observation here is that even qualitative knowledge can be effective in guiding an experiment. Thus, the development of general autonomous decision-making methods or protocols that can exploit a varying depth of domain knowledge would likely benefit a broad range of synchrotron users who study new materials or complex phenomena. In the context of GP-based decision algorithms, a few directions have been identified as promising paths toward achieving DK-aware decision-making, including advanced design of kernel functions and acquisition functions [370].

Integrating simulations and experiments into a real-time closed-loop feedback system. The field of Autonomous Experimentation continues to advance in the framework of closed-loop feedback systems. A crucial aspect of the closed-loop method involves a collaborative and iterative process where physics-based theory informs computational simulation, the results of the simulation inform experiments, and the experimental observations provide further guidance for theory [248]. The effectiveness of the Autonomous Experimentation concept is based on the system's ability to receive continuous feedback from its surroundings and use that information to make necessary adjustments. If a real-time closed-loop method is realized such that theory and/or computational simulation are updated in real-time to guide the experiment, the autonomous approach would become even more powerful.

Multimodal Autonomous Experiments by simultaneously driving multiple beamlines. Understanding materials often requires the use of multimodal characterization techniques that are offered at a set of instruments located at different beamlines. Typically, these measurements are carried out sequentially over a period of days to months. At NSLS-II, a novel approach has been developed, where unified multimodal measurements of a single sample library are conducted by autonomously coordinating experiments at distant instruments, utilizing a distributed network of agents. These agents use real-time analysis from each modality to inform the direction of the other [292]. This approach has demonstrated a high degree of coordination and efficiency, and will undoubtedly play a critical role in future experiments.

CHAPTER 11

Autonomous Infrared Absorption Spectroscopy

Hoi-Ying Holman

Biosciences Area, Lawrence Berkeley National Laboratory, Berkeley, California, USA

Steven Lee

UC Berkeley, Berkeley, California, USA

Liang Chen

Lawrence Berkeley National Laboratory, Berkeley, California, USA

Petrus H. Zwart

Molecular Biophysics & Integrated Bioimaging Division, Lawrence Berkeley National Laboratory, Berkeley, California, USA

Marcus M. Noack

Applied Mathematics and Computational Research Division, Lawrence Berkeley National Laboratory, Berkeley, California, USA

CONTENTS

O SCILLATIONS of dynamic dipole moments in molecules mostly have resonance frequencies within the narrow IR spectral region of 4,000-400 cm^{-1} wavenumber (or 2.5 - 25 μm wavelength), giving rise to characteristic absorption bands that are specific to the functional groups and structure of the molecules. Based on this principle, scanning synchrotron radiation-based Fourier transform Infrared absorption

DOI: 10.1201/9781003359593-11

spectroscopy (SR-FTIR) is a powerful non-invasive micro-probe technique that provides chemical information about biogeochemical and environmental samples without *a priori* knowledge. Data is commonly collected on a high-resolution Cartesian grid which often leads to the acquisition of redundant data and therefore inefficiencies in data storage, analysis, and experiment time—at each measurement point, high-dimensional spectra have to be collected. Autonomous collection of this data allows for faster data collection and effective focus on regions of interest.

11.1 SCANNING SYNCHROTRON RADIATION-BASED FOURIER TRANSFORM INFRARED ABSORPTION SPECTROSCOPY

When IR light shines on a molecule, infrared photons with frequencies close to the resonance frequencies are absorbed by the molecule's fundamental vibration modes if there is a net change in the oscillating dipole moment during vibration [103]. The larger the change in the transition dipole moment, the more intense the absorption will be. From a simplistic point of view, the bending or stretching of bonds involving atoms with different electronegativities (e.g., C=O, -N=O) will lead to intense absorption. All vibrations symmetric to the molecular center of symmetry (e.g., O=O, N≡N) are infrared-inactive because the dipole moment does not change. As the resonance condition is specific to the functional groups (or atomic groupings) of the molecules, the exact frequency of the absorbed light provides a characteristic signature of each molecular functional group. Since each material is a unique combination of atom groupings, the presence, position, and intensity of absorption bands create a specific absorption pattern representing a molecular fingerprint of the materials. Even though an infrared spectrum of a biogeochemical specimen is a sum of the contributions gathered from all biomolecules (i.e., proteins, amino acids, lipids, and nucleic acids, minerals), distinct absorption bands exist that can be related to known functional groups in biomolecules as well as minerals [61, 576, 46, 195, 103].

SR-FTIR data is high-dimensional in spectral space. While the FTIR approach collects all infrared wavelengths simultaneously, each data point in a spectrum is sampled at the interference fringe position using a He-Ne laser with an operating wavelength of 632.8 nm. The spectral resolution of an interferometer commonly ranges from 32 to 0.5 cm^{-1} and the number of data points ranges from 1,028 to 65,536. For solid-state matter like biogeochemical samples, a spectral resolution of 4 cm^{-1} is typically used which gives each spectrum 8,192 data points. For the prevalent grid-based SR-FTIR imaging approach, the number of grid points scales with the power of the dimensionality (>1,000) which makes SR-FTIR imaging computationally expensive.

In addition to high-dimensional data, SR-FTIR imaging experiments on biogeochemical samples often face challenges due to the sample's properties. Biogeochemical samples are heterogeneous with localized hotspots in biological, biogeochemical, or physical properties and high spatial gradients. One important consideration when making scanning SR-FTIR spectral measurements is the size of the targeted feature relative to the diffraction-limited infrared beam size which depends on the numerical aperture (NA) of the microscope objective and the instrumentation performance. At the ALS (LBNL, Berkeley, California), where Schwarzschild objectives with NA in the range of 0.4 to 0.7 are used, the beam size is $0.61 \times (\lambda/NA)$, which ranges from 0.5 λ to 1.2λ (i.e., 2 to 10 μm) where λ is the wavelength of the light [262, 263]. Depending on the vibrational frequencies of the targeted molecular functional groups, the beam size is smaller than eukaryotic cells like corallines algae, bigger than most of the prokaryotes or archaea, or is comparable to a small cluster of prokaryote or archaea cells in a microbial mat or biofilm. Example exceptions are a few mega-bacteria such as the sulfur bacterium Thiomargarita namibiensis which can reach a diameter of 700 μm or the thermophilic archaea Staphylothemus marinus which can grow to 15 μm in diameter. The challenge for the experimenter is to decide which data-acquisition strategy will lead to the most efficient exploration of the sample while honing in on eventual hotspots that are emerging.

11.2   A CHALLENGE OF EXPERIMENTAL DESIGN

Designing the appropriate pixel/grid size for experimentally optimal spatial resolution and information gain has been a challenge for decades. Unlike infrared microscopy with a conventional thermal source that is limited to using apertures (and appropriate pixel size) to achieve micrometer spatial resolution often with a drastic decrease in the signal-to-noise ratio, infrared microscopy with a bright synchrotron source has no requirement for apertures. When coupling the non-invasive SIR beam to an infrared microscope and using a single MCT detector element for detection, the photons can be focused on a spot with a size approaching the diffraction limit, or the highest spatial resolution. This improves spatial resolution with little loss in signal-to-noise ratio. However, it is important to match the mapping pixelation to both the NA and the point spread function (psf) of the microscope objective by using the appropriate pixel size during scanning. For a microscope objective with a numerical aperture of 0.65, an accurate diffraction-limited spatial resolution can only be achieved with a pixel spacing less than $\sim \lambda/4$ [346]. Although the accuracy at which a given feature is measured is greatly increased, because a synchrotron infrared beam allows the measurement to be taken only on the area of interest [195, 133], selecting a too-large pixel size leads to resolution loss. Conversely, selecting a too-small pixel size increases acquisition time with minimal improvement in spatial resolution. Since little information about the sample is known in advance, pixel spacing usually ranges from 0.5 to 5.0 μm, depending on the frequencies of the functional groups of the targeted molecules.

To address the challenges of computational costs linked to the number of Cartesian-grid measurement points and the high dimensionality of the spectra, traditional SR-FTIR experiments are conducted following a simple strategy: Measurements are performed at some initial points based on the practitioner's intuition. After acquiring these initial measurements, the practitioners identify features of interest in the data. These features are used to identify and then select candidate imaging regions. It is common to define a Cartesian grid over the imaging region with a pixel size ranging between 0.5 and 5.0 μm. Each region is limited in size by time constraints to fewer than about 2,500 pixels. This strategy has been employed to map and observe: microbes on a mineral surface reducing chromium(VI) to chromium(III) [197] or converting large recalcitrant polycyclic aromatic hydrocarbons into biomass [196], marine microbial communities in the ocean responding to an influx of petroleum [305, 198, 185, 36], the dynamics of cyanobacterial silicification [52, 53, 593], bacteria interaction with archaea in a cold sulfidic subsurface aquifer [414], and microbial diversity linking with the chemistry of their lithified precipitations [539].

For scanning SR-FTIR experiments on a dense Cartesian grid, the data-acquisition time for a single spectral image ranges from 6 to over 12 hours. This substantially limits our ability to measure the changing spatial distributions of transient biogeochemical processes temporally. In addition, while the design of the experiment and therefore the location of data points are chosen carefully—driven by the experimenter's intuition and experience—it is unlikely that the practitioner is able to optimally perform high-dimensional pattern analysis and decision-making in the SR-FTIR spectral spaces. Furthermore, experimenter bias can occur easily and skew the results; this will also affect the reproducibility of a particular experiment. In order to advance and optimize SR-FTIR data acquisition technology toward higher efficiency and unbiased real-time control, we employ two different autonomous grid-less data acquisition approaches. One is based on adaptive sampling that combines 2D barycentric linear interpolation with Voronoi tessellation (LIV) [194]. We reported that without human intervention, LIV was able to substantially decrease data-acquisition time while increasing measurement density in regions of higher physico-chemical gradients. The other method we employ is based on Gaussian process regression in the form of the openly available Python API *gpCAM*.

11.3 COMBINING 2D BARYCENTRIC LINEAR INTERPOLATION WITH VORONOI TESSELLATION (LIV)

Adaptive sampling that combines 2D barycentric linear interpolation with Voronoi tessellation (LIV)—as proposed by [194]—is based on leave-one-out cross-validation (LOOCV) to allow efficient and accurate approximations of the experimental model for predictive error calculations from which the algorithm rapidly identifies optimal regions for subsequent measurements. At the core of the concept is the idea that detectable phenomena arising from heterogeneity are primary regions of interest, information hot spots which should be spatiochemically resolved by increasing sample density in adjacent regions. The procedure starts by placing a number of random points within the domain. At each data-point location, leave-one-out error ϵ_{LOO} is calculated based on the Euclidean norm and normalized via the Voronoi predictive error. The next point is randomly chosen from within the region near the point of maximum ϵ_{LOO}. This loop is repeated until a break condition is fulfilled. The procedure is combined with standard interpolation techniques for model creation and Voronoi tessellation for uncertainty estimates. The area of the Voronoi cell is used to assess the relative importance of a suggested measurement. Principle component analysis is used for dimensionality reduction and the first five components were used for the adaptive exploration. This way, the speed of the data acquisition is comparable to traditional raster scanning while suggesting high-value measurements.

11.4   GAUSSIAN PROCESSES FOR AUTONOMOUS SR-FTIR

To enhance our effort, we consider a sampling technique based on Bayesian uncertainty quantification. For this, we employ the *gpCAM* algorithm as an additional tool. *gpCAM* is an implementation of the Gaussian process regression framework (see Chapter 4). We use the software's built-in uncertainty quantification to bypass the practitioner's bias while improving the SR-FTIR data-acquisition speed by focusing the sampling on areas of interest. In a pilot study, presented here and in [375], we began with a well-known microbialite sample from Bacalar, a karst coastal oligosaline lagoon system (Figure 11.2 [539]). We considered two problems for *gpCAM* to resolve: (1) a high-quality model of the microbialite sample purely based on model uncertainties, and (2) focusing on regions of interest in the microbialite sample. An overview of the autonomous loop is illustrated in Figure 11.1.

Exact Gaussian processes scale $O(N^3)$ with data set size; it is, therefore, infeasible to consider every spectral intensity as a data point. This motivated the consideration

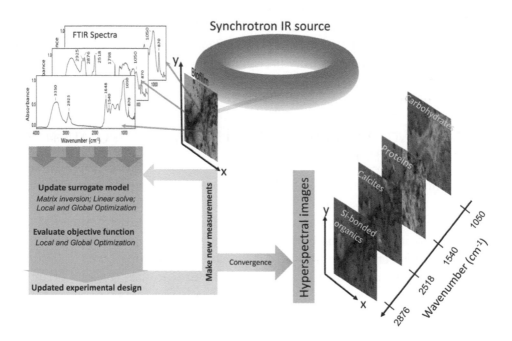

Figure 11.1   A schematic overview of the autonomous feedback loop that integrates Gaussian Process driven adaptive experimental designs and hardware that drives the acquisition of new data. At every measurement point a high-dimensional spectrum is collected and then either dimensionality-reduced or compared to a reference spectrum, depending on the experiment strategy. A surrogate model with associated uncertainties is then calculated via the GP which is then used to define an acquisition (or objective) function. The maxima of the objective function are then found and their locations are communicated to the instrument as new optimal measurement points.

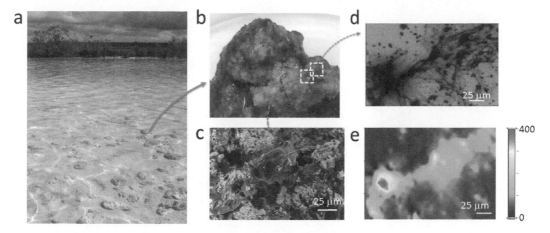

Figure 11.2 Bacalar lagoon microbialites used for the first autonomous SR-FTIR experiment study. Panel (a) shows the Bacalar lagoon in Mexico, which is the largest freshwater microbialite ecosystem in the world. Below the surface of the clear lake water, we see calcareous micobialite mounts. Panel (b) shows a microbiolite knob broken off from a microbialite mount and regions that were selected for sample preparation for SEM, bright-field visible, and IR absorption analysis. Panel (c) shows an SEM micrograph depicting mineral inclusions and possible organic matters. Panel (d) illustrates a bright field micrograph showing sample contrast from total visible light absorption. Panel (e) shows the associated infrared micrograph highlighting sample contrast from absorption of mid-infrared light. (From [539].)

of dimensionality reduction to address problem (1). A measured spectrum at each (x, y) point in the microbialite sample has thousands of intensities—it is a long vector corresponding to the intensity measurements at each sampled wavenumber—the typical dimensionality being 8,192. On-the-fly modeling, uncertainty quantification, and decision-making using this high-dimensional data directly require computational resources exceeding those available at most beamlines. Instead, we reduce the dimensionality of the data set while retaining the critical features of the spectra via Principle Component Analysis (PCA) [113]. PCA is a linear dimensionality-reduction technique that finds an orthonormal basis so that the basis vectors are linearly independent. Only the first few basis vectors are considered since they represent most of the variance of the data.

To address problem (2), we use *gpCAM* for feature finding (see Figure 11.3d). For this, we define a reference spectrum that is associated with the Si-Bonded organics we wish to identify (Figure 11.3d (left)). We define a so-called, acquisition function based on the correlation coefficient between the reference spectrum (Figure 11.3d) and the reconstructed spectrum

$$f_a(\mathbf{x}) = \tilde{\sigma}(\mathbf{x}) \left(\frac{1}{2} + \frac{1}{2} \tanh\left[\beta \left(\mathrm{r}\left(u(\mathbf{x}), v(\mathbf{x}) \right) - \alpha \right) \right] \right) \tag{11.1}$$

where β is the scaling factor, α is the correlation coefficient threshold, $u(\mathbf{x}_{\mathbf{ref}})$ is the reference spectrum, $v(\mathbf{x})$ is the reconstructed spectrum, $r(\mathbf{x}_{\mathbf{ref}}, \mathbf{x})$ is the correlation

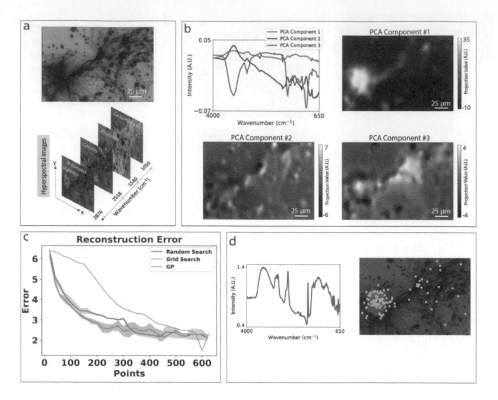

Figure 11.3 Illustration of all the key building blocks for the application of a Gaussian process (GP) for autonomous infrared mapping. (Panel a) Bacalar lagoon microbialites were used for the pilot study. (Panel a, top) A bright field micrograph (205 by 140 μmeters) illustrating a Bacalar lagoon microbialite used in the autonomous-experiment pilot study. (Panel a, bottom) Distribution maps of the Si-bonded organics, calcites, proteins, and carbohydrates that were extracted from a 1738-spectral-dimensions dataset (from [539]). (Panel b) SR-FTIR spectra were dimensionality-reduced from 1738 dimensions to three principal components via PCA while retaining 90% of the chemical information. The heat maps illustrate the spectral projections onto the first three PCA basis vectors (spectra), providing valuable information about the location of edges in the sample, the microstructure, and localized chemistry in the infrared domain. (Panel c) Reconstruction error for different sampling schemes. The interpolation in all cases was performed by a GP giving rise to small differences between random search and GP-guided exploration. In practice, random exploration would not naturally give rise to a GP model. (Panel d) Illustration of the feature-finding mode in the microbialite sample. This is accomplished by using a known reference spectrum associated with the Si-Bonded Organics (left) and including it in the acquisition function. The orange dots are the measurement locations, which are largely focused on the Si-Bonded organics "hotspot" in the ground truth map. The region of interest was identified and sufficiently well explored after circa 200 measurements, compared to approximately 10000 measurements needed for a full raster scan, leading to an estimated 50-fold improvement in this case. (Figure courtesy of [375].)

coefficient between the reference spectrum and the reconstructed spectrum, and $\tilde{\sigma}(\mathbf{x})$ is the posterior standard deviation of the predicted spectrum at $\mathbf{x}$. This acquisition function draws the focus of the data acquisition to regions where the correlation coefficient is above a user-determined threshold value α (usually $\alpha > 0.8$). The correlation coefficient is non-linearly transformed by a *tanh* function and multiplied by the posterior standard deviation—the posterior standard deviation is to avoid over-sampling. We first sampled 10 randomly chosen positions as initialization, followed by maximum-uncertainty-driven sampling using the $\sigma^2(\mathbf{x})$ acquisition function for 50 sampling points. That step was followed by applying the described feature-finding acquisition function for selecting the final 100 sampling points (Figure 11.3d (right)).

11.5 DATA ANALYSIS

Data coming from the detector are high-dimensional spectra which prevents us from making decisions based on the raw data directly. Therefore, we have to drastically dimensionality-reduce the measures spectra. While there are many methods to choose from, for instance, neural networks, clustering, and non-negative matrix factorization, traditionally, PCA has been shown to be effective for infrared absorption spectroscopy. As mentioned earlier, PCA is a linear dimensionality-reduction technique that attempts to find a new orthonormal basis in which all measured spectra can be presented as linear combinations of a few basis vectors. After these basis vectors are found, the projections of a new spectrum onto the first few base vectors (spectra) can be used as the function value for autonomous experimentation.

11.6 INFRASTRUCTURE AND INSTRUMENTATION

Scanning SR-FTIR measurements were performed on a Nicolet Nic-Plan IR microscope with a 32X, 0.65 numerical aperture objective and a Nicolet iS50 FTIR spectrometer bench (Thermo Fisher Scientific) using a KBr beamsplitter and HgCdTe (MCT) detector at Beamline 1.4.3 of the Advanced Light Source at Lawrence Berkeley National Laboratory. At each sampling point, an IR spectrum between 650 and 4000 cm^{-1} at 4 cm^{-1} resolution was measured with 6 co-added scans. The microscope stage and FTIR bench were directly controlled by the OMNIC 9.8 software (Thermo Fisher Scientific), and our program interfaced with OMNIC via the Dynamic Data Exchange (DDE) protocol to drive the microscope stage movement and acquire spectral data.

11.7 RESULTS

The results of the microbialite sample are depicted in Figure 11.3. After collecting spectra at a limited number of (x, y) positions, we performed PCA dimensionality reduction on the collected spectra. We found that the first three PCA components were necessary and sufficient for modeling and feature extraction; the first three components explained over 90% of the data variance. Projecting each of the spectra onto the three PCA base components (Figure 11.3b) reduced the dimensionality of the spectra from 8,192 to 3; these three projection values are provided to *gpCAM* in lieu of the entire spectrum at each measurement point. The posterior means of the three projection values are shown in Figure 11.3b. We can clearly see that the spatial distribution of the predictive mean associated with PCA component 1 is visually indiscernible to the dominant feature from the spatial distribution of the total visible light absorption in Figure 11.2d, e. Similarly, from the spatial distribution of their respective projection values, we can see that PCA components 2 and 3 encode the location of subsidiary biogeochemical components and their boundaries.

To quantify the success of the methodology, we calculate the Euclidean distance between the ground truth spectra and the reconstructed spectra. The reconstructed spectrum at each point is calculated by inverse PCA on the three interpolated PCA component projections. More specifically, the three projections are the posterior means of the Gaussian process. To make a fair comparison, the GP methodology was used to interpolate for all our tested sampling techniques. The results for all tested sampling schemes—random sampling, grid scanning, and GP sampling—are shown in Figure 11.3c. GP sampling means placing measurements where the predicted uncertainty (standard deviation) is a maximum. According to Figure 11.3c, using *gpCAM* for sampling leads to the fasted convergence compared to the other tested methodologies.

We use the same data to demonstrate the solution to problem (2)—feature finding (see Figure 11.3d). This silicate-containing-carbonate mineral sample has an unusually high content of organic carbons. We aim to efficiently resolve the spatial distribution of these silicate-containing carbonate minerals using *gpCAM*. The selected reference spectrum associated with the organic carbon-rich sample is shown in Figure 11.3 (d). Similar to all biogeochemical samples, microbialites are heterogeneous with localized hotspots in biogeochemical properties and high spatial gradients. Due to the complexity of the microbialite sample, it serves as an excellent test case for *gpCAM*, with the success criteria being that it must be able to determine the same localized region of the organic carbonate silicate that the expert has identified. To rephrase this feature-finding problem within the context of autonomous experiments, we require the automated process to sample more heavily where the desired biomolecule is located. We tackle this problem in *gpCAM* by defining a custom acquisition function (11.1) that considers the correlation coefficients between reference and reconstructed spectra. The reconstructed spectrum is an approximation of the original one, but the reconstruction error has negligible impact on the feature extraction as recognized in Figure 11.3. In Figure 11.3d, the measured points (the orange dots) are clearly clustered around the same localized region that the expert determined (the green region in the background).

11.8   SUMMARY

This pilot study has shown the potential benefit and success of autonomous SR-FTIR. We presented the accelerated mapping of a complex biogeochemical sample with improved efficiency. Far fewer measurement points had to be collected and analyzed reducing total experiment times. In addition, simple customizations allow us to target our search toward regions of interest, for instance, the likely occurrence of a specific molecule.

ACKNOWLEDGMENTS

The autonomous adaptive data acquisition for hyperspectral imaging was developed, implemented, and applied in two experimental cases with co-authors Yuan-sheng Fang, Liang Chen, Michael DeWeese, Hoi-Ying N. Holman, and Paul W. Sternberg of our published work. Special thanks for support from the Advanced Light Source (ALS), the Berkeley Synchrotron Infrared Structural Biology (BSISB) program, and Dr. Hans Bechtel [194].

CHAPTER 12

Autonomous Hyperspectral Scanning Tunneling Spectroscopy

Antonio Rossi

Molecular Foundry, Lawrence Berkeley National Laboratory, Berkeley, California, USA

Darian Smalley

Department of Physics, University of Central Florida, Orlando, Florida, USA

Masahiro Ishigami

Department of Physics, University of Central Florida, Orlando, Florida, USA

Eli Rotenberg

Advanced Light Source, Lawrence Berkeley National Laboratory, Berkeley, California, USA

Alexander Weber-Bargioni

Molecular Foundry, Lawrence Berkeley National Laboratory, Berkeley, California, USA

John C. Thomas

Molecular Foundry, Lawrence Berkeley National Laboratory, Berkeley, California, USA

CONTENTS

DOI: 10.1201/9781003359593-12

S CANNING PROBE MICROSCOPY is fundamental in materials exploration on the atomic scale in chemistry, physics, and materials science. Sample throughput and data production rate can, however, be low due to a) very long scanning times, specifically during dense spectroscopic acquisition, and b) the status of the atomically well-defined tip, and its preparation, requires time and expertise. Methods for experiments without any user bias in this field can better enable widespread accessibility to vested communities at large.

Material properties important for quantum communication, sensing, and computation require analysis at relevant length and time scales, which is at the length scale of single atoms and at the speed of molecular motion that is pertinent to applicable quasiparticles, such as electrons. Scanning tunneling microscopy holds the promise and application of both visualizing environments with sub-Ångstrom precision and, by coupling light in and out of the tunneling junction, measuring at the femtosecond timescale. The capability to characterize electronic properties above and below the Fermi level with atomic-scale spatial resolution is currently only possible with this system. However, it is extremely time-consuming due to complications with tip and sample, thermal drift, and skill of the user, to name a few issues that can arise during data acquisition. For example, scanning tunneling spectroscopy measurement acquisition times are on the order of seconds to minutes at each pixel, and thus collecting dense measurements of spectra can take weeks over one sample region. In order to increase tool accessibility and enable higher throughput, artificial intelligence and machine learning algorithms can provide a path for autonomous experiments without the need for any input from the scientist or engineer. Gaussian process regression has enabled a means for autonomous hyperspectral data collection on relevant 2D van der Waal materials, and can be subsequently classified by convolutional neural networks after a successful experiment which can be applied to any surface that is characterized by a scanning probe microscope. Here, we describe the implementation of a Python library built for such data acquisition, called gpSTS, through an open-access tutorial library, which can be run off test data to enable better access across a number of different toolsets available within the scanning probe field. We compare implementations of randomly acquired data with Gaussian process regression, making use of different available kernels and acquisition functions in gpCAM (*gpcam.lbl.gov*). In addition, we vary neural network hyperparameters during training with acquired imaging and hyperspectral data.

- 1D one-dimensional

- 2D two-dimensional

- 3D three-dimensional

- Å Ångstrom

- AI artificial intelligence

- ANN artificial neural network

- AFM atomic force microscopy

- CNN convolutional neural network

- DOS density of states

- eV electronvolt

- ExponentialCov standard exponential kernel using variance-covariance acquisition function

- ExponentialMIN standard exponential kernel using minimum of mean model function acquisition function

- fcc face-centered cubic

- FCN fully convolutional network

- GP Gaussian process

- hcp hexagonal close-packed

- IOU intersection over union

- LDOS local density of states

- LSTM long-short-term-memory

- ML machine learning

- Matérn2 Matérn kernel with second-order differentiability

- MatérnCov Matérn2 kernel using variance-covariance acquisition function

- MatérnMIN Matérn2 kernel using minimum of mean model function acquisition function

- mAP mean average precision

- nm nanometer

- PFM piezoresponse force microscopy

- ROC receiver-operator characteristic curve

- ReLU rectified linear unit

- SPM scanning probe microscopy

- STEM scanning transmission electron microscopy

- STM scanning tunneling microscopy

- STS scanning tunneling spectroscopy

- VAE variational autoencoder

- VGG visual geometry group

- WKB Wentzel–Kramers–Brillouin

12.1 INTRODUCTION

High-throughput hyperspectral measurements performed with an STM require deep knowledge of several interdisciplinary techniques. In this chapter, we delve into the fundamental concepts behind scanning tunneling spectroscopy and highlight machine learning methodologies that employ, e.g., neural networks and GPs to drive experimentation.

12.1.1 Scanning Tunneling Microscopy and Spectroscopy

The fundamental principle of quantum tunneling through a classically forbidden barrier in quantum mechanics paved the way for measuring local electronic and atomic structure at the picometer scale. Scanning tunneling microscopy, for which Binnig and Rohrer received the Nobel Prize in 1986 [59], makes use of this phenomenon by bringing an atomically sharp tip within tunneling distance to a surface of interest using an applied bias between the two conducting electrodes. The result is a tunneling current, I_T, that can be represented by the equation under the Bardeen [45, 515] formalism as

$$I_T = \frac{4\pi e}{\hbar} \int_0^{eV} \rho_S(E_F - eV + \varepsilon)\rho_T(E_F + \varepsilon)|M|^2 d\varepsilon \qquad (12.1)$$

where π is a mathematical constant, e is the elementary charge, $\hbar$ is the reduced Planck's constant, V is the bias voltage, E_F is the Fermi level, and ε is the energy variable for integration. Here, the change in the tunneling matrix element, $|M|$, can be considered negligible under certain considerations, such as in superconductors, and the tunneling current is then proportional to the convolution of the DOS of both the tip (ρ_T) and the sample (ρ_S). However, in many cases, the energy dependence of the tunneling matrix has to be taken into account, such as under the condition of elastic tunneling, where the gap region of the tip and sample can be represented as

$$\Psi_T(z) = \Psi_T(0)exp(-\kappa_T z) \qquad (12.2)$$

$$\Psi_S(z) = \Psi_S(s)exp(-\kappa_S(z - s)) \qquad (12.3)$$

and the two decay constants are then equal as

$$\kappa_S = \kappa_T = \kappa = \frac{\sqrt{2m\phi}}{\hbar} \tag{12.4}$$

where Ψ_T and Ψ_S are the wavefunctions of the tip and sample, z is the distance between electrodes that can be extended to 0 point reference and s distance, κ is a decay constant, m is the mass of an electron, and ϕ is the work function. Upon rewriting the tunneling current formula into a symmetric form and expanding the tunneling matrix element into its 1D form, it can be shown that the tunneling current can then be identified as

$$I_T = \frac{4\pi e}{\hbar} \int_{-\frac{1}{2}eV}^{\frac{1}{2}eV} \rho_S(E_F + \frac{1}{2}eV + \varepsilon)\rho_T(E_F - \frac{1}{2}eV + \varepsilon)|M(0)|^2 exp(\frac{\kappa \varepsilon s}{\phi})d\varepsilon \tag{12.5}$$

and in the case where s is large, the main contribution is due to a small energy near $\varepsilon \sim \frac{eV}{2}$. Here the differential of the current with respect to the voltage can be defined as

$$\frac{dI}{dV} \sim \rho_S(E_F + eV)\rho_T(E_F) \tag{12.6}$$

At positive sample bias, the unoccupied states of the sample can be probed and at negative sample bias the occupied states of the sample can be probed assuming a flat DOS of the tip, which is a fair approximation for commonly used metals. This is valid under the 1D WKB method [531, 243].

Additionally, the event of an inelastic tunneling event can be monitored under the second harmonic of I_T, and the tunneling matrix element can be rewritten in a time-dependent form and can indeed be considered constant such that only the molecule or adsorbate of interest is vibrating between the metal electrodes of both the tip and sample. The tunneling current, with respect to a vibrational state of energy $\hbar\omega$ can then be shown to be

$$I_T = \int_{E_F - eV + h\omega}^{E_F} f_1(E)f_2(E)dE = \left\{ \begin{array}{ll} 0, & \text{if } eV < \hbar\omega \\ eV - \hbar\omega, & \text{if } eV \geq \hbar\omega \end{array} \right\} \tag{12.7}$$

where ω is the angular frequency. This can be evaluated analytically such that the second derivative of I_T with respect to V obtains the line width of the vibrational excitation. This representation of the second harmonic is then

$$\frac{d^2 I}{dV^2} \propto c * exp(\frac{eV - \hbar\omega}{kT}) \frac{(\frac{eV - \hbar\omega}{kT} - 2)exp(\frac{eV - \hbar\omega}{kT}) + \frac{eV - \hbar\omega}{kT} + 2}{(exp(\frac{eV - \hbar\omega}{kT}) - 1)^3} \tag{12.8}$$

where k is the Boltzmann constant and T is temperature. Both the LDOS and vibrational spectrum driven by inelastic tunneling processes can be measured using a lock-in amplifier with an STM. These formalisms and capabilities inherent within SPM become exceedingly useful for understanding material systems at the nano and atomic limit.

"The coming nanometer age can, therefore, also be called the age of interdisciplinarity."

Heinrich Rohrer
Nobel Prize Recipient in Physics, 1986

12.1.2 Machine Learning Approaches in Scanning Tunneling Microscopy

Machine learning (ML) is defined as learning by computers without being explicitly programmed [464]. Supervised and unsupervised learning methods exist. In supervised learning, the input data is labeled (commonly referred to as the "ground truth") and used to train the machine learning model on specific tasks, while in unsupervised learning, the model learns patterns intrinsic to the unlabeled input data. ML has been found to be particularly useful for data classification, regression, natural language processing, anomaly detection, and clustering. Since ML is enhanced by maximizing the number of data with low noise that can be used for learning, its most recognized application has been computer vision. In chemistry and physics, ML has been used in areas that generate large sets of experimental and theoretical data such as high energy physics, materials property predictions and discovery, and condensed matter theory predictions [40, 97, 370, 399]. The challenge in the application of ML to scanning tunneling microscopy is the lack of large low-noise data sets [517]. In the following sections, the strategies to overcome this challenge specific to the data sets typical to scanning tunneling microscopy are discussed in detail.

12.1.2.1 Gaussian Processes

The definition of GPs can be classified as a generic supervised learning method designed to solve regression and probabilistic classification problems [314, 370]. An ideal GP would contain a collection of random variables, such that every finite collection of those random variables has a multivariate normal distribution. The distribution of a GP is the joint distribution of all those random variables. The main advantage of GP is that it offers a solution to the regression problem together with a range of confidence where the model is more likely to be precise and predictive. Finding a set of functions that well reproduce the collected data is a problem of finding suitable properties for the covariance matrix of the Gaussian distribution (kernel). Note, this gives us a model of the data, and characteristics which we can interpret with a physical meaning. GP has been widely used for a number of applications that span from astronomy to bioinformatics, and with applied use in materials science and materials discovery [148, 156, 125, 612, 224, 223, 413, 613, 356].

Despite its advantages, the application of GPs to SPM has been limited due to the inherent challenge associated with the image acquisition process. While high bandwidth techniques exist, scanning probe microscope images are acquired in minutes rather than seconds. Such long duration makes the acquisition process vulnerable to sample degradation via, e.g., surface contamination and thermal drift. These processes skew the data away from a normal distribution and prevent the application of GPs. Therefore, it becomes crucial to reduce the scanning time while preserving the data accuracy necessary to gain information about the physical properties under investigation.

Recently, GPs have been used for PFM, where it has been possible to significantly decrease the image acquisition time without loss in signal. Kelley et al. proposed a novel approach for fast scanning and identification of regions of interest using piezoresponse force microscopy [231]. They made use of GP regression reconstruction and extremely sparse spiral scanning to reduce the data collection time to identify a region of interest by almost a factor of 6. The GP algorithm is used to reconstruct the region of interest starting from a very sparse set of data. The algorithm has been proven to deliver an image deviating from a more dense data set with an error of 6%. Similarly, Ziatdinov et al. applied GP regression reconstruction to a multi-modal, or hyperspectral, version of PFM, i.e., band-excitation PFM, exciting and detecting multiple frequencies in parallel, and harmonic intermodulation methods that detect the mixing harmonics between the two excitation signals. This allowed for the reconstruction of spectroscopic 3D data [611]. Also, in this case, the amount of data necessary to reconstruct a reliable reproduction of the signal is a fraction of the original data set.

12.1.2.2 Neural Networks

A supervised machine learning method that loosely models the action of neurons and synapses in animal brains in the form of nodes connected by directed and weighted edges can be used to define an ANN [245]. Each node represents an artificial neuron that receives input signals and can activate to output a signal to other connected neurons. The weight of each edge dictates the strength of the received signal. Signal propagation occurs when a neuron is active, where activation is computed by a non-linear function of the sum of the weighted inputs. A collection of these neurons form a hidden layer. This layer is then connected to input and output layers to construct the neural network. In a feed-forward network, each sample in the labeled input data is input to the network, and the error in the output, compared to the expected result, is computed via some loss function. The network learns by iteratively updating neuron connection weights to minimize the loss function. This is accomplished using a generalization of the least mean squares algorithm called backpropogation. Remarkably, it was shown that just one hidden layer multilayer feed-forward neural network can approximate any function [202].

This network construction requires input data with a 1D shape, so applying an ANN to SPM, which produces 2D data, would require flattening it into a vector. Flattening an image into a long vector removes inherent spatial context from the data. Instead, neural networks can be extended to 2D data and computer vision problems by utilizing the convolution operation. The convolution slides an array of numbers, called a filter/kernel, over each pixel and computes the linear combination of the filter elements with the overlapping input pixels. Over one pixel, this produces a single number. When swept over every pixel, a new image is formed called a feature map. The convolution extracts relevant image features using kernels that are learned during training. This is the basis of the CNN, the backbone of most recent machine learning approaches to computer vision.

While CNNs have seen widespread application in many fields, such as medical imaging, satellite imaging, social media filters, and many others, they were not applied to STM until 2018 by Rashidi and Wolkow [109, 385, 4, 429, 511]. They trained a shallow, binary CNN to predict a "sharp" tip or a "double" tip by classifying 3500 images, 28×28 pixel, of isolated surface dangling bonds of a hydrogen-terminated silicon surface, H:Si{100}. Their model was able to achieve 97% accuracy initially and up to 99% accuracy after including majority voting [429]. In 2019, Gordon et al. bolstered the work done by Rashidi by expanding the problem from binary to multi-class. They trained and evaluated a selection of models on 128x128 pixel images of both metallic Cu{111} and Au{111}, and semiconducting H:Si{100} surfaces. Gordon reported that CNN architecture performance was surface dependent, and found that VGG-like networks produced the best results for Au{111} and Cu{111} [174].

In 2020, both Gordon and Rashidi advanced their own techniques in different directions. Rashidi et al. expanded the previously used H:Si(100) dataset by adding pixel-wise labels to 28, 1024×1024 pixel, images. Those images were then divided into 64, 128×128 pixel, smaller images that were artificially expanded with rotations and reflections to a total of about 15,000 images. They trained an encoder-decoder network, like SegNet [33], to enable semantic segmentation of the images to provide localization and classification of 6 defect classes. Using this model, the authors were able to perform automated hydrogen lithography [428]. Gordon also advanced work done on the H:Si(100) surface by exploring classification of partial scans and investigated treating the problem with video content recognition techniques. They found that a model including a secondary LSTM network just before the final, dense CNN layer to make predictions over a rolling, partial scan window could achieve similar or better performance as previously reported results [176].

Other microscopists sought automation as a means to continue experimentation away from the lab. Here, we explore automated tip conditioning possibilities on the Au{111} surface using CNNs to optimize the tip for spatial resolution through classification of herringbone reconstruction in STM images. In 2021, Wang et al. demonstrated optimization for spectroscopy resolution by classification of Au{111} STS spectra until the surface state was visible [555]. Similarly, we used a 1D-CNN to classify STS spectra from the face-centered cubic and hexagonally close-packed phases of the herringbone reconstruction along with pristine and defective regions of WS_2 [517].

The Unet architecture has seen widespread usage in object detection for a variety of imaging datasets [449, 144]. In CNN architectures involving multiple convolutional layers, the convolved feature maps are typically pooled to reduce the map size. This results in a compression of feature representation within the network which is typically encoded on to a small vector. This network design is referred to as an encoder, and the encoded vector is said to exist in the latent space of the model. The opposite is true of a decoder, which learns to decode an encoded vector into an image. The combination of these two forms an encoder-decoder model which is the basis of Unet. In STEM, Kalinin and Ziatdinov et al. trained a FCN model on images of a simple lattice of Gaussians, i.e., synthetic STEM images, and corresponding masks to atomically segment experimental STEM images of silicon-doped graphene [610]. In a follow-up, they were able to use this method to track electron beam damage to the lattice in real-time and explore order parameters of the lattice learned by a VAE trained on the segmentation maps produced by a CNN [222]. In this case, the architecture of the VAE was modified to include invariance to rotation by explicitly adding a latent space variable to encode a rotation angle. This approach was shown to be quite effective on STEM images; the restriction of the use of one variable as rotation angle enabled the rest of the latent space vector to disentangle a more subtle representation of the atomic segmentation maps, representing the crystallinity of the lattice. In STM, CNNs have been used to detect gold nanoparticles, identifying self-organized structures, and de-noise images [31, 381, 175, 219].

All of the above successes depend upon the ability of the neural network to generalize and handle shifts between the original training data and new target data encountered after model deployment. This is referred to as a data distribution shift and is a point of failure for supervised learning systems. Distribution shifts can happen for all sorts of reasons, from training data class imbalance to a change in the manual labeling scheme, to the difference in noise profiles between STM systems. So, it is important to measure the effect of these shifts on the predictive performance of the network. Supervised models are commonly measured during training by one or more user-defined accuracy metric(s) and post-training through several statistical methods. These statistics are measured by applying unseen test data to a trained model. In the case of multi-class classification, the confusion matrix, precision, recall, and ROC can provide information on prediction bias, false positive rate, and generalizability. The mAP is a quantity used to compare model performance on a particular dataset using a single number that incorporates the results of the previous metrics. In the case of image output, the IOU is computed between the test image and the predicted image, and used to measure model performance. The shift in these metrics over time or between deployments reveals the effect of the distribution shift. For example, when models were trained for classifying STM images of the Au{111} surface, it was found from the confusion matrix that a class imbalance in the training data dominated by the herringbone class resulted in a bias for the model to produce herringbone false positives. The bias was corrected by carefully augmenting the training data to compensate for the dominant class and balance the class distribution.

One of the drawbacks of supervised learning is the need to manually label large training datasets. It is a time-consuming process in which labeling bias inevitably permeates the dataset and is learned by models during training. Generating synthetic training data can help overcome these challenges.

Additionally, when constructing a neural network there are several parameters of the network's design that must be chosen. An example would be the number of nodes in a hidden layer. Since there is no general theory of neural networks there is no clear best choice for the number. This value is part of a collection called hyperparameters which dictate some aspect of the network and must be optimized.

12.2 HYPERSPECTRAL SCANNING TUNNELING SPECTROSCOPY DATA COLLECTION

GP models have been used in sparse data collection, piezoforce microscopy, and others, where the model can be defined over a given dataset, $D = x_i, y_i$, and the regression model can then be defined as $y(x) = f(x) + \epsilon(x)$, where x are the positions in some input or parameter space, y is the associated noisy function evaluation, and $\epsilon(x)$ represents the noise term. The kernel functions $k(x_i, x_j; \phi)$ define the variance-covariance matrix of the prior Gaussian probability distribution, where ϕ is a set of hyperparameters that are found by maximizing the log marginal likelihood of the data. Here, we explore both the Matérn kernel with second-order differentiability and the standard exponential kernel to visualize performance on a test dataset with two different acquisition functions [517], where a predictive mean and variance can then be defined given a Gaussian probability distribution from a chosen kernel function. We make use of both the covariance acquisition function and the mean function during point acquisition (Figure 12.1).

12.2.1 gpSTS: Using a Tutorial Library for Autonomous Scanning Tunneling Spectroscopy

In order to properly introduce the library, gpSTS [517], we begin by forming a tutorial library that implements gpCAM [357] on a test dataset of hyperspectral data collected

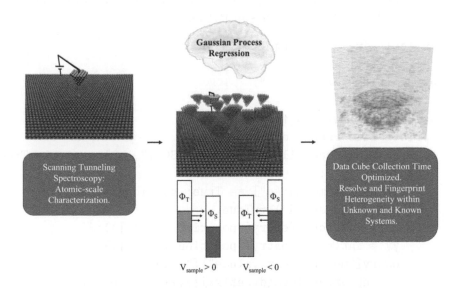

Figure 12.1 A generalized schematic of obtaining an optimized hyperspectral datacube with gpSTS, where the tip is directed autonomously by a Gaussian process for point scanning tunneling spectroscopy. Here, the tip height is held constant and the bias is swept to observe states above and below the Fermi level, and this is continued until the experiment is exited. Modified after Thomas et al. (2022) [517].

on a chalcogen vacancy within the TMD, WS_2 [517]. With all the defined requirements met, please install the tutorial library by

```
git clone -b tutorial https://www.github.com/jthomas03/gpSTS
```

for the necessary libraries discussed in the following sections. For the purposes of driving a GP-driven experiment, we make use of both the covariance-driven and the maximum acquisition functions. This can be optimized by returning the negative value of the mean model function and by returning the positive value of the covariance, respectively, which can be obtained through the description below.

```
### Acquisition Function 1 ###
def my_ac_func(x,obj):
  mean = obj.posterior_mean(x)["f(x)"]
  imin = -mean
  return imin

### Acquisition Function 2 ###
def my_ac_func(x,obj):
  cov  = obj.posterior_covariance(x)["v(x)"]
  return cov
```

For the purposes of this tutorial, we introduce an experiment that is randomly driven and that can be compared to a GP-driven experiment. First, a data class is instantiated to create an object that can be passed to the tool or to a tool simulator, which can be described by the following lines in Python.

```
### Begin Tool Simulation ###
while True:
    print("waiting for file...")
    time.sleep(2)
    if os.path.isfile(read_file):
        print("file received")
        a = np.load(read_file, encoding="ASCII", allow_pickle=True)
        for entry in a:
            if entry["measured"] == True: continue
            entry["measured"] = True
            xx = int(round(entry['position']['x1']))
            yy = int(round(entry['position']['x2']))
            entry["measurement values"]["values"] =
                np.array([sum(dtout[xx][yy][lpix:upix])])
            entry["measurement values"]["value positions"] =
                np.array([0])
            entry['measured'] = True
        os.remove(read_file)
        np.save(write_file, a)
        print("results written")
```

This enables the GP interface to connect to the tool (or tool simulator) via a command file, which can also be modified for a more optimized TCP/IP connection as performed in libraries such as deepSPM [246]. The command file directs the next point of measurement, which is collected and read back to the interface to update the model. In order to direct the tool to perform a purely random measurement, a class implementation of both data output (command file) and a control can be used. In order to avoid redundancy, each visited point is stored, and random points are created in each command output.

```python
def write_command(self,location):
    com = []
    def get_command():
        def get_random(lower_limit,upper_limit):
            out = random.uniform(lower_limit, upper_limit)
            return out
        idx = 0
        com = []
        com.append({})
        com[idx]["position"] = {}
        for gp_idx in self.conf.random_process.keys():
            dim = \
                self.conf.random_process[gp_idx]
                ["dimensionality of return"]
            num = \
                self.conf.random_process[gp_idx]["number of returns"]
            for para_name in self.conf.parameters:
                lower_limit =
                    self.conf.parameters[para_name]
                    ["element interval"][0]
                upper_limit =
                    self.conf.parameters[para_name]
                    ["element interval"][1]
                com[idx]["position"][para_name] =
                    get_random(lower_limit,upper_limit)
            com[idx]["cost"] = None
            com[idx]["measurement values"] = {}
            com[idx]["measurement values"]["values"] =
                np.zeros((num))
            com[idx]["measurement values"]["value positions"] =
                np.zeros((num, dim))
            com[idx]["time stamp"] = time.time()
            com[idx]["date time"] =
                datetime.datetime.now().strftime("%d/%m/%Y_%H:%M%S")
            com[idx]["measured"] = False
            com[idx]["id"] = str(uuid.uuid4())
```

```
            return com, idx
    write = False
    while write == False:
        if len(self.data_set) == 0:
            com, idx = get_command()
            self.visited.append(com[idx]["position"])
            np.save(location,com)
            write = True
        else:
            com, idx = get_command()
            if com[idx]["position"] not in self.visited:
                self.visited.append(com[idx]["position"])
                np.save(location,com)
                write = True
            else:
                write = False
```

Initial parameters are defined in `Config.py` and a new dictionary can be added in the configuration file to define additional parameters needed for randomly-directed experiments. Here the only information required is the number of returns, the dimensionality of the returns, and the plot function, as to plot the model and variance output after each collected data point.

```
random_process = {
    "rand": {
        "number of returns": 1,
        "dimensional of return": 1,
        "plot function": plot_2drand_function,
    },
}
```

The data class can be instantiated in the main loop for random data collection. Here, we remain within the maximum iterations defined in our configuration file, read the data provided by the tool, use the data class to define the next random point that has yet to be visited, and remain in the loop until the process has been fully completed.

```
def random_main():
    ### Begin Random collection loop ###
    data = RandData(conf)
    data.write_command(write_file)
    test = 0
    while test < conf.likelihood_optimization_max_iter:
        print("Waiting for experiment device to read and
            subsequently delete last command.")
        time.sleep(2)
        if os.path.isfile(read_file):
```

```
            print("result received")
            a = np.load(read_file, encoding="ASCII", allow_pickle=True)
            data.read_data(a)
            data.update_file(dvispath)
            os.remove(read_file)
            data.write_command(write_file)
            out = data.get_dlen()
            print(out)
            print("command written")
            test += 1
    out = data.get_data()
```

The tool simulator can interface between both the GP and the randomly driven main loop by defining this in the Config.py file. Here a run parameter can be either defined as a random process or by Gaussian processes.

```
run_param = "random_process" | "gaussian_processes"
```

Additionally, we can add the exponential kernel, in comparison to the Matérn kernel with second-order differentiability, in order to contrast results from the provided acquisition functions to both kernels.

```
def exponential_kernel(x1,x2,hyperparameters,obj):
    ### exponential anisotropic kernel in an input space with 12 ###
    hps = hyperparameters
    distance_matrix = np.zeros(((len(x1),len(x2))))
    for i in range(len(hps)-1):
        distance_matrix +=
            abs(np.subtract.outer(x1[:,i],x2[:,i])/hps[1+i])**2
    distance_matrix = np.sqrt(distance_matrix)
    return   hps[0] *  obj.exponential_kernel(distance_matrix,1)


def kernel_12(x1,x2,hyper_parameters,obj):
    ### Matern anisotropic kernel in an input space with 12 ###
    hps = hyper_parameters
    distance_matrix = np.zeros(((len(x1),len(x2))))
    for i in range(len(hps)-1):
        distance_matrix +=
            abs(np.subtract.outer(x1[:,i],x2[:,i])/hps[1+i])**2
    distance_matrix = np.sqrt(distance_matrix)
    return   hps[0] *  obj.matern_kernel_diff2(distance_matrix,1)
```

With this implementation, we can begin to run both random and GP-driven experiments under the provided kernels, which define the covariance [370]. For an initial test, the randomized run can be set for 200 iterations, and, in addition, we can run a subsequent GP run with an equal maximum number of iterations to visualize the test dataset, which is described more explicitly in Thomas et al. and Noack et al.

N = 10   N = 20   N = 40   N = 80   N = 160

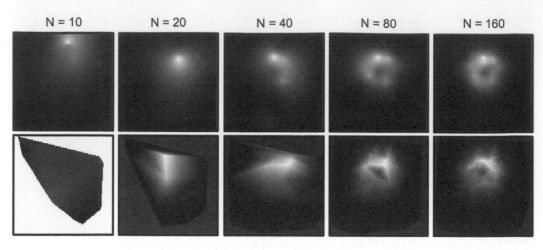

Figure 12.2   A GP-driven process run under an exponential kernel (top) compared to randomly driven point acquisition (bottom) evolving from 10 to 160 points. Ground truth data used for analysis is from Thomas et al. (2022) [517].

[370, 517]. In Figure 12.2, a GP-driven toy experiment is shown over a given number of iterations that can be directly compared to a randomly driven experiment.

12.2.2   Image Correlation

In order to cross-correlate data output from each experiment we can define a new class to compute correlation across each iteration, which is adapted from the literature [518, 517].

```python
class Corr(object):
    def __init__(self, im1, im2):
        ### Constructor ###
        assert im1.shape == im2.shape
        self.im1 = im1
        self.im2 = im2
        self.corrlen = ((self.im1.shape[0]*im1.shape[1])+1)
        self.corrs = []
        self.ind = []

    def get_corr(self,ind):
        def corr_vec(xx, yy, ll):
            return np.sum(((xx-np.mean(xx, axis=0))*
                (yy-np.mean(yy, axis=0))),axis=0)/
                ((ll-1)*np.std(xx, axis=0)*np.std(yy, axis=0))
        def specnorm(data):
            dmin = np.min(data)
            dmax = np.max(data)
            out = np.zeros(data.shape[0])
```

```
for i in range(0,data.shape[0]):
    out[i] = (data[i] - dmin)/(dmax-dmin)
return out
im1_corr = specnorm(self.im1.flatten())
im2_corr = specnorm(self.im2.flatten())
corrd = corr_vec(im1_corr,im2_corr,self.corrlen)
self.corrs.append(corrd)
self.ind.append(ind)
```

Additional lines provide the capability to update images, get correlation data, and plot the correlation as a function of step size. Implementation of this class can be instantiated through the provided `Run_correlation.py` script, which can be run after a dataset is obtained through `Run_gpSTS.py`. The tutorial library can be individually tuned, and one can modify both acquisition functions and kernels in order to optimize the GP to best fit a given substrate or sample of interest. In Figure 12.3, we compare the results of random point acquisition with that of combining the Matérn2, with second-order differentiability, and its calculated variance-covariance matrix for point acquisition. Here, we reach 95% correlation in only 59 points using Matérn2 and reach the same benchmark at 144 points with the random method.

In order to better determine the performance of the presented kernels and acquisition functions, we can perform multiple experimental runs and perform a oneway ANOVA analysis on the resulting data using the Scipy package available in Python [187, 549]. For our purposes, the point where greater than 95% correlation can be used to represent how effective each kernel and acquisition function combination, which is also compared to multiple randomized acquisition runs, are to one another (Figure 12.4). Assumptions important in this test are as follows: samples are independent, each sample is from a normally distributed population, and population

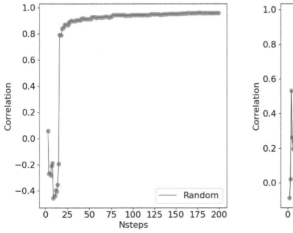
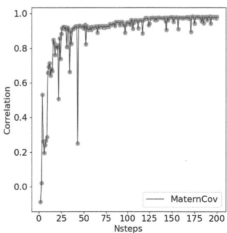

Figure 12.3 Correlation between ground truth data over a given number of points acquired, or steps, for both Random (left) and covariance-driven Matérn2 (right). Ground truth data used for analysis is from Thomas et al. (2022) [517].

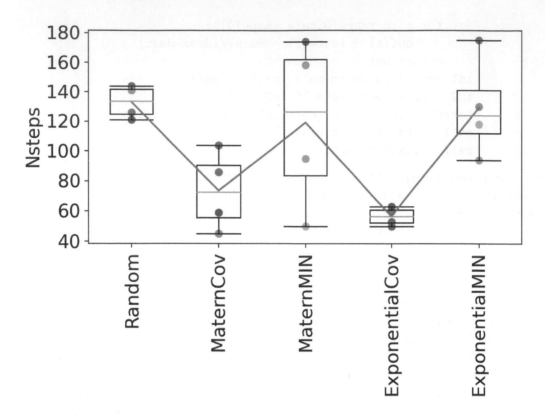

Figure 12.4 Statistical analysis of two kernels run with two different acquisition functions, that is then compared to a random-driven process, with the mean value of each group of runs connected by the red line.

standard deviations of groups are equal, or show homoscedasticity. We then test the null hypothesis, where two or more groups have the same population mean.

Oneway ANOVA analysis between random, MatérnCov, MatérnMIN, ExponentialCov, and ExponentialMIN yields a p_{value} of 0.013, where the means are statistically unequal ($p_{value} < 0.05$). However, if we separate the mean values and compare random, MaternMIN, and ExponentialMIN, we obtain a p_{value} that is equal to 0.88, showing that the mean values are statistically equivalent. Additionally, upon comparison of MatérnCov and ExponentialCov alone, we obtain a p_{value} that is also well above the commonly accepted value of 0.05 ($p_{value} = 0.26$). This information can be interpreted such that Random, MatérnMIN, and ExponentialMIN are consistently elevated with no clear top performer among the three processes. However, both MatérnCov and ExponentialCov perform better with ExponentialCov showing a tighter distribution with less high fliers ($Random_{iterations} = 133.0 \pm 9.7$, $MaternCov_{iterations} = 73.5 \pm 23.0$, $MaternMIN_{iterations} = 119.3 \pm 49.7$, $ExponentialCov_{iterations} = 56.5 \pm 5.22$, $ExponentialMIN_{iterations} = 129.3 \pm 29.4$). The data only represent 4 runs in each category, so more data is certainly required for meaningful results. However, it does reveal a path to take when installing gpSTS onto a new scanning probe platform. A number of different kernel functions and acquisition function choices are available within gpCAM [357], and can be subsequently implemented as required by an individual user.

12.2.3 Application of Convolutional Neural Networks in 1D

A number of papers have shown usage of CNN in SPM, such as in tip state identification, automated lithography, and even in non-contact AFM [429, 428, 246, 16]. As mentioned earlier, hyperparameter tuning can be non trivial and there are a number of reports and libraries dedicated to such optimization as it relates to ML and a variety of learning structures such as Neural Networks [145, 601, 271, 169, 54]. Even so, gpSTS uses the Python library PyTorch [395], where a number of parameters are available for optimization within the context of what is presented for gpSTS and, as it is open-source software, the package can be modified to use a large number of available structures, e.g., within the field of Neural Networks. We use the dataset that is made available in Thomas et al., which is the spectroscopic dataset composed of Au$\{111\}$ hcp, Au$\{111\}$ fcc, pristine WS$_2$, and a V$_S$ within pristine WS$_2$ [517]. The dataset consists of 424 Au$\{111\}$ fcc, 709 Au$\{111\}$ hcp, 158 V$_S$, and 191 WS$_2$ spectra for a total of 1482 scanning tunneling spectra.

With the available dataset, hyperparameter optimization is possible with the four spectroscopic classes identified. The neural network architecture used is an extension of those used on datasets obtained from the Airborne Visible/Infrared Imaging Spectrometer [289], where two convolution layers are implemented with a 1×3 kernel (stride 1, padding 1), batch normalization, max pooling, and ReLU activation [317, 385]. Additionally, dropout is applied to help overcome overfitting. By setting the outputs of some neurons to zero during dropout, in a random fashion during training, the neurons are not included in the forward pass and subsequently also not in back-propagation. ReLU is a nonlinear operation that accepts input if positive or returns a 0 if the input is negative [162]. By combining both ReLU and dropout, sparse-regularization is achieved which solves the overfitting problem [94]. As the model progresses through training epochs, the Adam algorithm minimizes computed cross-entropy loss and uses momentum and adaptive learning rates to converge faster with less hyperparameter tuning [234].

A number of hyperparameters are available for optimization, such as the learning rate and the number of convolutional features produced at each layer with a given kernel size, to list only a few parameters. The effect of the learning rate, taken from `Config.py` on our dataset is shown in Figure 12.5, where learning rates of 0.001 and 0.0001 show the same accuracy on test data after 10 training epochs. Learning rates greater than 0.01 show low accuracy scores across classes. Choosing 0.001 as the learning rate for our CNN, the effect of the number of filters can be probed. This can be modified in the `cnn1dimport.py` file, which is shown below.

```
class Conv1d(nn.Module):
    def __init__(self,num_classes=4):
        super(Conv1d, self).__init__()
        self.layer1 = nn.Sequential(
            nn.Conv1d(1, 8, kernel_size=3, stride=1, padding=1),
```

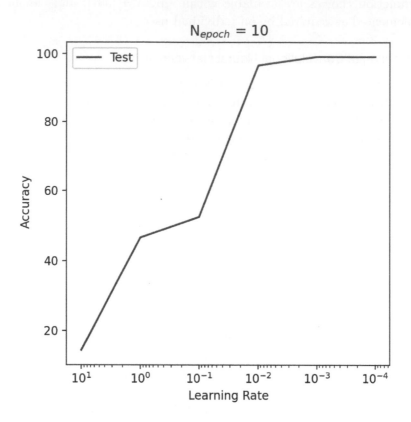

Figure 12.5 Accuracy as a function learning rate on test data. Accuracy is shown to reach 99% with a learning rate of 0.001. Test data are from Thomas et al. (2022) [517].

```
        nn.BatchNorm1d(8),
        nn.ReLU(),
        nn.MaxPool1d(2, stride=1))
    self.layer2 = nn.Sequential(
        nn.Conv1d(8, 16, kernel_size=3, stride=1, padding=1),
        nn.BatchNorm1d(16),
        nn.ReLU(),
        nn.MaxPool1d(2, stride=1))
    self.layer3 = nn.Dropout(p=0.2)
    self.fc = nn.Linear(self.getinput(), num_classes)

def size_postopt(self, x):
    out = self.layer1(x)
    out = self.layer2(out)
    out = self.layer3(out)
    return out.size()
```

```
def getinput(self):
    size = self.size_postopt(torch.rand(1,1,
        conf.gpsts_config['Experiment_Settings']
        ['NumSpectralPoints']))
    m = 1
    for i in size:
        m *= i
    return int(m)

def forward(self, x):
    out = self.layer1(x)
    out = self.layer2(out)
    out = self.layer3(out)
    out = out.reshape(out.size(0), -1)
    out = self.fc(out)
    return out
```

Here, the first convolutional layer takes 1 input channel and outputs 8 convolutional features with a defined kernel. The second layer then takes 8 input features and outputs 16 features. This can be varied, as one example, to extract as many meaningful features as required to obtain a model with low loss and high accuracy. Looking at Figure 12.6, it can be shown, with up to 30 training epochs, how a different number of channels at each layer affects the quality of the produced model. In the case of 8-channel output in layer 1 and 16-channel output in layer 2 compared to the 16-channel output in layer 1 and 32-channel output in layer 2, there is not much gained in terms of loss minimization and accuracy increase. In fact, in this case, the overall performance at 30 epochs is reduced from 98.3% accuracy to 97.5%. However, in the case of 32-channel output in layer 1 and 64-channel output in layer 2, a closer match between training and validation predictions is seen in both loss and accuracy, where the test dataset predictions show off-diagonal elements in the confusion matrix, which represent type i, false positive, or type ii, false negative, error. However, moving to the case of 64 channel output in layer 1 and 128 channel output in layer 2, all off-diagonal elements in the test dataset are minimized at only 8 epochs, training accuracy and loss scores reach at most 99.5% and 0.0%, and validation accuracy and loss scores are above 98% and below 0.1%, respectively. In order to verify the model of choice, we can move to even more output channels totaling 256, where the model fails to perform as well in the prior case. It is worthwhile to note that maintaining the same train/validation/test split while exploring hyperparameter space can be beneficial when comparing performance, however, having a well-balanced dataset that is standardized and normalized has a substantial influence in how well a Neural Network learns and subsequently predicts or classifies a given spectrum that it has not seen during training.

The results presented in this section serve as only an introduction to optimizing specific hyperparameters on provided hyperspectral STS data. There are multiple

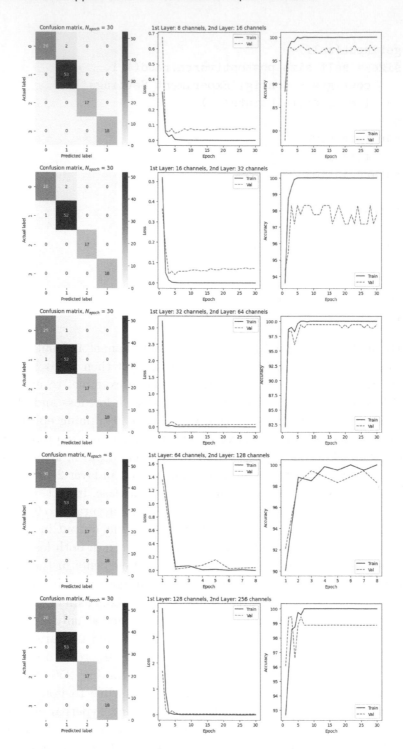

Figure 12.6 Confusion matrix, loss, and accuracy as a function of the number of input and output channels for two convolutional layers over a given number of training epochs. Training, validation, and test data are from Thomas et al. (2022) [517].

available routes for such optimization with many more parameters that were not discussed. As such, gpSTS was designed to be highly modifiable to meet the demands of a variety of substrates and samples for interrogation. Where a trained CNN model on Au{111} fcc and Au{111} hcp can serve well to identify the quality of an STM tip and the capability to produce reliable spectra, another route for tip optimization is using the as-obtained scanning tunneling micrograph over Au{111}.

12.3 SCANNING TUNNELING MICROSCOPY: CONVOLUTIONAL NEURAL NETWORKS IN 2D

In STM, atomic resolution is achieved due to the exponential dependency of the tunnel current on the tip-sample distance. The tunneling current is also related to the integrated LDOS between the Fermi level of the system and the relative tip-sample bias. Images represent a convolution of the topographic features of the sample and the LDOS contributions from both the tip and the sample in the region of interest [90]. As such, atomic scale configuration, including the composition, of the tip sensitively affects resulting images and, subsequently, the quality of spectra that are obtained during autonomous hyperspectral data collection. Therefore, the STM tip must be sufficiently prepared and conditioned to yield stable tunneling behavior and well-known electronic characteristics before performing any experiments. STM tip preparation typically involves a human operator manually manipulating the tip with respect to an atomically flat metallic substrate in such a way that an ordered layer of metal atoms forms at the apex of the tip. Au{111} is the material of choice when conditioning STM tips for many STM groups, since it imparts well-known electronic properties to the tip. It has the added benefit of hosting a surface reconstruction, which exhibits a distinctive "herringbone" pattern that is used as the accepted standard to determine that the tip is in a good condition for imaging surfaces [93]. Thus, the condition of the tip may be both assessed and improved using a single sample.

Unfortunately, this tip conditioning process is stochastic and usually extremely time-consuming, requiring hours to even days depending on the state and cleanliness of the tip. Even after conditioning, during experiments, the tip will interact with surface contaminants and adsorbates which degrade the geometry of the tips apex and reduce scan resolution. Thus, STM tips require frequent re-conditioning. This fact severely limits the throughput of STM systems and consumes valuable time that an STM operator could otherwise spend performing more meaningful tasks. Additionally, since the conditioning process is stochastic, there is no universally agreed-upon technique to produce a well-conditioned tip. This leads to a large variation in the quality of the conditioned tip, making a direct comparison between different STM operators and STM systems non-trivial.

This section describes our efforts toward implementing a semi-automated tip preparation tool to improve the productivity of STMs. Such a tool will ultimately be able (1) to autonomously condition the tip and (2) to recognize good images once the tip is conditioned. Here, we present initial work on the second aspect of the problem: recognizing good images taken from a well-conditioned tip. To do this we created and

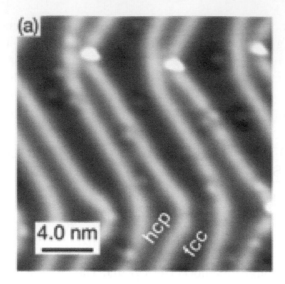

Figure 12.7 Constant current height map of the Au{111} surface showing both fcc and hcp regions of the herringbone reconstruction at a temperature of 4.5 K, ($I_T =$ 100 pA, $V_{sample} = 1$ V).

labeled a new dataset of STM images. Classification results obtained from applying a VGG-like convolutional neural network to recognize features commonly observed in images of Au{111} during the tip conditioning and evaluation process are presented. The motivation of this work is to eventually increase the throughput of the STM by training a CNN to classify images of the Au{111} surface to detect the presence of the herringbone reconstruction and thereby identify a sufficiently conditioned probe tip automatically.

12.3.1 STM Image Dataset

STM tip conditioning efforts can utilize Au{111} for the purpose of assessing and improving the condition of STM tips. Before being used for preparation and assessment, the Au{111} is cleaned by repeated cycles of thermal annealing and Argon ion sputtering in ultrahigh vacuum ($\leq 10^{-10}$ Torr). This removes surface contaminants such as water, adventitious carbon, and other adsorbate species. After being cleaned, Au{111} shows a surface reconstruction in which alternating regions of hcp and fcc crystal structures form a pattern known as the "herringbone" reconstruction (shown in Figure 12.7). We perform tip conditioning and assessment in ultra-high vacuum and with the sample and tip cooled to cryogenic temperatures using either liquid nitrogen (77 K) or liquid helium (4 K).

All STM-acquired data were taken on Au{111} and labeled to facilitate the use of a fully supervised training regimen. The labeling process involved using a Python script to present a trained STM user with a single STM image which had been flattened using a first order plane-fit and then asking the user to classify the image by picking a single label from a list of predetermined options. Figure 12.8 shows example

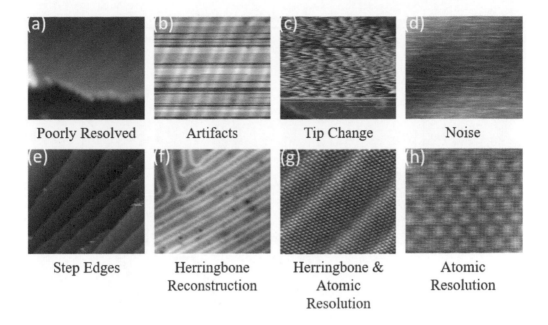

Figure 12.8 Examples of unconditioned (a-d) and conditioned (e-h) STM images of Au{111}.

images from the obtained dataset, where (a-d) presents images commonly observed when a tip has not been fully conditioned, as indicated by the degradation of the quality of the image, and (e-h) shows images of a tip which has had sufficient conditioning. These images clearly show the features one would expect from a well-resolved reconstructed Au{111} surface, namely: step edges, herringbone reconstruction, and even the atomic lattice.

The Au{111} dataset presented shows an uneven distribution of observations across defined classes, with a bias toward the herringbone class. To compensate for this class imbalance, per-class weights were applied and less represented classes were oversampled with random combinations of rotation, zoom, sheer, and translation augmentations to artificially balance the dataset. After augmentation, thirty percent of the labeled data was set aside, separate from the train/validation split, and used to test the classifier after training was completed.

12.3.2 Neural Network Model

The CNN used was inspired by the VGG2019 model with some modifications to better suit the Au{111} dataset (Figure 12.9). It was found that employing the VGG2019 architecture directly produced poor accuracy, in the 40-60% range, and that performance improved as the number of parameters in the network was reduced. This was attributed to the model being over-parameterized ($\sim$ 40 million parameters) relative to the small size of the dataset.

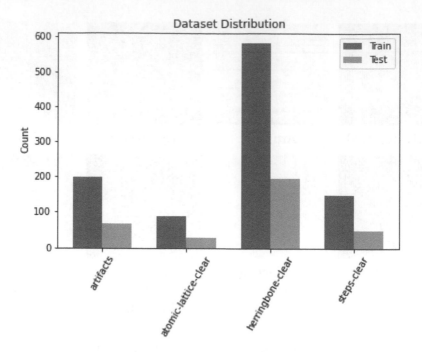

Figure 12.9 Au{111} dataset distribution.

The first layer was then changed to use 7×7 kernels to reduce the parameter count and account for the how the herringbone reconstruction pattern usually extends across the entire image and is not localized to one area. The number of convolution layers and kernels used per convolution layer were tuned to obtain best performance. The ReLU activation function was used for all except the final layer where softmax was used instead due to the multiclass nature of this problem. Batch normalization was applied to the output of the activation layers followed by down-sampling via 2×2 max pooling to form a set of feature maps. There are five blocks of convolutional, batch norm, and max pooling layers before the feature maps were passed to a fully connected neural network. Categorical cross-entropy was used as the loss function, which is minimized during training to achieve better model performance, along with the Adam optimizer.

Figure 12.10 shows results from the best model. On the left of Figure 12.10 is a graph of the training and validation accuracy, and on the right is a graph of training and validation loss. The x-axis in both cases is the number of training epochs. While the best validation accuracy achieved by the best model tested was 90%, noticeable instability is present in the validation curves. This indicates that, while the model is learning, it is also attempting to randomly guess the correct classification and shows that the accuracy metric can be misleading on imbalanced datasets such as this one.

As mentioned previously, a portion of the training data was set aside for testing purposes. After the model was trained, it was tasked with making predictions on this test dataset. A typical example produced by the model is shown in Figure 12.11. Here, off-diagonal elements in the confusion matrix indicate both type i and type

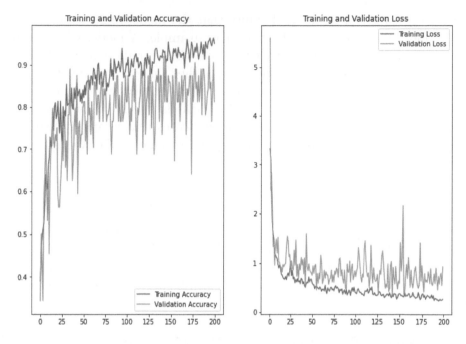

Figure 12.10   Training and validation performance of the VGG-like model.

ii errors but also show that the herringbone class dominates the correctly labeled predictions. While not ideal, this is still acceptable as the primary interest lies in recognizing when herringbone is present in an STM image.

Additionally, the ROC curve was calculated per class and is shown on the right in Figure 12.11, as well as the area under the ROC curve. An ROC curve measures the performance of a classification model at all classification thresholds while the

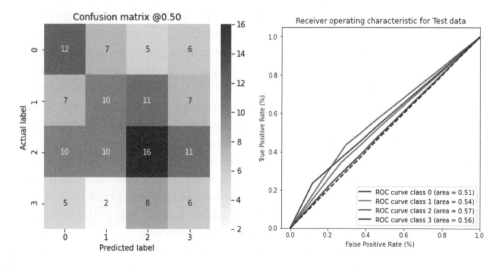

Figure 12.11   Confusion matrix (left) and ROC curve (right).

AUC can be thought of as the probability that the model ranks a random positive example more highly than a random negative example. A model whose predictions are 100% wrong has an AUC of 0.0 while one whose predictions are 100% correct has an AUC of 1.0. In this regard, the black, dashed line on y=x equates to an AUC of 0.5, which would imply an uninformative classifier that was randomly guessing labels. So, while none of the AUC values for each class were 0.5 or lower, the highest was 0.57 for the herringbone class. This indicates that relatively high accuracy ($\sim$ 85%) on validation data does not correspond to high predictive performance on test data.

12.4 OUTLOOK & PROSPECTS

The tutorial and results presented provide a path toward better enabling a highly complex and difficult toolset for widespread usage. AI/ML have enabled a computational renaissance within the field of SPM, and new developments further lower the boundaries of what may be perceived as difficult workflows for scientists and engineers.

12.4.1 Autonomous Discovery in Scanning Probe Microscopy

Autonomous experimentation, which requires no user input, is especially useful for tasks such as collecting dense hyperspectral maps over regions of interest across an atomically smooth substrate. The combination of GPs and a 1D CNN enables not only the collection of an optimized hyperspectral grid but also the subsequent classification of obtained spectra. STS data collection can take on the order of 2-5 minutes for high-resolution voltage spectra, and, as such, GPs are well-poised for point acquisition optimization and tool direction. Additionally, mapping heterogeneity across substrates is of substantial interest to the fields of semiconducting devices, quantum information science, and two-dimensional materials, where defects that contribute to pristine disorder in the system present both complications and routes toward enabling new physical phenomena to be understood and applied. As current imaging tunneling spectroscopy and LDOS mapping can enable scanning probe measurements of phenomena such as Majorana modes, atomic-scale quantum emission, and even mapping quantum coherence transport [517, 159, 472, 474], boosts in data collection capability should enable acquisition with higher throughput and greater reproducibility. Additionally, the modification of both the Kernel and the acquisition function provides routes toward operating in exploration mode, where locations are suggested to improve the statistical model, in exploitation, with the goal of locating the global optimum, or in some combination, which are possible and can be readily adapted to any scenario within the available software. One complication that naturally arises with a supervised machine learning scheme, such as a 1D CNN, is the availability of properly labeled datasets. As the field continues to develop and more groups make their data available in open-source platforms, training on larger datasets will allow a greater number of network architectures to compete for top performance. The authors hope that packages such as gpSTS and others will help break a number of limitations within the field of SPM.

12.4.2 Applied Neural Networks

This analysis shows that a relatively basic CNN can delineate between common tip condition states on an unbalanced dataset. By utilizing confusion matrices and ROC curves, the per-class performance metrics revealed some predictive accuracy on her-ringbone reconstruction. The relatively poor performance on unseen test data, espe-cially compared to recent work in the field, can be attributed the size of the training dataset. Current top performance results were obtained on datasets with sizes from 5,000 to 15,000 images, which is an order of magnitude larger than the dataset pre-sented. In order to achieve the best possible model performance, a large Au{111} dataset is perhaps required. To this end, generative adversarial networks can be ex-plored in order to perform style transfers of STM images of Au{111} taken under different data acquisition conditions to greatly expand the amount of overall training data. Additionally, since most STM images can be convoluted by topographic and electronic contributions, a multi-labeled approach can be explored. The inclusion of residual bottleneck layers and concept whitening should be investigated to improve model performance and interpretability. Detecting and tracking specific atomic de-fect sites using autoencoders for image segmentation is another area of interest within the field. A fully open-source, experimental STM image database, like the materials project, would be fruitful for STM groups attempting to utilize computer vision.

12.4.3 Concluding Remarks

Overall, we present a tutorial workflow for autonomous data acquisition based on the library gpSTS, which implements the autonomous discovery software gpCAM. This is paired with exploring different options for training a 1D CNN, and results that may be obtained. The logical progression of extending this into 2D image recognition is shown over Au{111} systems. Additional techniques concerning defect detection and classification, atom manipulation, and autonomous data collection workflows, we predict, will continue to receive attention throughout the field and literature.

CHAPTER 13

Autonomous Control and Analysis of Fabricated Ecosystems

Peter Andeer

Environmental Genomics and Systems Biology Division, Lawrence Berkeley National Laboratory, Berkeley, California, USA

Trent R. Northen

Environmental Genomics and Systems Biology Division, Lawrence Berkeley National Laboratory, Berkeley, California, USA

Marcus M. Noack

Applied Mathematics and Computational Research Division, Lawrence Berkeley National Laboratory, Berkeley, California, USA

Petrus H. Zwart

Molecular Biophysics & Integrated Bioimaging Division, Lawrence Berkeley National Laboratory, Berkeley, California, USA

Daniela Ushizima

Applied Mathematics and Computational Research Division, Lawrence Berkeley National Laboratory, Berkeley, California, USA

CONTENTS

R ESEARCH in the optimization of environmental conditions has used imaging data from plant roots and leaves to monitor and maximize biofuel productivity. The morphology of plants, such as *Brachypodium distachyon*, can be correlated to agricultural management techniques, but little is known about the impact of environmental stress, such as drought, and how to quantify plant variation under different

DOI: 10.1201/9781003359593-13

climate conditions. By exploring Lawrence Berkeley National Laboratory (LBNL)'s EcoFABs, new analytical pipelines can be created for plant surveillance. EcoFABs are substrates that mimic soil conditions and enable spatially and temporally resolved root/leaf mappings and the potential for self-driving labs for plants.

In this chapter, we explore the obstacles and discuss opportunities associated with fostering a low-carbon economy and advancing bioenergy businesses. We will delve into the challenges that arise when implementing analysis protocols and conducting plant research with EcoFABs. The goal of these new capabilities is to empower scientists in the field to study plant morphology in greater depth and to fine-tune the Autonomous Experiment system for automated management of plants using EcoFABs.

13.1   LOW-CARBON ECONOMY AND BIOENERGY

There are several reasons why we need to advance our understanding of improved agriculture and bioenergy. Firstly, with the world's population projected to reach 9.7 billion by 2050, there will be an increasing demand for food and energy. Agriculture and bioenergy are key sectors that can contribute to meeting this demand sustainably. Secondly, environmental degradation poses significant demands to food and energy security, and improving the efficiency and sustainability of agriculture and bioenergy research can help address these challenges. Thirdly, bioenergy is an important component of the transition to a low-carbon economy, and understanding how to optimize its production can help reduce greenhouse gas emissions and mitigate climate change. Finally, improving agriculture and bioenergy can also create new economic opportunities and improve livelihoods, particularly in rural areas where these sectors are important sources of income.

Climate change has made it difficult to predict yields of food and bioenergy crops due to extreme weather events, making it necessary to develop strategies to increase crop resistance to unanticipated changes in temperature and precipitation. These strategies should aim to reduce atmospheric carbon emissions and increase carbon uptake by soils. Bioenergy crops like switchgrass and sorghum have the potential to improve soil quality and can be grown in areas unsuitable for most food crops, contributing to efforts to improve crop yields and decrease soil degradation (Figure 13.1).

Conventional agricultural methods use inorganic fertilizers like nitrogen, potassium, and phosphate, which is unsustainable as it causes soil carbon leaching, water pollution and requires precious resources. Organic fertilizers like biochar, compost, and biosolids are seen as a more sustainable option as they not only provide necessary nutrients for healthy crop production but can rehabilitate poor soils over time. The composition of organic fertilizers and their impact on performance over time needs further understanding. For example, microorganisms play a crucial role in soil health and plant growth by exchanging nutrients and bioactive compounds, impacting biogeochemical cycles, and improving soil properties like water retention and nutrient composition.

The rhizosphere, or the soil surrounding plant roots, receives up to 20% of plant photosynthates in the form of diverse molecules. These molecules serve as a primary growth substrate for microbial communities, which benefit plants through improved

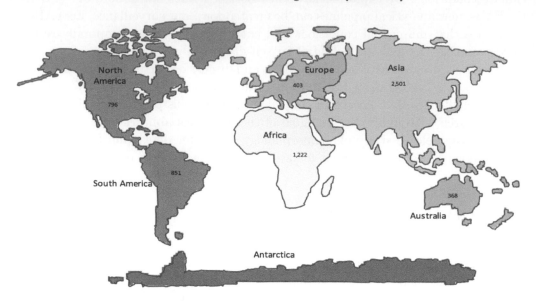

Continental scale estimates of degradation (million ha)

Figure 13.1 Land degradation estimates in million hectares: one-fifth of the Earths land area affected, with global trends indicating increased human-induced process, such as desertification, deforestation, improper soil management, cropland expansion and urbanization [163, 347].

nutrient acquisition, pathogen exclusion, and drought resistance. Microbial communities also significantly impact the fate of fixed CO_2 in soils, as plants take up and metabolize a range of organics from the environment. The relationship between microbes and plants is crucial for maximizing plant growth and understanding soil property changes over time.

Plant microbial interactions, most notably the association of nitrogen-fixing bacteria with legumes have been known for decades yet our understanding of the relationship between microbial communities, plants, and soil health remains poorly understood. The reasons for this can be attributed to several factors: (1) soils are heterogeneous in nature, (2) soil microbial communities are large, diverse, and often uncultured, (3) environments around roots are dynamic, and (4) a shortage of tools and methods to study microbial communities in these contexts.

The soil's heterogeneity has long been acknowledged, and the complexity of soil microbial communities has been recognized since the development of environmental molecular techniques. However, there is a need for methods that can predict not only which microbes to use for plant pairing but also how environmental conditions influence these interactions.

There are laborious steps in studying plant-microbial-chemical interactions and designing interventions. Traditional methods for studying plant impacts on ecosystem function involve ecologically driven experiments that seek correlations between field observations with little control, or highly structured experiments with a single plant and/or single microbe. Correlation analyses can link variables, but they cannot determine causality. On the other hand, experiments with single species can establish causality but may lack real-world relevance.

Testing grounds that incorporate specific elements of natural environments can address these issues. However, microbial communities are particularly challenging to study because methods become increasingly complex and hard to reproduce. Furthermore, the poorly defined nature of microbial communities, including which bacteria to use, their ratios, and inoculant sizes, can affect results, depending on the type of study. Methods are limited and may not apply to a given system. Depending on the hypotheses/objectives, methods may not be transferable, particularly if boutique environments are being used. Sharing of microbial communities can be expensive via culture collections or time-consuming and prone to error from lab to lab. Reproduction and comparison of results without appropriate benchmarks can be burdensome.

13.2 FABRICATING ECOSYSTEMS WITH ECOFABS

The vision driving EcoFABs, illustrated in Figures 13.2–13.4, revolves around the aspiration of unraveling the intricacies of complex microbial communities and unearthing a vast realm of scientific knowledge. By developing and standardizing these advanced ecosystem fabrication systems, scientists hope to accelerate their understanding of these complex communities through inter-laboratory studies that build upon one another. With standardized protocols and experiments that incorporate different environmental parameters, comparative experiments, and reproducible research are expedited. From individual microbes to genes, each study can serve as a touchstone to one another, ultimately leading to a more cohesive model of these systems. And with the inclusion of temporal and spatial analyses, scientists have the opportunity to advance their hypotheses and unlock the secrets of these communities.

Figure 13.2 *Brachypodium distachyon* grown in EcoFAB 2.0—photograph by Thor Swift, LBNL.

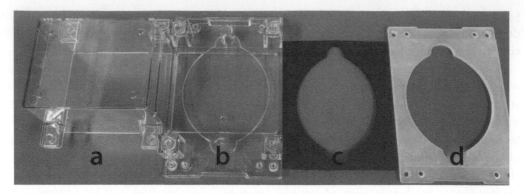

Figure 13.3 EcoFAB 2.0 design: 4 custom components that comprise: (a) EcoFAB plant chamber, (b) EcoFAB 2.0 base, (c) gasket that forms the root growth area, (d) backing plate.

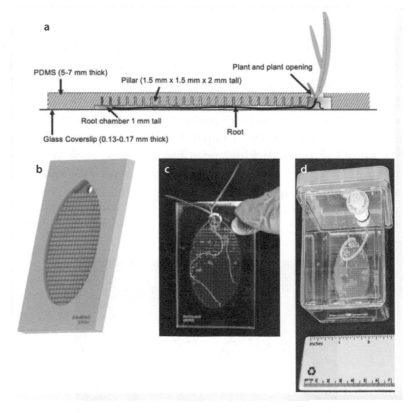

Figure 13.4 EcoFAB (a) Side-view configuration. The incorporation of pillars within the device facilitates the smooth flow of media while promoting the flat growth of roots against the coverslip. (b) A representation of the PDMS component of the device showcases the emergence of shoots from the top port. (c) EcoFAB device is presented in conjunction with a plant at a developmental stage of 20 days following germination. (d) EcoFAB device, housed under sterile conditions, exhibits the presence of a plant at a growth stage of 14 days postgermination, also cultivated hydroponically. Source: Jabusch et al. [207].

After a series of scientific meetings and workshops that were held between 2016 and 2019 [136], microbial and plant ecologists, biologists, and modelers combined expertise to determine the necessary features for model fabricated ecosystems that could bridge the gap between the study of cultures of unicellular organisms and entire ecosystems. Based on the outcomes of these meetings, LBNL developed EcoFABs (short for Ecosystem Fabrication), which are single-plant-scale systems. Specifically, EcoFABs were designed to fulfill the following requirements: (1) enable easy experimentation under sterile conditions, (2) be compatible with spatial analysis of root zones, including microscopy, and (3) be suitable for metabolomics analyses.

In order for EcoFAB systems to be suitable for these types of interlaboratory experiments it is important to have materials and methods that cover all aspects of benchmark experiments that have been validated across multiple researchers, ideally working in different locations [513]. With this in mind, a multi-laboratory study was conducted using sterile, axenic *Brachypodium distachyon* (Figure 13.2) plants, EcoFABs made of polydimethylsiloxane (PDMS), and three different growth media. Researchers in 4 laboratories located across the world then conducted parallel experiments and compared phenotypic plant data including plant root and shoot fresh weights, phosphorous and metabolite contents in plant tissues, and plant exudation profiles. The results demonstrated that differences between plants were reproducible and were driven by treatments, not by the person conducting the experiment. A follow-on experiment will be conducted using a community of isolated bacteria to assess how these methods extend to more complex systems.

13.3 AUTOMATION AND AUTONOMOUS DISCOVERY IN PLANT SCIENCES

Automation can be extremely beneficial in studying biological systems plate readers have long been used to track microbial growth in multi-well plates which greatly increases the throughput of microbial growth assays compared with individual tubes that either need to be subsampled or hand delivered to a spectrophotometer for measurements. For studying plant growth in general and particularly in the context of microbial interactions, automation can drastically accelerate processing times and provide enriched datasets that can lead to more effective and efficient experimentation.

For plant systems, a lot of time has been invested in developing automated methods for data analysis, particularly for analyzing images of plants. Image analyses are powerful tools for plant studies as images do not necessarily require destructive sampling, i.e., removal of the plant from the soil. One of the more popular programs is PlantCV [142], a software package developed by Fahlgren et al., that has over 100 citations at the time of this writing. This open-source software aids in the processing of multiple plant image types from single plants taken with a single camera to imaging densely packed plants imaged using multi-camera arrays. PlantCV is the foundation of other software packages including RhizoVision [478] which was designed to process pictures and scans of roots in order to get biomass information as well as phenotypic profiling of roots. These types of image segmentation processes greatly accelerate the processing of time compared with manual methods, e.g., weighing and measuring lengths manually.

Automated systems also exist for the maintenance, sampling, harvesting and imaging plants. These automated systems typically range from dedicated rooms to entire field sites. Indoor and greenhouse facilities have begun to incorporate automated imaging and scanning systems where planted containers are transported to various imaging systems through conveyor belts or through the use of cameras and other scanning equipment on motorized gantry that traverse the growing areas to acquire data on the plants during growth. Additional capabilities can include sensing equipment (e.g., pH, soil moisture) and weight measurements.

Advanced imaging and sensing have also been deployed across entire fields. Drones equipped with hyperspectral cameras can routinely photograph crops and there are even field sites equipped with motorized scanners that can image an entire field. These capabilities when linked to sensors deployed throughout the field to capture soil parameters can be of great benefit for understanding how soil properties impact plant growth, health and yields.

13.4   PLANT MONITORING USING ECOFAB

Original EcoFABs were made for custom plant growth and analyses. PDMS was selected as it is a common polymer in microfluidics and for culturing bacteria. Because PDMS devices are made using soft lithography, designs can be altered and customized simply by changing the mold based on the experiments and systems being conducted. PDMS can be bonded to glass using a plasma cleaner and then autoclaved to sterilize. These PDMS "chips" which can range from 1 to 20 mls are then placed in sterile containers for growth. The EcoFAB design targets high-resolution confocal imaging of small plant roots, as illustrated in Figure 13.5 but also works well for relatively inexpensive scanners as shown in Figure 13.6. Imaging microbial plant interactions was determined to be crucial, but burdensome because ideal microscopic methods require plant roots to be flat against the bottom surface so the root does not traverse in and out of the path length of the microscope objective. Moreover, this can present a problem for plants actively growing in these devices because the liquid is displaced by the plant via transpiration which can create pockets of air enmeshed within their roots. This issue tends to exacerbate as the plant ages and becomes more productive. The lack of vertical space between the roots, cover slip, and PDMS layer hampers refilling the devices. To combat this problem, imaging EcoFABs are cast in a special mold that contains pillars that are 2/3 the height of the root zone with channels for air and liquid to flow through them. Because these pillars are 1 mm short of the EcoFAB coverslip, roots that are 1 mm thick or less are able to grow through the device but are kept close to the slide bottom while liquid and gas can travel through the channels above the root.

EcoFAB 2.0 was designed to expand the user base and facilitate automated imaging and handling [377]. The EcoFAB 2.0 base has the same footprint that the Society for Laboratory Automation and Screening (SLAS, formerly SBS) laid out in the definition of microtiter plates. This allows these EcoFABs to be handled by robotic

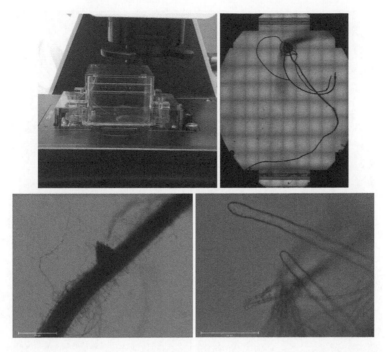

Figure 13.5 Images of a non-sterile *B. distachyon* grown in EcoFAB 2.0 after 3 weeks, scanned with an EVOS Auto Fl 2.0 inverted scanning microscope. An EcoFAB 2.0 resting in the stage of the microscope (top left), a stitched scan of the plant roots taken in an EcoFAB 2.0 using a 2X objective lens (top right). A single image of part of the plant root with fungal hyphae in contact, taken with a 4X objective lens (bottom left), and close-up images of the root hairs with interwoven fungal hyphae taken with a 60X objective lens (monochrome).

platforms built to hold, transport, and otherwise manipulate microtiter plates. To avoid the secondary containers required with previous EcoFAB iterations, EcoFAB 2.0 has the plant growth chamber affixed to the device base with sample ports that directly access the root zone chamber of the devices located on either side. The distance between these sample ports and the plant chamber was selected to allow large (5 ml) robotic liquid handling tips to freely access these sample ports from above even if the EcoFAB is tilted 45 degrees to aggregate liquid to one side.

Therefore, changing of media (Figure 13.8), sampling, or adding specific microorganisms or chemical compounds can be performed automatically using a liquid-handling robot. While EcoFABs constructed out of PDMS allow for customized solutions that can be tailored for specific applications, i.e., the imaging EcoFAB, however, device construction is time and labor-intensive and requires equipment to bond the PDMS to glass which limits production. To make EcoFAB 2.0 scalable, a need for high-throughput automation experiments, the primary components in EcoFAB 2.0 can be mass-produced by manufacturers with approval from LBNL. There are four primary components fabricated for the device that are assembled along with a large glass slide or other suitable viewing material with screws and washers. The EcoFAB

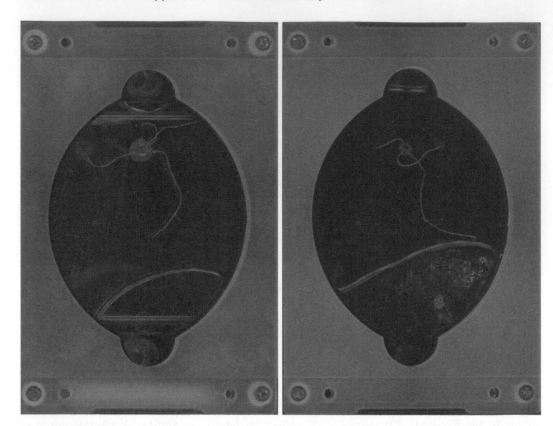

Figure 13.6 Images of *B. distachyon* root zones taken from EcoFAB with a transparent (left) or opaque (right) base. Images were collected with an EPSON Perfection V850 Pro document scanner.

2.0 plant growth chamber and base, which form the top of the device are made from polycarbonate via injection molding and the root zone is formed by a silicone gasket and slide held against the bottom of the base using a backing plate made with injection molded polycarbonate with glass fiber inserted to add stiffness. Since EcoFAB 2.0 is made out of injection-molded polycarbonate, the color (and transparency) of various parts can be altered by adding dyes during manufacturing. This can include

Figure 13.7 RGB images taken of a 3-week-old *B. distachyon* plant collected from three different angles using a Specim IQ camera.

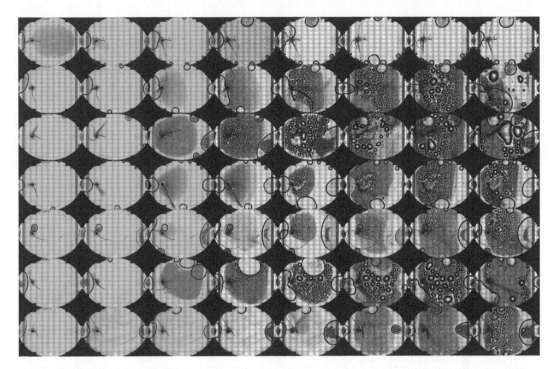

Figure 13.8 Experiments using EcoFAB with EVOS-scanned images to capture root growth under different media: rows show seven time series, with columns illustrating the root development at approximately 8 equally spaced points in times.

the addition of visually opaque dyes that are transparent to near-infrared light which allows the imaging of roots through the EcoFAB when lighting and camera sensors expand over 900 nm.

The ability to conduct experiments in the EcoFAB 2.0 under sterile conditions in order to control the microbial communities introduced into each plant was at the forefront of its design. Since all of the components are made of glass, polycarbonate, silicone, or metal, the entire device can be partially assembled and sterilized in an autoclave before final assembly and sowed with a seed/seedling within a sterile environment such as a laminar flow hood or biosafety cabinet. As mentioned, the plant shoot growth chamber is screwed into the EcoFAB base and since the top chamber is clear it does not have to be opened until harvest and will remain uncontaminated. The two sampling ports can be sealed using gas-tight septa, microbial-resistant membranes, or tape or robotic-friendly caps so that sampling or the addition of microbial or nutrients can be performed in a sterile environment but does not contaminate the device during growth. Imaging the entire plant, both roots and shoots, during growth, is one of the most important factors when EcoFABs were designed. Since the entire growth area at the bottom of the EcoFAB 2.0 is visible through the glass slide and rests flat the entire root chamber can be imaged from below using a number of different methods, provided that the entire device can fit within the imaging devices.

The top growth chamber of the EcoFAB is transparent on all four sides and above which allows the plant shoots and leaves to be captured from multiple angles.

One way that plant roots and the surrounding environment can be imaged is with a flatbed scanner. Because scanners use reflective illumination it is not necessary for light to penetrate through the EcoFAB device and thus transparent or opaque parts work. The versatility of many scanners allows for a wide range of resolutions (e.g., 300–12,800 dpi) and output file formats (e.g., .png, .tiff, .jpg).

Using automated scans of the entire EcoFAB 2.0 viewing area images clearly show root details like root hair density. Fungal hyphae can also be observed at these magnifications though up to 60X (with a long-working distance objective) might be needed. Individual microbial cells can often be observed, in particular, if they express fluorescent proteins at high magnification (40 - 60X objective lenses); though quantification can be challenging and needs to be further investigated.

The default height of the EcoFAB 2.0 root zone was designed to be 3 mm thick to allow liquid to move freely over roots when refilling media, however, this thickness is determined by the gasket thickness so thinner root zones with more in-plane root images can be made with 1–2 mm gaskets. Unlike the smaller PDMS EcoFABs, Eco-FAB 2.0 can not accommodate 0.1 mm cover slides as they are too fragile, however, 0.55 mm gorilla glass or similarly treated glass are possible. This should help with higher resolution objectives (e.g., 60X) where even with long-working distances their vertical range can be less than 3 mm.

The chamber top is made with optically clear polycarbonate to allow visual images from all angles including from above. This is particularly important with grasses like *Brachypodium distachyon* where prior studies have found that imaging from three angles is ideal for converting imaging data to size (mass) estimates of shoots. Figure 13.7 shows an example of RGB images taken using a hyperspectral camera from three separate vantage points (above, side and from in front).

13.5   CHALLENGES IN DEVELOPING AUTOMATED AND AUTONOMOUS EXPERIMENTS

While automated data collection can generate data more densely and quickly than manual experimentation, autonomous processing can make for smarter and more productive experimentation. Whether studying at the single plant scale or across acres of land, the number of parameters to study in plant-soil-microbial systems make designing experimentation suitable for predictive models extremely complex. This can be particularly difficult in plant systems where each iteration can take weeks to months and the natural variability of plant growth require more replicates than work with most unicellular organisms. Machine learning and AI-driven experimental design can help explore this space much more efficiently through the assessment of variable importance, reproducibility, and sensitivity.

Implementing automation in the study of plant and soil microbiomes requires finding a balance between functionality and relevance to natural environments. While indoor pots and planters have been used for greenhouse plant growth, they fail to replicate the dynamic nature of soils. Non-microscopic components are not designed with automation in mind, which means that measurement methods need to be adapted to existing infrastructure or custom-made to fit specific needs. Researchers must decide on an approach based on their budget, available space, the organisms being studied, and the questions being asked.

Although automation platforms are commercially available and can be integrated into greenhouse designs, there are too many use cases for them to always meet researchers' needs. Oftentimes, it is crucial to understand the experimental requirements of a research question and balance the trade-off between automating and capturing all relevant variables. Additionally, integrating equipment that is not designed to work with one another requires experts, such as biologists, computer scientists, engineers, and statisticians.

Plant phenotypic profiling poses laborious analysis when automating root data from underground, which are difficult to track over time in natural systems. The next section discusses recent research toward inferring root structures from plants in the environment.

13.6 MATH AND ML FOR FABRICATED ECOSYSTEMS

As raw data sets leave the acquisition apparatus, which can be the microscope, spectrometer, and other modalities available for EcoFABs, collected images such as illustrated in Figure 13.8 have to be transformed into representations amenable to AI schemes so that characterization can be performed, followed by dimensionality reduction before traits can be used for decision-making. Building new mathematics and algorithms for plant data (e.g., signals, images, text) analysis enables the execution of the following tasks: segmentation, characterization, pattern detection and/or classification, querying and retrieval, as well as experiment recommendation.

In the initial stages of developing Computer Vision pipelines, the focus was on utilizing EVOS-acquired images (Figure 13.5). These images were generated by stitching together multiple scanned tiles, resulting in a final image size of 10,337 pixels in width and 7,697 pixels in height. While this high-resolution approach offered advantages such as clear visualization of root hair, the challenges associated with tile stitching, as well as issues related to condensation and bubble formation, introduced unnecessary storage and computational time requirements.

The subsequent experiment involved shifting to Epson-acquired scans, as depicted in Figure 13.6, and developing a fully automated system for root segmentation. This transition aimed to streamline the root biomass estimation and reduce the computational burden associated with the previous approach.

The current proof-of-concept demonstration of *B. distachyon* root scans analysis has successfully utilized a computer vision framework to process an entire experiment that consists of capturing root images for plants under different nutritional media enclosed in different EcoFABs for a month (Figure 13.9) that incorporates three main modules: image annotation, semantic segmentation, and root biomass estimation. The image annotation module [536] provides an interactive interface for partially annotating the root scans, allowing for the labeling of regions corresponding to both foreground (root areas) and background (non-root areas). It is important to note

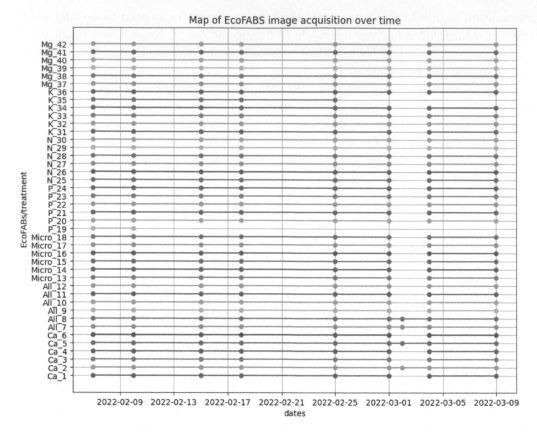

Figure 13.9 Experimental map and time series: summary of all the root scans (points) acquired for each EcoFAB (y-axis) over time (x-axis) given different nutritions.

that only a small percentage of the data undergoes annotation, and even within the annotated subset, only a fraction of the pixels are labeled. The annotated images play a crucial role in developing segmentation algorithms and assessing their performance.

The subsequent module takes both annotated and unlabeled data and automatically distinguishes the foreground from the background. This is achieved by employing a deep neural network, built with PyTorch, that utilizes an enhanced version of the Residual U-Net [204] architecture, incorporating information about the EcoFAB's substrate geometry. The Residual U-Net approach addresses challenges related to training difficulties and slow convergence by introducing residual connections. These connections facilitate the direct flow of information from one layer to another without alteration, thereby mitigating the vanishing gradient problem and enabling effective gradient propagation across deeper layers. This technique ensures reliable model creation. Chapter 7 provides further details about neural networks.

Lastly, the module responsible for root biomass estimation refines the semantic segmentation results from the previous step by integrating them with geometric priors derived from the EcoFAB design [377]. By incorporating these priors, the module calculates the final measurements of root biomass and tracks root growth over time, providing a comprehensive analysis of root development.

In addition to root images, hyperspectral imaging of plant leaves (Figure 13.7) has emerged as a powerful tool for monitoring and understanding plant health and development. By capturing images across a range of wavelengths, hyperspectral cameras can provide detailed information about the biochemical and physiological properties of plants, enabling researchers to detect subtle changes and track plant growth over time. However, analyzing hyperspectral data can be a daunting task due to the high-dimensional nature of the data. In order to analyze hyperspectral data, we have been exploring pipelines to extract information about plant growth and development and use it as feedback during autonomous experiments.

The analysis pipeline consists of three main steps, which are performed for each of the three vantage points. The first step involves image registration, which is necessary to correct for camera shifts that occurred during hardware maintenance. By performing a simple Fourier-based template matching method, we are able to register images within a few pixels of a known standard, which allows us to have more precise location information about the image. At the same time, this step also allows us to check the quality of the image and potentially identify possible hardware problems that occurred during data collection.

The second step in the pipeline involves semantic segmentation of the images into plant and non-plant pixels. This is performed on the hyperspectral data using data from a manually selected band-pass range, using neural networks that are trained on hand-labeled images. The final classification is passed through a size and location-based filter that removes pixels that are likely to be incorrectly classified. This step is critical for extracting meaningful information from the data, as it allows us to separate the plant pixels from the background noise and focus on the biochemical and physiological properties of the plant itself.

The final step in the pipeline involves the extraction and analysis of spectra from the plant pixels. The spectra are first normalized by their population standard deviation and then dimension-reduced. By embedding the spectra in a two-dimensional manifold, we are able to visualize the biochemical and physiological properties of the plant in a way that is easy to interpret and analyze. This step is particularly useful for identifying patterns and trends in the data, and for comparing the spectra from different plants and time points.

Upon integrating efficient algorithms that combine Computer Vision and Machine Learning, such as Deep Neural Networks for transforming images into feature vectors, the application of probabilistic frameworks, specifically Gaussian Processes (GPs), can greatly enhance statistical inference and interpretability.

By leveraging GPs, the extraction of common patterns across various domains becomes feasible, enabling the acquisition of knowledge from diverse high-resolution and multidimensional datasets. As described, computer vision modules facilitate feature extraction using convolutional neural networks, while also enabling data reduction to allow uncertainty quantification and automated, intelligent processes. The GP

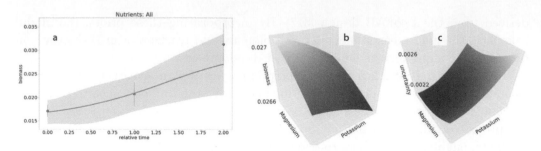

Figure 13.10 An example of a GP regression model of the biomass as a function of time (a) and nutrient concentration (b-c). (a) shows the surrogate and the uncertainty in measurements and predictions in one dimension (time), at a particular point in the space of nutrients. (b) and (c) show the last time step over a two-dimensional slice through the parameter space of nutrients. (b) shows the surrogate, and (c) shows the uncertainty in the form of the posterior variance. This information is available everywhere in the parameter space and can be combined with function optimization to steer the data acquisition and create models for interpretation.

uncertainty estimates can be used to aid the determination of whether and where additional data is required to improve the model. Consequently, the system can selectively gather more data in areas where discrepancies arise between experimental data and expert models, or when there is a potential for maximizing the improvement of information entropy. An example GP model over a parameter space of nutrients is shown in Figure 13.10.

GP-driven uncertainty quantification and decision-making are involved and computationally costly processes that can only be realized on dimensionality-reduced data. By using dimensionality-reduction techniques, we can extract a smaller number of important quantities of interest (QoIs) from the raw data (for instance biomass). These QoIs are then used as inputs for Gaussian process regression (Chapter 4) to create a surrogate model over some space, often representing plant-growth factors such as soil bacteria composition or nutrient content. Gaussian processes are characterized by a prior normal distribution over functions, and the hyperparameters of the kernel function are optimized to maximize the log marginal likelihood of the observations. Once optimized, the prior distribution can be conditioned on the observations, resulting in a stochastic view of the model function, given the collected data—the posterior probability distribution. This distribution is defined over the entire domain and provides a probability density function for the QoI everywhere in the space, including unmeasured regions. This posterior distribution can then be used in concert with function optimization, allowing for optimal conditions to be identified quickly and decisions about future data acquisition to be made.

Figure 13.10 illustrates the result of running a GP, using root features coming from plants imaged during the three initial dates of plant development (see Figure 13.9 for the full experiment). For more information on Gaussian processes, please refer to Chapter 4.

13.7 CONCLUSION

In conclusion, this chapter has highlighted the significance of research focused on optimizing environmental conditions which can potentially enhance biofuel productivity. The utilization of imaging data obtained from plant roots and leaves has proven valuable in monitoring plant development and growth. While there is an understanding of how agricultural management techniques impact the morphology of plants, such as *Brachypodium distachyon*, little is known about the effects of environmental stress, specifically plant nutrition, and how to accurately quantify plant variation under varying climate conditions.

The Lawrence Berkeley National Laboratory's (LBNL) EcoFABs have emerged as a valuable tool for the development of analytical pipelines and plant surveillance. These EcoFABs replicate soil conditions, providing a platform for detailed spatial and temporal mapping of roots and leaves, opening avenues for the potential establishment of self-driving labs dedicated to plant research.

Throughout this chapter, we have explored the challenges and opportunities associated with advancing the bioenergy sector and fostering a low-carbon economy. We have discussed the obstacles encountered when implementing analysis protocols and conducting plant research using EcoFABs. The ultimate goal of these advancements is to empower scientists in the field to delve deeper into the study of plant morphology and to refine the Autonomous Experiment system for automated plant management using EcoFABs.

By combining cutting-edge technologies, such as imaging, analytical pipelines, and the EcoFAB platform, researchers can gain deeper insights into plant behavior and fine-tune their approaches to maximize bioenergy production, contributing to the broader objective of sustainable and efficient bioenergy businesses.

ACKNOWLEDGMENTS

This chapter acknowledges the contributions of all those who participated in this study, particularly LBNL Twins and CAMERA research teams and former students who are listed in alphabetical order: K. Johnson, L. Teran, M. Soe, Z.Sordo. This work was supported by the Office of Science, of the U.S. Department of Energy (DOE) through the Advanced Scientific Computing Research and Biological and Environmental Research program under Contract No. DE-AC02- 05CH11231, and the LBNL Workforce Development & Education program. Any opinions, findings, and conclusions or recommendations expressed in this material are those of the authors and do not necessarily reflect the views of DOE or the University of California.

CHAPTER 14

Autonomous Neutron Experiments

Martin Boehm

Institut Laue-Langevin, Grenoble, France

David E. Perryman

Institut Laue-Langevin, Grenoble, France

Alessio De Francesco

Consiglio Nazionale delle Ricerche, Istituto Officina dei Materiali, Grenoble, France

Luisa Scaccia

Dipartimento di Economia e Diritto, Università di Macerata, Macerata, Italy

Alessandro Cunsolo

University of Wisconsin-Madison, Madison, Wisconsin, USA

Tobias Weber

Institut Laue-Langevin, Grenoble, France

Yannick LeGoc

Institut Laue-Langevin, Grenoble, France

Paolo Mutti

Institut Laue-Langevin, Grenoble, France

CONTENTS

DOI: 10.1201/9781003359593-14

FOLLOWING Reyes and Maruyama [444], the three major steps of an Autonomous Experiment consist of a *(Bayesian) belief*, which is based on the response after a limited number of experiments, the *decision policy*, which selects experiments based on the potential information yield and, finally the *experimental results* themselves, which are fed back to update the belief. Classical triple-axis spectroscopy (TAS) measurements follow the same steps where, up to now, belief and decision policy rely entirely on the scientists. Because the data acquisition with TAS is sequential and on the order of minutes, after each measured point, a decision is made about what the next point to measure should be. The experimental challenge is that a single measurement detects signals only in a small subset of the entire space, which is sparsely distributed within the space the measurements explore. TAS steering is thus similar to gold veins in a large geological massif where TAS scientists, similar to geologists, need to probe the space point-by-point for signals and base their decision of where to go next on very few measured data points. Extending recent advancements and algorithms to assist scientists during experiments is an exciting and (to the authors) overdue question, along with how best to measure the efficiency of an autonomous data acquisition loop compared to traditional data acquisition strategies. In the 1950s Daniel G. Krige, a South African geologist, sought to estimate the most likely distribution of gold from a few boreholes in the Witwatersrand reef complex in South Africa. He developed the Kriging algorithm named after him [107], which became Gaussian Process Regression in the following years.

About the same time as Krige explored the Witwatersrand reef for gold, Bertram Brockhouse developed the foundations of experimental Inelastic Neutron Scattering in pioneering work at the Chalk River reactor, which earned him the Nobel prize in Physics together with Clifford Shull in 1994. Today it is a standard method in solid-state physics and other areas of condensed matter research to investigate the dynamics of large atomic or molecular ensembles on the atomic scale with a time resolution ranging from a few picoseconds on triple-axis and time-of-flight (TOF) instruments to nanoseconds, on back-scattering or spin-echo spectrometers. While first experiments were still limited to large samples of several grams, continuous progress in neutron optics achieved flux gains by several orders of magnitudes and pushed the technique down to nowadays milligram samples for some applications. Large position-sensitive detectors, especially on TOF and back-scattering instruments, replaced single counters with tremendous gain in pixel power and, hence, sensitivity. The neutron instrumentation progressed considerably in the following seven decades, but the data acquisition strategies on TAS remained surprisingly unchanged. For TOF instruments with large angular detector coverage a noteworthy Bayesian approach was proposed by Sivia and co-workers in the 1990s for the interpretation of quasi-elastic signals [486, 485], which found resonance also in other neutron techniques recently [115, 116, 199, 308, 312, 324, 410]. Butler and al. proposed to accelerate the interpretation of single-crystal inelastic neutron data obtained by TOF instruments with the help of deep neural networks [80]. A similar approach was successfully applied for the extraction of interaction parameters in single-crystal experiments [462, 463]. For TAS experiments, on the other hand, linear 1-D scans along selected paths in the feature space, very similar to grid scanning, remained the standard since the times of Brockhouse. How many individual measuring points are actually needed to acquire sufficient information is an often ignored question, both in the completely agnostic case, where no prior physical information on the systems to be studied is at hand, and in the alternative case, where pre-established theoretical models need to be confirmed with sufficiently precise model parameters.

In this chapter, we use the concept of Autonomous Experiments (AE) as an alternative and more efficient way of data acquisition in inelastic neutron scattering. With AE we entrust the statistical analysis of the acquired data and the best guess for the future data points to an algorithm, while scientists control the progress of the experiment by defining acquisition and cost functions, potential models, priors, or the type of kernels. We propose three different examples, which apply to different types of spectrometers, either in real experiments or in simulations. While all cases use the cyclic concept of Autonomous Experimentation, the algorithms, and steering processes differ from case to case. Specifically:

1. We start with sequential data acquisition for an agnostic case, i.e., under the assumption that no (or very little) prior physical information on the system is available. Typically, in this case, neutron scientists tend to use instruments that provide the mapping of large sections of the feature space with simultaneous energy transfer measurements, such as TOFs and back-scattering instruments. The sequential TAS acquisition is considered more efficient only for restricted, previously identified, sections in this space, especially in combination with complex sample environments, due to the inherent higher flux. In this case, the question is to which extent or for which applications, an autonomous TAS approach makes sense. We use Gaussian Process Regression as our algorithm [430].

2. The second example uses Bayesian tools to infer the posterior probability distribution of free model parameters in TOF measurements. This approach is no longer agnostic but uses a very general formalism of the expected neutron intensity distribution in the likelihood expression. Some initial guesses on the values of the model parameter might be made by adapting the prior probability distributions. Applications of Bayesian data evaluation in inelastic neutron scattering have been published by A. De Francesco and co-workers for Spin Echo spectra and Brillouin neutron scattering, a technique adapted to low scattering angles [115, 116].

3. Finally, in the third example, we provide an outlook on physics-informed sequential explorations in the four-dimensional feature space. We develop the parametric Bayesian concept of the second example further and construct a showcase for an AE based on a known magnetic exchange model for a single

crystal. This approach is unexplored, to our knowledge, but we believe that it has the strongest potential to increase the data acquisition efficiency of the instruments.

We start this chapter with a very brief introduction to the various inelastic neutron techniques, which are relevant to the three different cases. We also briefly describe the nature of the measured signals and the convention for presenting them in the form of a scattering function, which will be a multidimensional surface $S(\mathbf{Q}, E)$ in the feature space under investigation. We limit the feature or data space to four dimensions[1], with the three components of the momentum transfer, $\mathbf{Q} = (Q_h, Q_k, Q_l)$, equivalent to the reciprocal space in crystallography up to the Planck constant h, and the energy transfer E.

14.1 A BRIEF INTRODUCTION TO INELASTIC NEUTRON SCATTERING

14.1.1 The Feature Space and the Problem

Inelastic neutron scattering probes the collective dynamics of atoms and electrons (see e.g., Ref. [488] for a general introduction). In crystalline samples, neutrons interact with the atomic lattice via the strong force, and with electrons via a dipole-dipole interaction between the neutron magnetic moment and the spin of the electrons. Restricting the discussion to the latter, the best-known example for collective magnetic excitations are spin-wave motions of magnetic moments in magnetically ordered systems (see e.g., [528] for application in inelastic neutron scattering). In classical magnets, this order sets in below the Curie-temperature T_C for ferromagnetic or the Néel-temperature T_N for antiferromagnetic interactions, where the magnetic exchange interactions between electrons overcome competing thermal fluctuations. In the quantum case, many fundamentally interesting phenomena arise due to the competition between quantum fluctuations—favored by small magnetic moments and magnetic frustration—temperature, and the influence of the surrounding lattice. Well-known examples are the dynamic magnetic responses in high-T_C materials, known as "resonances" [482, 529], which seem to be linked to the complex electron pairing mechanism in these materials, or spin liquid systems in frustrated magnets [41], showing spinon excitations in inelastic neutron spectra [326].

The energy of collective magnetic excitations is usually of the order of 1 to 10^2 meV, which matches the kinetic energy of the incident neutrons on the sample. In the scattering process, the neutrons transfer (gain) momentum and energy to (from) the spin-wave and thereby probe the excited states with a resolution of only several percent of the incoming energy.

The detector count-rate I is proportional to the sample volume V and to the double differential scattering cross-section $\frac{d^2\sigma}{d\Omega dE}$, i.e., the probability of scattering neutrons from the sample into the solid angle $d\Omega$ with a given energy interval E. The scattering cross-section can be further expressed in the form of the *scattering*

---

[1]Here the terms *feature* and *data* space are equivalent. It is the space, where the data (features) are measured. The *parameter* space, in contrast, is the multidimensional space defined by the number of model parameters.

function $S(\mathbf{Q}, E)$, which is a function of the momentum and energy transfer

$$I \propto V \cdot \frac{d^2\sigma}{d\Omega dE} \cdot \Delta\Omega\Delta E \propto S(\mathbf{Q}, E). \tag{14.1}$$

Introduced by Van Hove, $S(\mathbf{Q}, E)$ is the Fourier transform of the time-dependent pair correlation function $G(\mathbf{r}, t)$, which describes the correlation of the spin density at the position $\mathbf{r_0}$ at the time $t=0$ and the spin density at $\mathbf{r_0} + \mathbf{r}$ at time t. As the momentum transfer $\mathbf{Q}$ is a three-dimensional vector, the energy transfer E and the three components of the momentum transfer (Q_h, Q_k, Q_l) span a four-dimensional feature space, in which $S(\mathbf{Q}, E)$ forms 4D surfaces or regions. Non-zero values (or non-zero scattering probability) are restricted to points $(\mathbf{Q_0}, E_0)$, which fulfill the conditions of the dispersion relation

$$E_0 = E_0(\mathbf{Q_0}, \mathbf{J}, \mathbf{D}) = E_0(Q_{h,0}, Q_{k,0}, Q_{l,0}, J_1, .., J_N, D_1, .., D_N). \tag{14.2}$$

The $\mathbf{J} = (J_1, .., J_N)$ are the magnetic exchange integrals, which express the coupling strengths between magnetic moments, and $\mathbf{D} = (D_1, .., D_N)$ are magnetic anisotropy parameters. While the values on the surface $S(\mathbf{Q}, E)$ vary according to the scattering probability, Equation 14.1, its shape is given by the exchange and anisotropy parameters via Equation 14.2. It is the goal of any inelastic neutron experiment to track $S(\mathbf{Q}, E)$ in the feature space in the most efficient way, and to deduce from the measured $S(\mathbf{Q}, E)$ the exchange parameters and, hence, the model.

14.1.2 Measuring $S(\mathbf{Q}, E)$

Various inelastic neutron techniques have been developed over the last decades. They differ in the way they define the incoming E_i and analyze the final neutron energy E_f, and how they cover the solid angle with detectors.

The scattering process can be illustrated in the feature space by the *scattering triangle* (see Figure 14.1a), which represents the momentum, Equation 14.3, and energy, Equation 14.4, conservation laws of a scattering event

$$\mathbf{Q} = \hbar\mathbf{k_i} - \hbar\mathbf{k_f} \tag{14.3}$$

$$E = \frac{\hbar^2}{2m}(k_i^2 - k_f^2), \tag{14.4}$$

where $\mathbf{k_i}$ and $\mathbf{k_f}$ are vectors representing the direction and magnitude of the incident and scattered neutrons, respectively, while $\hbar$ is the reduced Planck constant. The space coordinates for the three momentum components are conveniently expressed in reciprocal units, defined by the lattice parameters of the crystalline samples.

Frequently, single crystals are not available, but micro-sized crystals can be combined into powder samples with sufficient volume V. For these systems, and for liquid or glassy ones, a quasi-infinite number of reciprocal spaces are superposed, which leads to a loss of directional information, but information can still be obtained by expressing the scattering function as a function of the momentum amplitude $|\mathbf{Q}|$, which

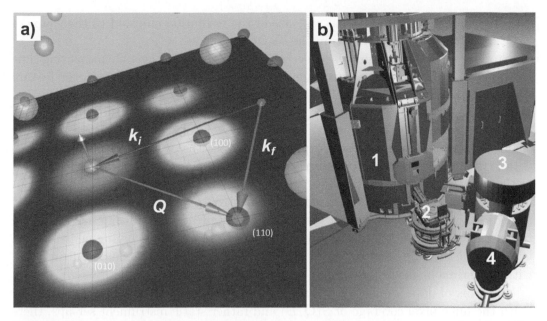

Figure 14.1 (a) Schematic representation of a section of a 3D reciprocal space. The blue shaded plane represents the selected 2D *scattering plane* during the measurement, and yellow-green colors represent various intensity values of $S(\mathbf{Q}, E)$. Here $\mathbf{k_i}$: incoming scattering vector, $\mathbf{k_f}$: outgoing scattering vector, $\mathbf{Q}$: measuring position in the reciprocal space. Numbers in brackets indicate Miller indices in the reciprocal space. (b) Snapshot of the digital twin of the triple-axis spectrometer ThALES at the Institut Laue-Langevin [64]. 1: monochromator drum containing the monochromator, which defines $\mathbf{k_i}$ (1st axis). 2: sample table carrying the sample environment and sample (2nd axis). 3: analyser drum containing the analyser (3rd axis), which measures $\mathbf{k_f}$. 4: detector shielding.

reduces the feature space to two dimensions, and allows us to express the scattering surface as $S(|\mathbf{Q}|, E)$. The technical limitations of every spectrometer result in small deviations for $\mathbf{k_i}$ and $\mathbf{k_f}$, in angle and magnitude, around their mean values. This results in a 4D resolution ellipsoid, which represents the restricted detection area of one single detector in the feature space [106, 409, 135]. The detector count-rate P at a given $(\mathbf{Q}, E)$ is in fact the convolution of the signal $S(\mathbf{Q}, E)$ with the volume of the resolution ellipsoid.

Classical TAS instruments (see Figure 14.1b) explore $S(\mathbf{Q}, E)$ point-by-point. The shape of the triangle depends in real space on six rotational movements, where always two share a common rotation axis, leading to the name of a *triple-axis* spectrometer [481]. The vector $\mathbf{k_i}$ is related to the monochromator (axis), $\mathbf{k_f}$ to the analyzer (axis), and finally, the sample reciprocal space has a rotational degree of freedom around the user-defined sample axis. The very same point $(\mathbf{Q}, E)$ can be accessed by various combinations of $\mathbf{k_i}$ and $\mathbf{k_f}$, which all have a slightly different shape and size of the resolution volume. This experimental flexibility is used to tune the best signal-to-noise ratio by adapting the orientation and size of the resolution ellipsoid to the shape of $S(\mathbf{Q}, E)$ at the various positions in the scattering plane.

Standard TAS experiments last several days. Assuming an average measuring time per point of several minutes, this leads to the order of 10^3 data points in the feature space per measurement. Up to 10^4 to 10^6 data points (pixels) are measured in parallel on *time-of-flight spectrometers*, which cover huge parts of the momentum space by position-sensitive detector units. The measurement of the various energy transfer channels is almost simultaneous, exploiting the fact that variations of neutron (kinetic) energies of sub-meV order correspond to variations in the neutron *time-of-flight* of the order of microseconds, which can be easily resolved by the detector electronics. For TOF instruments, separated pulsed neutron packages are needed to trigger periodically the starting point of the TOF measurement. The pulsing has a price in the incoming neutron flux, cutting out a large fraction of the incoming neutrons and lowering the incoming flux by three to four orders of magnitude compared to a TAS with a continuous neutron beam.

In terms of data acquisition strategies, these two techniques are complementary. With the huge pixel coverage of $S(\mathbf{Q}, E)$ and rather static instrument set-ups, there are fewer decisions to take during a TOF measurement. The available measuring time sets the limits for the counting statistics in the individual channels. The data volume at the end of the measurement is large, of the order of Gigabytes, which requires suitable software tools to sift through and analyze the data *a posteriori*. Traditional TAS strategies, in contrast, are not unlike those of a battleship game, where one tracks a signal $S(\mathbf{Q}, E)$ in a two-dimensional plane with a sequence of equidistant points. In a more systematic way, grid-scanning over restricted areas in the feature space might be applied, if the remaining measuring time permits. The efficiency of the TAS data acquisition depends on the prior physical information at hand, especially on the shape of $S(\mathbf{Q}, E)$. Often, TAS instruments explore details of TOF-measured scattering functions in strongly restricted areas of the feature space. In the absence of experimental information, theoretically calculated scattering functions could guide the search, with strong uncertainties in the model and the values of the parameters. In any case, a certain level of *a priori* examination of the expected values is necessary to increase the chances of success.

The availability of Machine Learning algorithms and the increase in computing power could become a game changer if automated data acquisition policies base their decisions on available experimental and theoretical information, and if computing times for the prediction of the next points remain below the typical measuring time of a TAS measurement. Below, very recent examples of applying such algorithms with varying prior information content are presented.

14.2   AGNOSTIC SCANNING WITH GPR

In this section, we present a summary of Autonomous Experiments performed on TAS spectrometers assuming no prior information on $S(\mathbf{Q}, E)$. The main goal is the collection of a sufficient number of relevant data points in the feature space with a minimum amount of experimental budget, i.e., measuring time. Ideally, these data

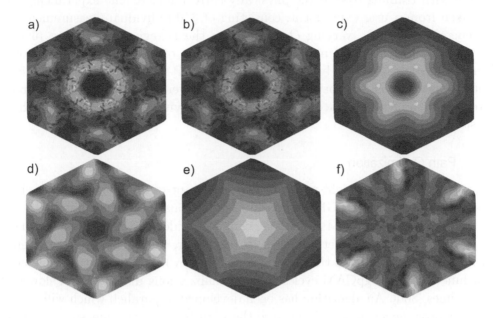

Figure 14.3 Comparison of GPR results trained with experimental data (320 data points) for (a) a non-symmetric (standard Matérn) kernel, and symmetric kernels using either the (b) average, (c) product, (d) maximum, (e) minimum, or (f) median of the covariance between all pairs of equivalent points.

scattering function $S(\mathbf{Q}, E)$ has to obey the symmetry conditions of the (here magnetic) point group of the system. The best strategy thus far is to use the symmetry information for identifying the smallest fundamental domain before the experiment and to apply the Gaussian Process (GP) only in this reduced domain. Alternatively, points generated in the complete domain by the autonomous loop could be mapped into the fundamental domain with the help of the symmetry elements. As a further alternative, we tested the performance of symmetry invariant kernels [366, 165]. There are several ways to construct a symmetry invariant kernel, e.g., using the average, product, or median of the covariance between all pairs of equivalent points. Figure 14.3 shows a comparison of a non-symmetric kernel with various symmetric kernels used within gpCAM on experimental data. Although every symmetric kernel is a mathematically valid construction, one has to pay attention to the usefulness of the results. Choosing the maximum, minimum, or median covariance is intuitively like choosing the furthest, shortest, or median distance between all the distances of symmetry-equivalent points. When using the product, Figure 14.3c, and minimum distance, Figure 14.3e, it seems that points close to the symmetry axis are favored, as all symmetry-equivalent points are close to each other. Mapping the original points to a subdomain, Figure 14.3a, or using the average covariance, Figure 14.3b, are the only ones that include the feature of the rings toward the border of the region and converge with training toward the physically correct image. The experimental data were taken from a user experiment, consisting of 320 individual measurements at a fixed energy E and Q_l, varying Q_h and Q_k in the hexagonal scattering plane. In practice, assumptions about magnetic symmetry can often turn out to be incorrect, causing symmetry kernels to become too biased for GP scanning. Therefore, in experiments, despite the satisfactory performance of the averaged kernel, we rather reduce the region in the feature space to the smallest non-symmetric unit and run the GP with standard kernels.

14.2.2 Path Optimization

A prior reflection on the right region is also important to minimize the time elapsed between two points in the GPR sequences on TAS instruments. Compared to grid scanning, the total path in GPR scanning can be considerably longer, which can drastically worsen the overall performance. In particular, movements along the energy transfer axis are slow (up to the order of minutes/meV). We, therefore, implemented a cost function in the gpCAM cycles, which penalizes long distances and slow motor driving times [366]. An algorithm has been developed in parallel, which will synchronize the motor movements and optimize the path between two measuring points to keep the instrument at a safe distance from any obstacle [563, 564].

14.2.3 Metrics

The efficiency of our Autonomous Experiments was defined as the maximum information obtained in the feature space for a minimum experimental budget. Increasing efficiency is the obvious motivation for an optimized agnostic Autonomous Experiment. However, defining the best metrics to measure efficiency is not evident in this case. In agnostic scanning with a limited amount of experimental budget, one can only make the best guess that the chosen acquisition strategy steers the instrument into those regions of interest, which will be relevant for the post-processing and parameter estimation. In the examples shown above we decided to use a quite arbitrary division into ROIs and defined success as the yield obtained inside these ROIs. Once the hyperparameters are trained, the success rate depends on the acquisition function rather than the GP itself. In Ref. [514] the authors suggest a formal procedure for such metrics allowing the comparison between different algorithms and between different experiments. In test runs with simulated data, obtained with a given model, one can use this very same model for different metrics, which evaluate the information content of the data points directly from the fit quality to the model parameters. We have tested two further approaches: the first metric is the squared bias plus variance of a nonlinear least squares fit to the experimental data, $\sum(\beta_i - \hat{\beta}_i)^2 + \sigma^2(\beta_i)$. The second metric for the Autonomous Experiments uses a Bayesian framework, where prior distributions are placed over the physical parameters, and a Markov Chain Monte Carlo method is run to produce a posterior distribution for the parameters.

14.3   AUTONOMOUS TOF EXPERIMENTS

This section makes a quick detour from sequential data acquisition to introduce a technique for Autonomous Experimentation on time-of-flight instruments. As mentioned in Section 14.1.2 the positions in $(\mathbf{Q}, E)$ and the total number of data points (orders of magnitudes higher compared to TAS) are given and fixed by the instrument geometry. This makes the application of an algorithm for the optimization of these positions and numbers superfluous. Nevertheless, careful instrument time management is also here essential, pondering the benefit of a longer acquisition against the penalty it imposes on the measurement plan. Assigning an unreasonably short time slot to a given measurement could result in very noisy spectra and model parameters correspondingly affected by large error bars. Even worse, ulterior background corrections or other manipulation of data sets may render the final counting statistics too weak to be credibly modeled without the risk of over-parameterization and parameter sloppiness. On the other hand, an excessively long measurement could make the spectral acquisition redundant, harmfully reducing the time available for further meaningful measurements.

For a given model, what finally matters is the statistical accuracy of the data points yielding the expected target uncertainty of the model parameters. We, therefore, intend to infer the values and target uncertainties of the model parameters at any given time before or during data acquisition, given the experimental information and any possible prior information on the parameters at that moment. With this Bayesian approach, we aim for on-the-fly model evaluation. The approach is cyclical, subdividing the global measuring time into a number of time windows and using the posterior distribution at time n as the prior at the acquisition $n + 1$. This approach starts with an assumption on $S(|\mathbf{Q}|, E)$, or Bayesian belief, which includes a quite general model with an unspecified, although limited, number of parameters at the start of the experiment. Their relevance and precision are determined by the posterior probability distributions. The process either stops when a user-defined precision of the parameters is reached or when the global measurement time is exceeded. No decision policy is required, which is one of the three major steps of the Autonomous Experimentation loop, as stated in the introduction. However, due to the cyclic data acquisition and the Bayesian evaluation of data, we believe that this application is of interest in the context of Autonomous Experiments with inelastic neutron scattering.

So far, this Bayesian approach has been tested with simulated data (see example below) and successfully applied in a posteriori data treatments [115, 116]. The extension to the here suggested cyclical "Measurement Integration Time Optimizer" (MITO) awaits the proof-of-concept on the instruments [150]. Bayesian inference has not yet become a standard tool in the evaluation of neutron data. Noteworthy exceptions are the book on neutron data analysis by Sivia and Collins [485], the application of the analysis of neutron diffraction data [199], the determination of protein conformational ensembles from a limited amount of neutron data [410], neutron reflectometry analysis [308] or a careful analysis of low counting statistics on neutron spectrometers [253]. We first give a brief introduction to our Bayesian approach in Section 14.3.1, followed by examples given in Section 14.3.2.

14.3.1 Measurement Integration Time Optimization

In the following, we assume poly-crystalline, powder, or liquid samples. In this case, the scattering function $S(|\mathbf{Q}|, E)$ is reduced to a two-dimensional feature space, where only the size of the momentum transfer $|\mathbf{Q}|$ can be determined, as the directional information is averaged out. Often, the intensity I in Equation 14.1 is integrated over the complete $|\mathbf{Q}|$ range and the data structure of $I(E)$ can be considered as a one-dimensional feature vector $\mathbf{I} = \{I_1(E_1), .., I_n(E_n)\}$ for the n equidistant energy channels E_i. We also introduce the vector ensemble $\theta = \{\theta_1, .., \theta_m\}$, whose m components are the parameters θ_m of the model adopted to approximate a measured spectral shape.

According to Bayes' theorem, the probability $P(\theta|\mathbf{I})$ of the model parameter set θ, conditional on the measured data-set $\mathbf{I}$ is the product of two factors: (1)$P(\mathbf{I}|\theta)$, which expresses the likelihood of the data measurement conditional on the model parameters, and (2) $P(\theta)$ representing the best guess over the parameter range, as elicited by the researcher, based on the prior knowledge of the model

$$P(\theta|\mathbf{I}) = \frac{P(\theta)P(\mathbf{I}|\theta)}{P(\mathbf{I})}. \tag{14.5}$$

Our experiments aim at measuring a spectral shape as precisely as possible to identify all its relevant features ultimately unveiling the microscopic dynamic events inside the sample. Once the spectrum is collected, we typically need a model, determined by the parameter set θ, to approximate the measured spectral shapes $\mathbf{I}$ and establish a correspondence between spectral features and microscopic degrees of freedom. Following an inferential procedure described in [150, 115, 149], the spectral shape acquired after an initial counting time can be used to draw a preliminary θ posterior $P(\theta|\mathbf{I})$. This posterior distribution will then be fed into the Bayes equation as an "adjourned prior" for the next measuring cycle. This next run $\mathbf{I}'$ will draw more refined posterior and spectral shape renderings. This procedure will be repeated until convergence is reached on the parameter set. Using the multiplication rule of probability and assuming that the data are mutually independent, Bayes' theorem gives

$$P(\theta|\mathbf{I}, \mathbf{I}') = \frac{P(\mathbf{I}'|\theta, \mathbf{I})P(\theta|\mathbf{I})}{P(\mathbf{I}'|\mathbf{I})} = \frac{P(\mathbf{I}'|\theta)P(\mathbf{I}|\theta)P(\theta)}{P(\mathbf{I}')P(\mathbf{I})} = \frac{P(\mathbf{I}, \mathbf{I}'|\theta)P(\theta)}{P(\mathbf{I}, \mathbf{I}')}. \tag{14.6}$$

This is equivalent to using the initial prior multiplied by the likelihood of the combined measurements but allows for the periodic update at every run.

14.3.2 Examples of the Measurement Integration Time Optimizer

We follow the example given in Ref. [115] assuming a physical model of damped harmonic oscillators (DHO) [147], to describe the spectral shape of $\mathbf{I}$ coming from Brillouin scattering measurements:

$$I \approx R(E) \otimes S(E) = R(E) \otimes \left\{ A_e(Q)\delta(E) + [n(E)+1] \sum_{k=1}^{2} \frac{2}{\pi} \frac{A_k(Q)E_k^2(Q)\Gamma_k(Q)}{[E^2 - E_k^2(Q)]^2 + 4[\Gamma_k(Q)E]^2} \right\} \tag{14.7}$$

The parameter set is thus given by $\theta = \{A_e, A_1, A_2, E_1, E_2, \Gamma_1, \Gamma_2\}$ where A_e represents the area of the elastic peak and A_k, E_k and Γ_k are the areas, the undamped frequency and the damping coefficient of the j-th inelastic mode. Notice that in this case the number of excitations k was imposed to be $k=2$. In the general approach, proposed by De Francesco and collaborators, the number of DHO excitations k is a priori unknown and itself part of the parameter set. [n(E)+1] is the temperature-dependent Bose factor, which is not part of the parameter set here. The $|\mathbf{Q}|$ dependence is integrated out so that the scattering function $S(E)$ only depends on the energy transfer E. As mentioned above, $S(E)$ needs to be convoluted with the instrumental resolution function $R(E)$, which is Gaussian in shape.

The likelihood term in Equation 14.6 can now be expressed with the help of Equation 14.7, in order to calculate the posterior distribution $P(\theta|\mathbf{I}, \mathbf{I}')$. Usually, the posterior distribution is drawn up to the normalization constant $P(\mathbf{I}, \mathbf{I}')$, which we are unable to compute analytically. For this reason, such a posterior is commonly calculated by resorting to Markov Chain Monte Carlo methods [115, 164, 522] which simulate the sought-for posterior distribution. Once the full joint posterior distribution of the model parameters is obtained, we can focus on the posterior distribution of one specific parameter, chosen to be E_1 here. This estimate is simply obtained by marginalization, i.e., by integrating the posterior over all the other parameters

$$P(E_1|\mathbf{I}, \mathbf{I}') = \int_{\theta - E_1} P(\theta|\mathbf{I}, \mathbf{I}')d\theta_{-E_1}. \tag{14.8}$$

Figure 14.4 shows the evolution of the posterior distribution of the parameter E_1. The data are simulated, but similar to spectra typically obtained by a ToF spectrometer like BRISP [11, 10], an instrument once installed at the High Flux Reactor of the Institut Laue Langevin (Grenoble, France). The posterior distribution evolves with the data acquisition time, leading to a shift of the posterior mean value and a sharper distribution, hence, higher precision of E_1. As these are simulated data, the hours indicated in the figure caption are an estimate for the required measuring time, based on the comparison with previously measured experimental data. As the acquisition time increases, the distribution becomes increasingly well-shaped and sharp, thus delivering a more precise estimate of the optimal parameter values. The Bayesian approach on 1D TOF spectra without the recursive condition has already been successfully applied to Brillouin spectra, as just shown above, and neutron-spin-echo data [115, 116] and could be applied to any data with similar structure, such as TOF crystal-electric field measurements. Sivia et al. already considered this approach for quasi-elastic neutron scattering data [486]. However, quasi-elastic scattering necessarily has the same $E = 0$ center for various superimposed peaks, which cannot give any meaningful statistical answer beyond $k = 2$, without any further physical information from elsewhere.

The full potential of this Bayesian approach comes in by considering the number k of peaks in the model as part of the parameter set θ. The estimation of the correct number k is experimentally often the most difficult part, especially when the resolution does not allow for unambiguously distinguishing between neighboring peaks. It has been shown that Bayesian statistics provides reliable, quantitative results on the number k of inelastic components contributing to the spectrum, estimated conditionally on available data [150].

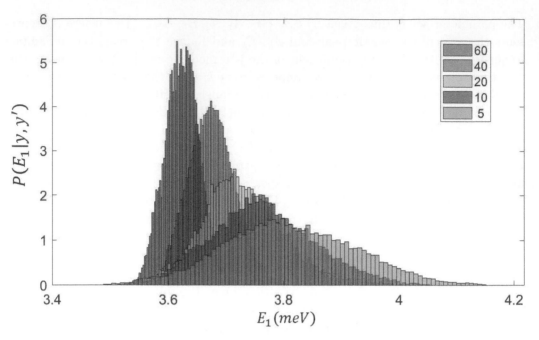

Figure 14.4 Evolution of the posterior distribution of the model parameter E_1 as estimated marginalization of the full joint posterior distribution according to Equation 14.8. E_1 is the energy of a collective excitation in a fluid measured on the ToF instrument BRISP as given in Equation 14.7. The analysis was done with simulated data, numbers indicate imaginary hours of sample exposure.

14.4 PHYSICS-AWARE TRIPLE AXIS SPECTROMETER STEERING

Any additional information on the system will increase the efficiency of steering for sequential data acquisition. Before starting a neutron scattering experiment, typically, some physical or chemical details of the samples are known, such as the crystalline and magnetic structure, magnetic susceptibility, specific heat data, or data from previous neutron experiments on the same sample.

14.4.1 Including a Simple Spin-Wave Model into the Autonomous Loop

We propose a simple example of how to integrate physical knowledge into the process of Autonomous Experimentation on a TAS. We assume that the magnetic structure is already known from previous neutron diffraction experiments. We further believe that the collective magnetic excitations are described by a dispersion relation $E(\mathbf{Q}, J)$, Equation 14.2, and an intensity distribution $S(\mathbf{Q}, E, J)$, which can be approximated by a classical spin-wave calculation. Using a concrete example from the literature we take the magnetically ordered system of CuB_2O_4. The four magnetic Cu^{2+} ions in the unit cell are related by one exchange parameter J, which is of equal strength between all nearest neighbors [63]. An algorithmic expression for $E(\mathbf{Q}, J)$ was obtained by semi-classical spin-wave theory. Figure 14.5 shows two-dimensional

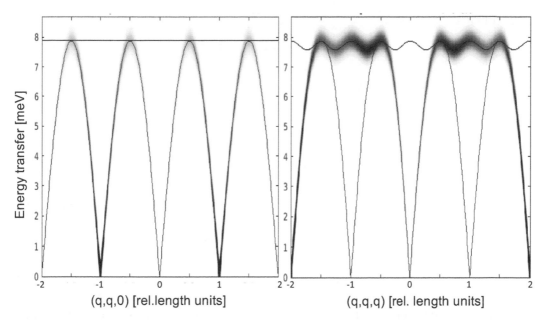

Figure 14.5 Two-dimensional E vs. $\mathbf{Q}$ cuts along two high symmetry directions in the reciprocal space of the magnetically ordered system CuB_2O_4.

E vs. $\mathbf{Q}$ cuts along the high symmetry directions $[Q_h,Q_h,0]$ and $[Q_h,Q_h,Q_l]$ with the literature values of $J_0 = 3.93$ meV and $D \approx 0$. $S(\mathbf{Q}, E)$ is a convolution of a spectral weight function $A(\mathbf{Q}, E, J)$, containing the model relevant parameters, and the equally known multivariate Gaussian resolution function $R(\mathbf{Q}, E, \eta)$, containing instrument parameters η, which define the thickness of the S surface in the feature space

$$S(\mathbf{Q}, E, J, \eta) = A(\mathbf{Q}, E, J) \otimes R(\mathbf{Q}, E, \eta). \qquad (14.9)$$

Finally, the neutron counts in the detector, Equation 14.1, follow Poisson statistics

$$I(\mathbf{Q}, E, J, \eta) \sim \mathrm{Pois}(S(\mathbf{Q}, E, J, \eta) + Bg(\mathbf{Q}, E)). \qquad (14.10)$$

In Equation 14.10 we added a feature space-dependent background function, which is not known before the experiment.

The expectation of a successful Autonomous Experiment is the convergence of the predicted mean toward the ground-truth value J_0 with a user-defined, acceptable uncertainty σ_J after N measured neutron intensities $\{I_1, .., I_N\}$. Thus, a narrow posterior probability distribution $P(J|I_N(\mathbf{Q}, E))$ around the mean J_0. The decision policy should be such that the number of chosen points N is minimal for reaching the expected σ_J.

As mentioned in the introduction, the three major steps of an Autonomous Experiment consist of (Bayesian) belief, the decision policy, and the experimental results [444]. The Bayesian belief is expressed by the likelihood, which in this example is the joint probability of the Poisson distributions of N observed data given the parameter

J. Explicitly

$$P(\mathbf{Q}, E | J) = \prod_{i=1}^{N} I(\mathbf{Q}, E, J, \eta), \tag{14.11}$$

with I from Equation 14.10.

The range of J is not completely arbitrary and can usually be restricted before the start of the experiment. Here, we impose a Gaussian prior around a reasonable starting value J_0 with a sigma of $\sigma_{J_0} = |J_0|$

$$P(J) \sim \mathcal{N}(J_0, \sigma_{J_0}). \tag{14.12}$$

We have deliberately chosen offsets to the literature value of J to test convergence to the ground truth with the autonomous loop. The posterior distribution is again obtained by Monte Carlo sampling, in this case with Hamiltonian Markov Chain Monte Carlo (HMCMC).

We kept the same decision policy as in the example of the agnostic Autonomous Experiment in Section 14.2, i.e., an upper confidence bound criteria. Figure 14.6 shows the performance of this physics-informed Autonomous Experiment as a function of the measuring points. We compare this performance with a random choice of points and a Gaussian process over the same domain size in feature space. The explored domain is four-dimensional with a size of about 4 Å^{-1} in each reciprocal space dimension and about 6 meV in the energy transfer. Considering the instrumental resolution, about 1000 steps in every direction would be required in a grid scan mode, leading to the hypothetical value of about $(10^3)^4 = 10^{12}$ voxels to map out the complete domain. The signal density within this domain is below 20%, as estimated from the hitting percentage in the Region of Interest of a random run after 500 points (see Figure 14.6a). Measured by the percentage of measuring points inside the ROIs, the GPR is not performing significantly better than the random run. It seems that in the higher dimensional case, the typical number of measuring points during a TAS measurement is too low to establish a meaningful correlation between them. The parametric Bayesian method, by contrast, rapidly converges toward the correct value of the exchange model (see the Bayesian expectation value for the exchange parameter Figure 14.6b), which increases considerably the hit percentage into the ROIs after the first hundred points. Figures 14.6c and d show the evaluation of the mean value J and the standard deviation σ as obtained by least-square fits as a function of the measured points. In this simple example with only one unknown parameter value all three methods provide a good estimate of the parameter, even with a low number of points measured in meaningful intensity regions for the random and GPR case.

14.4.2 Future Directions

We believe that the physics-informed Bayesian method has considerable potential to evaluate a model with multiple parameters during the sequential data acquisition of an autonomous TAS experiment. More generally, one could imagine an Autonomous Experiment evaluating several competing models, each one with its own, distinguished set of parameters $\mathbf{J}$. Taking the example of spin waves, the number of exchange (and

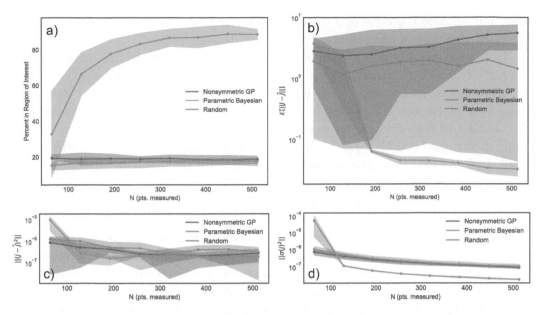

Figure 14.6 Metrics for simulated experiments on a CuB_2O_4 copper sublattice TAS experiment. The Gaussian process method is using the variance acquisition function. The figure on the top left shows the % of points in a region where the signal is at least 25% of the intensity. The figure on the top right shows the mean of the absolute difference between the true J value and a sample from the posterior of J, given the data. The figures on the bottom left and right show the bias and variance of a maximum likelihood estimate of the parameters. Confidence regions are estimated by several simulated experiments.

anisotropy) parameters is often ambiguous before a neutron measurement, leading to varying numbers of possible exchange terms in a model. Their relative impact on the global model could be evaluated in parallel from a probabilistic viewpoint. The most interesting questions in a physics-informed Autonomous Experiment arise around the right decision policy. A decision policy needs to work out the most likely among the various models with a minimum amount of invested measuring points, selected by the policy. While the Autonomous Experiment selects measuring points $(\mathbf{Q}, E)$ in the feature space (and gets back the intensities $S(\mathbf{Q}, E)$ from the feature space), the success of the policy will depend on the extent to which the points will contribute to the rapid convergence of the parameters toward the ground truth region in the parameter space. One can suppose that not all points in $(\mathbf{Q}, E)$ contribute with equal weight to the posterior distribution $P(\mathbf{J}|S_N(\mathbf{Q}, E))$, and their choice and sequences should be guided by the theoretical available information. The impact of physical information on a promising policy can be demonstrated on the basis of the previous example. The surface $S(\mathbf{Q}, E)$ in Figure 14.5a has maxima at half-integer values, e.g., $\mathbf{Q_{max}} = (0.5, 0.5, 0)$, regardless of the value of the a-priori unknown J. Identifying the location of the maximum $(\mathbf{Q_{max}}, E_{max})$ among the manifold of points $\{\mathbf{Q_{max}}, E\}$

fixes immediately the value of J. Investing some points along the E direction around $\mathbf{Q_{max}}$ is, therefore, a good starting point for a decision policy.

In a considerably more complex example with no algorithmic expression of $S(\mathbf{Q}, E)$ at hand, Samarakoon et al. extracted five exchange parameters of a spin-1/2 Heisenberg Hamiltonian with Kitaev anisotropy. The experimental data were images over selected 2D regions of the static structure factor $S(\mathbf{Q})$, measured at the CORELLI beamline, and of dynamic structure factor $S(\mathbf{Q}, E)$, measured on the TOF spectrometer SEQUOIA, both located at the Spallation Neutron Source, SNS [463]. In this work, two autoencoders were trained for the static and dynamic structure factors with simulated data of $S(\mathbf{Q}, E)$, which were generated from (Metropolis) Monte Carlo and Landau Lifshitz dynamics (MC/LLD) for random sets of exchange parameter. A so-called iterative mapping algorithm (IMA) evaluated a mean square distance estimate between experimental and simulated data and generated the next set of parameters for the MC/LLD solver. Applying neural network architectures in the context of autonomous TAS experiments, one could imagine training networks to identify those points $(\mathbf{Q}, E)$, where e.g. the variance between different structure factors $(S_i(\mathbf{Q}, E, \mathbf{J_i})$ is a maximum to allow a rapid distinction between various parameter sets $\mathbf{J_i}$.

We conclude with a rather general statement of including probabilistic statistics in the experimental workflow of neutron experiments. The Bayesian approach supposes *some* knowledge before the experiment. This approach presents the scientist with two problems. First, the experiments feel biased. Second, in order to bias the experiment, physical information must be at hand and correctly implemented, a priori, to the experiment. But, the typical expectation of a scientist visiting a neutron facility is that the collected data are sufficiently rich that regardless of the level of prior physical knowledge, some conclusions can always be drawn in a-posteriori data processing. As the period between the formulation of an experimental idea and the actual experiment is typically a year, this position is understandable. This leads to the natural approach of *the more, the better*, supposing that data storage and procession remain manageable.

As shown above, the sequential data acquisition on a TAS forces scientists to come up with some a priori information. The Bayesian approach is only the logical continuation in order to combine the experimentally available information in a, statistically founded, most meaningful way. Even by continuing a simple grid acquisition strategy, a statistical correlation of grid data in the form of a Gaussian process might have some predictive character, although it is not exploited for an acquisition policy. The other way around, if the data acquisition strategy cannot determine a model, the obtained experimental data are still valid for any other posterior data treatment. One has to admit that the data presentation looks different with a seemingly randomly distributed data set compared to a grid scan, although the information content might be very similar (or even richer in the first case).

Further, data on their own contain rarely sufficient information for publishing as such. Data need interpretation. The Autonomous Experiment on a TAS or TOF just reverses the order of the workflow. The interpretation is formulated as a hypothesis before the experiment and confirmed with a certain probability. This is different from the up-to-now usual case, where the data are fitted to a model after the experiment.

As argued throughout this chapter, we believe that the autonomous approach is more efficient, as data are only collected up to a level where they confirm the hypotheses with a user-defined precision. It also minimizes the risk that collected data turns out to be useless during the fitting process and experiments need to be repeated.

CHAPTER 15

Material Discovery in Poorly Explored High-Dimensional Targeted Spaces

Suchismita Sarker

Cornell High Energy Synchrotron Source, Ithaca, New York, USA

Apurva Mehta

SLAC National Accelerator Laboratory, Menlo Park, California, USA

CONTENTS

T HE RAPIDLY EMERGING fields of Artificial Intelligence (AI) and Machine Learning (ML) initiated an exciting new direction and elevated expectations for materials science in the past decade to unveil the holy grail of material design, discovery, and development which can provide greener energy, affordable manufacturing alternatives in technological applications, and societal development for a sustainable future. Traditionally, new technology deployment from the laboratory to commercial products often takes more than 15 to 20 years [466, 281]. Several important discoveries in the past have occurred by serendipity or through brute-force trial-and-error experimentation. Both these approaches are expensive and time-consuming as they depend on empirical rules and hypotheses of synthesis pathways to a chemical and structural state and associated properties, shown in Figure 15.1. However, there is no mechanism to avoid brute-force random permutation of chemistry and synthesis parameters, if the structure of the complex material, and the associated properties, begin to deviate from the empirical rules. With the advancement of technology, data acquisition speed and storage capacity have increased significantly which opens up the unprecedented possibility of real-time course corrections and an accelerated exploration of high-dimensional design spaces, rapid optimization of properties, and a

DOI: 10.1201/9781003359593-15

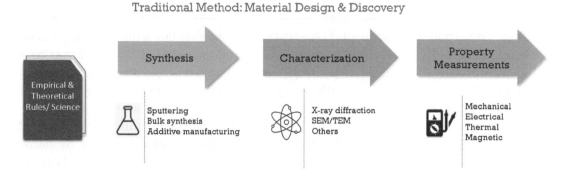

Figure 15.1   The traditional approach for material design and discovery: starting from a hypothesis, empirical rules, or theoretical science to select the material for synthesis, characterization, and property measurements.

deeper understanding of fundamental science through data-driven approaches and high-throughput experimentation.

15.1   CURRENT CHALLENGES

One example of a complex target materials space is the poorly explored phase space of compositionally complex alloys (CCA) [526]. Complex alloys containing three to seven elements in comparable concentrations promise a range of unique and desirable alloy properties, from high-temperature creep resistance to superior thermoelectric performance. The challenge, however, is that the CCA target space is vast and complex. Traditionally, domain experts knowledge and theoretical rules help to design new materials. These approaches work fairly well in well-defined comparatively low-dimensional targeted spaces but fail for CCA. These approaches fail for CCA for several reasons. Although computational methods, such as density functional theory (DFT) or finite element method (FEM), and molecular dynamics (MD), have advanced significantly over the last decade, they are still too slow and expensive when it comes to computing the ground states of solid solutions of multiple elements [184, 591]. Furthermore, as the compositional complexity grows, local heterogeneity and defects become more common and more critical in stabilizing metastable structures and influencing properties. Different processing paths for nominally identical compositions arrive at different metastable configurations and properties. Physio-chemical theories of these vast spaces, even when they exist, deviate from predictions of the synthesized realization of the CCA. These deviations, sometimes subtle, but often critical, are naturally captured experimentally. Learning from experimental observation is essential for CCA, and the vastness of the target space and the expense of experiments puts a high premium on selecting experiments with the highest new information, and hence the learning potential.

15.2 STEPS TOWARD MACHINE LEARNING AND MATERIALS DISCOVERY

The realization of experimental observations in the mid-90s launched the pioneering work of data-driven machine learning in materials science [281] to predict corrosion behavior [440] and investigate melting points [99] and damage tolerance in composite materials [424], creating the path to exploring material properties. A lot more focus on computational materials science emerged after that. However, a truly visionary effort was made more than a decade ago by the Material Genome Initiative (MGI) that revolutionized accelerated material design, development, and discovery by harnessing data and different simulation approaches together with experimentation [118]. Strategies for rapidly distilling the information from previous experiments, guided by current physiochemical theories, and using it to drive future exploration (research) is driving advancements in autonomy, autonomous research, and self-driving laboratories. This data-driven revolution and, in particular, Bayesian-inspired material discovery and synthesis optimization, are poised to transform materials science and technologies evolving from it.

15.3 HIGH-THROUGHPUT EXPERIMENTATION AND DATA-DRIVEN MACHINE LEARNING

The direction materials science envisions to accelerate new materials discovery is through data-driven machine learning and high-throughput experimentation as research complexity increases. Here, we summarize different strategies and challenges of different approaches in material discovery for poorly explored high-dimensional targeted spaces. To expedite the acceleration of materials discovery the traditional process started with existing experimental databases. One of the critical challenges of these purely data-driven approaches is the sufficient quantity and diverse quality of datasets for predicting and optimizing material properties or decision-making for new material synthesis. Numerous efforts have been made to reshape these material intelligence ecosystems by creating machine-readable material, data repositories, and materials informatics [48, 203]. Human efforts to extract data from more reliable material datasets [228, 153] and clean the experimental data [339], or machine learning-assisted material discovery from failed data become an interesting endeavor [418]. Recent guidelines of FAIR (findable, accessible, interoperable, and reusable) principles address some of the challenges of the digital ecosystem for scientific data-rich research management and stewardship [568].

After creating clean training datasets, the next steps are to generate predictions and uncertainty estimates of different desired models and machine-guided material recommendations for parallel synthesis, high-throughput structural, and property characterizations, closing this feedback loop by adding new experimental datasets to the model and improving predictions with different iterations to accelerate new materials discovery shown in Figure 15.2 [184, 48, 455].

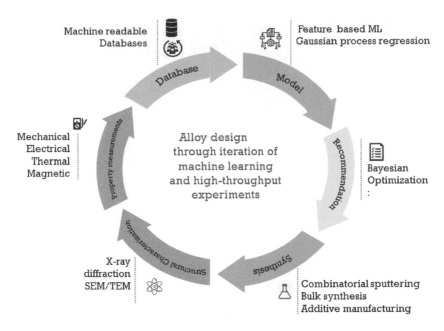

Figure 15.2 Data Driven Approach for material design and discovery. Step I: creating databases. Step II: model generation. Step III: recommendations. Step IV: high-throughput synthesis. Step V: structural characterization, and Step VI: property measurements and complete the feedback loop.

In the past decades, these data-driven high-throughput experimental approaches have been taken by researchers to design new materials such as high-entropy alloys [425], metallic glasses [443], superconductors [491], thermoelectric materials [173], shape memory alloys [172], batteries [276], and solar materials [512]. Others explored modeling ahead of experiments to predict and understand the composition-structure-property relationship. For example, recently, we explored metallic glass that is known to pose superior strength, hardness, wear resistance, and corrosion resistance compared to its crystalline counterpart due to the absence of the traditional deformation pathways. Finding metallic glasses with desired properties has been a constant pursuit for numerous experiments in the last six decades. However, the lack of established theoretical rules creates an immense challenge to finding desired glass-forming alloys in high-dimensional space. Finding metallic glasses with desired properties makes the search space even more challenging. Therefore, a data-driven machine learning approach has been taken to explore this vast design space both for glass-formability and mechanical properties. In general, robust feature-based supervised machine learning depends on suitable training datasets and physicochemical-based descriptors also known as features. The extraction and quantification of alloys are described as composition-based physically relevant feature vectors generated from the properties of the elements such as positions in the periodic table, melting point, Mendeleev number, electronegativity, and many more. There are several open-source packages [208] that help to extract these features from materials raw data and

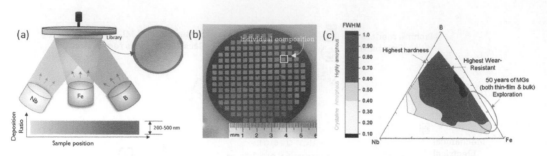

Figure 15.3 (a) Schematic diagram of thin-film deposition with Fe, Nb, and B. (b) Combinatorial libraries of 177 spots where each spot represents individual compositions of the ternary diagram. (c) Experimentally measured glass formability with the highest hardness and wear-resistance composition.

combine them as mathematical functions before implementing machine-learning algorithms like the data-science community uses. These feature extractions generate a relationship between input and output variables which is one way to explore domain science. Therefore, reliable feature-based machine learning requires designing physically meaningful parameters and selecting the relevant features from a large pool [200]. The Matminer package is used here to optimize these features and explore the glass-formability and mechanical properties of metallic glasses and recommend the desired material for synthesis in FeNbB ternary [560]. To explore this large ternary space, high-throughput synthesis is performed by thin-film deposition i.e., combinatorial libraries [287] that can create large datasets to verify the feasibility of hypotheses and rapidly validate any computational models. These libraries are fabricated by combining magnetron sputtering and creating a composition gradient to cover a large fraction of the ternary. The composition of the co-sputtered samples is determined by high-throughput wavelength dispersive spectroscopy (WDS). A rapid and non-destructive technique of synchrotron high-throughput x-ray diffraction (HiTp-XRD) is used to measure amorphous (glass) region shown in Figure 15.3. After identifying the metallic glasses, dynamic nanoindentation equipped with a Berkovich-geometry diamond indenter is used to map elastic modulus and hardness, correlate properties with glass formability, and validate the model, and we were able to discover several of the new alloys exhibit hardness greater than 25 GPa, which over three times harder than hardened stainless steel and only surpassed by diamond and diamond-like carbon [469].

15.4 AUTONOMOUS CLOSED-LOOP ALLOY DESIGN

In the prior example, the recommendation from the model of the entire FeNbB ternary synthesis for metallic glass has been synthesized before the experiment based on iterative results of machine learning and high-throughput experimentations. However, as the dimensionality of the CCA rises, the high-throughput (HiTp) synthesis tools become increasingly difficult to scale to synthesize the entire target space or characterize it as well. The goal has to be to guide our searches to desired regions of the

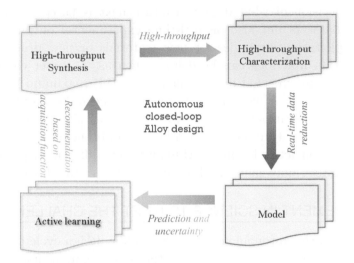

Figure 15.4 Autonomous closed-loop experimentation starting from high-throughput synthesis, using high-throughput characterization and real-time data reduction for modeling to generate predictions and quantify uncertainty, and perform active learning and Bayesian optimization to choose either the next characterization datapoint or next material for synthesis.

target space. The search takes on an iterative aspect. Start with an HiTp search of a lower-dimensional or simplified target space, use the lower-dimensional results to estimate a promising next search direction or a set of experiments using a Bayesian approach, and continue redirecting the search to a more complex space. The closed-loop materials discovery and synthesis optimization using the Bayesian approach is the future. These autonomous closed-loop material discovery ecosystems must try to reduce the cost of each iteration and perform real-time information extraction for smart decision-making under uncertain conditions. To make this process of circling the loop faster and cost-effective, several approaches have been taken from high-throughput parallel synthesis (mostly combinatorial thin-film) to high-throughput characterization (such as synchrotron x-ray [443], optical method [499], or electrical measurements [614, 588]).

The four parts that make up the closed-loop cycle are illustrated in Figure 15.4: high-throughput synthesis, high-throughput characterization, model predictions, and active learning. One of the successful ways to integrate this high-throughput experimentation is by Bayesian active learning driven by Gaussian Process Regression (GPR) which is a nonparametric probabilistic machine learning model for truly creating an autonomous material discovery ecosystem [360]. In GPR, the surrogate function is created with preliminary data and can adapt the model complexity as more data arrives. In addition, the physics knowledge can be integrated either by prior-mean functions or kernel designs which become advantageous for decision-making of material synthesis in high-dimensional search spaces [248]. GPR generates predictions and quantifies the uncertainty of the observations and predictions.

Therefore, the decision of what data to capture next is based not only on previously collected datasets but also on real-time experimental output. Based on exploration or exploitation strategies for new materials or new process parameters, different acquisition functions can be defined such as (i) upper confidence bound (UCB), (ii) maximum expected improvement (EI), (iii) maximum likelihood of improvement (MLI) or others [68]. There are several successful examples of the closed-loop Autonomous Experiment for structure predictions [596, 280], phase mapping [248, 552, 20], microstructure optimization [379], and other property measurements. Bayesian-optimization active learning techniques using different acquisition functions truly create an autonomous material discovery ecosystem to alloy design.

15.5 ADVANCEMENT IN SCIENCE TO MEET THE CHALLENGES

The slowest step in the autonomous closed-loop alloy design cycle determines its efficiency and demand on the performance of the other components. The synthesis is the bottleneck in such cycles, followed by the latency in shipping synthesized samples to the characterization labs. Therefore, the challenge ahead for the community is to eliminate the latency between the recommendation and faster high-throughput synthesis capabilities. The dynamic range of processing conditions in the multiple principal element alloy spaces makes this scenario difficult. In the prior example, although the data-driven machine learning and high-throughput characterization approach can discover several hardest known wear-resistance thin-film metallic glasses, the challenge is to use different synthesis procedures and understand their effect on the properties to design the new materials as well. In metallic glasses, one of the parameters for glass formability is the critical cooling rate of the molten metal where the atoms dont have enough time or energy to rearrange themselves for crystal nucleation and the liquid reaches the glass-transition temperature to solidify. The critical cooling rate for sputtered thin-film metallic glasses is 10^8 K/s, which is very different than the cooling rate of melt-spun ribbon 10^6 K/s [278] and bulk metallic glasses 10^5 K/s [468]. The properties discovered in thin-film metallic glasses cannot be directly extrapolated to melt-spun or bulk metallic glasses. Therefore, understanding the effect of different manufacturing conditions needs either high-throughput synthesis procedures for each one or steps to control the processing parameters of the cooling rate. In the last decade, there have been several isolated efforts toward a high-throughput parallel synthesis of both thin-film [287, 290] and bulk materials [386]. However, there is no straightforward way to do that. Therefore, the absence of capturing these processing parameters for the metastable state makes it extremely difficult, expensive, and time-consuming to find new materials and optimize properties for different manufacturing conditions. To explore this dynamic range of processing conditions, the community needs an approach for rapid high-throughput infrastructure not only for combinatorial libraries but also for bulk sample synthesis and optimizing the properties with different affordable manufacturing processes. Recently, several efforts have been made to integrate high-throughput synthesis with high-throughput characterization to explore multi-dimensional parameter spaces to map out the processing and properties of the metastable state through laser annealing or flash annealing.

In addition, real-time data processing in several characterization processes is still very challenging, and finally, any type of human intervention during decision-making needs to be addressed as well. We have an exciting time ahead for automation and autonomy to explore material design rapidly and this has far-reaching implications for the development of a sustainable future.

CHAPTER 16

Autonomous Optical Microscopy for Exploring Nucleation and Growth of DNA Crystals

Aaron N. Michelson

Brookhaven National Laboratory, Upton, New York, USA

CONTENTS

I THIS CHAPTER, we will demonstrate an application of autonomous investigation using Gaussian process regression. We will cover the development of autonomous microscopy experiments to explore the nucleation and growth of DNA crystals via DNA origami nanotechnology. First, we will give a short background on the problem being investigated, namely, the optimization of crystal growth. Second, we will describe how we set up the autonomous loop for microscopy. In the last section, we will demonstrate some results from the experiment highlighting where things went well and where things went awry, and where we will go in the future.

16.1 DNA NANOTECHNOLOGY

The field of DNA nanotechnology was kickstarted by Ned Seeman [477], with the notion of using DNA for structural assembly leveraging the precise A,G,T,C base pairs to create by design material to act as a scaffold for protein crystallography. A branch of this technology was further developed by Paul Rothemund [452] with the introduction of a method to create nanoscale objects by folding a long strand of DNA with many smaller strands of DNA, called "DNA-origami". Leveraging DNA further, these DNA-origami were modified to interact with each other with DNA overhanging at vertices allowing for the self-assembly of these nano-objects into micron-scale

DOI: 10.1201/9781003359593-16

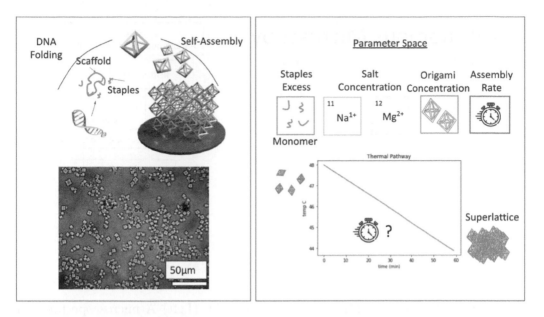

Figure 16.1 (Left) DNA Origami synthesis and assembly along with optical microscopy of DNA lattices in solution. (Right) Parameters for DNA assembly, including purity (excess staples for folding the long strand), salt concentration in buffer, origami concentration, and assembly/annealing rate. The plot schematically shows the thermal annealing of monomers into superlattices following some linear thermal path.

material built entirely from DNA [521]. The self-assembly of these DNA origami nanoscale building blocks tends to be described by classical nucleation and growth kinetics, typified by a single energy barrier between the transformation between the melt and solid [120]. The assembly of DNA crystals is a large parameter space and exploring how to optimize the assembly process would involve tuning properties such as purity of monomers, salt concentration (needed for DNA stability and rigidity), the concentration of monomer, and the thermal annealing rate. The annealing step is one of the most critical for creating well-ordered materials and is challenging to optimize given the wide array of annealing histories that can be applied to a sample. Please see Figure 16.1 for an overview [316]. A redeeming quality, however, is that DNA crystals grow large enough to be clearly visible with basic microscopy, which creates the potential for Autonomous Experimentation. If the investigator can evaluate the crystals after an annealing pathway and suggest new pathways in order to optimize some parameters, we would be able to move on to more and more complex architectures.

16.2 AUTONOMOUS MICROSCOPY

The first step to starting autonomous microscopy was outlining the components needed to close the loop for investigation. For this to work we need a chamber and

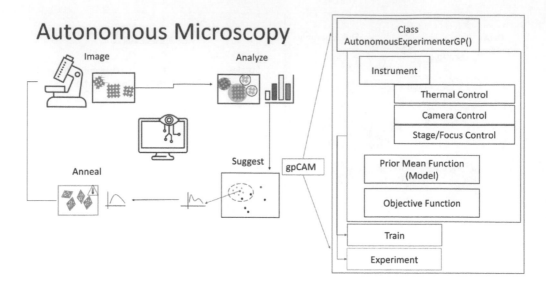

Figure 16.2 Autonomous Microscopy and gpCAM. (Left) A microscope takes images of the sample, which is then analyzed using ML or image processing to yield a figure of merit to the investigator which can then suggest a new thermal annealing pathway. (Right) gpCAM setup incorporates the instrument (thermal annealing and camera controls) along with the model we propose and the objective function for the suggestions.

thermal annealing apparatus, along with a camera, microscope, stage to move around the sample, image analysis software, and a brain to suggest new data points. See Figure 16.2 for the schematic drawing outlining the experimental procedure. We engineered a thermal stage and microscope/camera which could be controlled directly in Python. Depending on the need, an additional motorized stage could be deployed to add scanning to the microscopic image collection. After annealing and image collection, images need to be processed reducing the 2D images to data that can be used as inputs for the autonomous algorithm. The algorithm would then attempt to suggest new data points.

What is wonderful is that there are a number of packages that have been put together which can quickly enable us to go from being a biologist to a superpowered biologist in the lab. Much credit can go to the engineers and scientists who innovated to create open packages in Python and Java for the control of cameras, motorized stages, and a healthy biotech industry that could be easily co-opted from live cell imaging to our problem of DNA crystallization. For our example, we will use an implementation of Gaussian process regression (gpCAM, https://gpcam.lbl.gov/) which introduces an Autonomous Experimenter (AE) class that we will use to act as our brain for suggesting new pathways. The code is flexible to attach to our problem. Since we separately built simple functions to control operations like "snap an image", "set a temperature", and "analyze this image", the AE class will only need a few arguments to get started. The first will be to define the "instrument" and in this

```
my_ae = AutonomousExperimenterGP(
    parameters,
    hyperparameters,
    hyperparameter_bounds,
    instrument_func=instrument,
    init_dataset_size = 1, dataset = 'localdata.npy',
    acq_func = 'maximum',
    kernel_func = kernel_l2_single_task,
    prior_mean_func = priormean1,
    training_dask_client = None,
    acq_func_opt_dask_client = None)
```

```
def instrument(data, instrument_dict={ }):
    for entry in data:
        #Run the Thermal Path
        t, T = thermalpath(entry['position'][0], entry['position'][1],
                           entry['position'][2], entry['position'][3],
                           entry['position'][4], 60, camera=True)
        #Simulation from Model for the given path
        Total, path, cummulative = grainsize(t, T, g0=48, n0=46, path=True)
        entry["Total"]= Total
        #Direct Evaluation from Microscopy using YOLO
        a = YOLOSIMPLE(im)
        entry["value"] = a
        entry["variance"] = 0.001
    return data
```

```
def priormean1(gp_obj,x,hyperparameters):
    y = []
    for i in x:
        temp = graphingpathgrowth(i[0], i[1], i[2], i[3], i[4])
        y.append(temp)
    y = np.asarray(y)
    return(y)
```

Figure 16.3   Code excerpt.

case, we additionally provide a prior model for the experiment which is like giving a taste of the physics to the brain to save time and avoid exploring obvious pitfall regions. The instrument function collects the simple functions we need to 1. Control the thermal stage, 2. Take microscope images, 3. Evaluate the images with a call to a machine learning algorithm. Multiple suggested pathways can be obtained before coming back to the brain to ask for more suggestions.

It is important at this stage to discuss the details of how we carefully define our problem and how we will interrogate the problem with the autonomous experimenter. We encapsulate these details in the parameterization of our problem, see Figure 16.4. The thermal annealing pathway of the nano-objects will start as monomers in solution; at some time point during the thermal annealing, the monomers will bond to each other and reach a critical nucleus size, persist, and grow. We could ask a number of questions about this process, for instance, what is the maximum size obtained? What is the fastest time to see nucleates form? And is there a critical pathway that needs to be traversed for the crystals to form? These questions help us formulate the objective we will give to the experimenter. As you may have seen in the code excerpt above, we provided the acquisition function (acq_function) a "maximum", so in this example, we attempt to maximize the crystal size. So we perform measurements through an annealing protocol and feed into the algorithm the histogram or maximum crystal from the distribution to judge the quality of the experiment we performed.

At this point, we need to have some tools to evaluate the images to be able to give the AE some way to evaluate the microscope images. For this, we developed a You-Only-Look-Once algorithm trained from a few (10-100) images to detect and draw circles around each object in order to evaluate their size. For an example of how we might have looked at the evaluation of the transformation of the crystals over time, we could use the structural similarity index metric to detect changes in

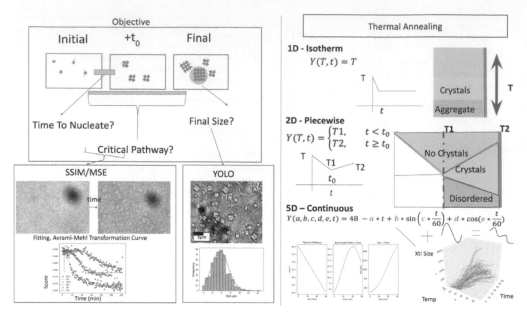

Figure 16.4 Parameterization (Left) schematic of nucleation and growth along with variables of interest along with two methods to attempt to quantify crystals in solution with either the structural similarity index measure, mean square error for time transformations, and a You-only-look-once (YOLO) algorithm for identifying crystals in the image with an accompanying histogram of the data. (Right) Attempts to parameterize the thermal annealing pathway comparing simple 1D, 2D, and 5-dimensional spaces with an abstraction of the interoperability within these examples.

the field of view over time, fitting the result of all the autocorrelations to some model of material transformation (shown in Figure 16.4 bottom left).

Once we have the tools to evaluate the crystal size, we have an additional challenge to think about how and when to probe the sample. This additional level of parameterization is crucial for the experimental design. The simplest thermal path is an isothermal path where we simply provide a single temperature and wait until the end of the time window we set. It would then map out the existence of the crystal across our variable: "Temperature". The experimenter would thus learn how to relate the temperature to the size and hone in on an optimal pathway for assembly. However even in this one dimension with respect to temperature, we still have to contend with the time discretization of the experiment, how long we let the experiment run, or how often we make decisions (after every input? after every 5? 10?).

If we attempt to add complexity to the thermal pathway by adding two temperature points in the thermal path we can potentially map out the value of the starting and final temperature; this is also useful because the space of constant functions fits into this model as well since f(T1,T2) collapses to f(T) for T1=T2. The model would then tell us how the initial vs. final temperature of the assembly relate and can be visualized on a 2D map. The more time/temp steps we add to the piecewise function the higher-dimensional the parameter space becomes. In a simple case of one experiment per hour, if we asked the AE to suggest a temperature setpoint every second this would be 3600 dimensions for it to work through. Adding more and more points for the model to suggest on the thermal path has added complexity to the problem to an extent where we would not know how to effectively visualize the problem anymore. Even worse, it may not be possible to explore this space with enough experiments to converge on a solution.

To address the problem of creating more complex pathways while also retaining the ability to interpret the result of the model, we leverage our domain knowledge about the superlattice assembly process. Knowing that we mostly need slow cooling, we combine a linear function and sinusoidal function to potentially allow for oscillations about a temperature ramp or force the function to spend more time around the optimal temperature in the thermal path. This still leaves us with a 5-dimensional problem to attempt to interpret.

To address the 5D problem, we turn to our domain knowledge of self-assembly in general in which we can make a simple model that estimates a growth and nucleation rate as a function of time and temperature, this is shown in the plots of Figure 16.4. This function, which we call a prior mean function, evaluates the thermal pathway to yield a "crystal size" from the time and temperature parameters. The thermal pathway function now becomes a facsimile for optimization of the nucleation and growth rate. An optimal route would find the best time/temperature to spend in the nucleation window of the crystal and spend the rest of the experiment time growing those crystals. This not only provides us with a means to evaluate the 5D parameters but also lets us directly feed this model to the AE. The AE suggests thermal paths and checks later against the prior model to see how the real sample compares. Combined with the acquisition function, the number of experiments becomes far reduced. For an example see Figure 16.5 wherein we compare the grain size with and without using the prior mean function. The simulated results show that with the inclusion of a prior mean function, we can achieve "crystals" which are evaluated as 1400 within a few iterations of the experiment, validating the model. When using this same approach for the real sample, we can see crystals are improved in the second experiment. It's important to note that the prior model didnt have meaningful units for the actual experiment, it just describes success and failure on a spectrum of about 1400 points, this could easily have been scaled to 0-100 or any metric for a direct evaluation of each thermal curve.

16.3 CLOSING REMARKS, FUTURE

We will now make good on our introductory remarks and discuss where things went well and where things went awry, and talk about plans for the future of this experimental platform. We hope you can feel your way through the way we attacked the problem and how a number of different options emerged. These problems can be termed intrinsic and extrinsic domain challenges. Intrinsically, for time-constrained problems, we can not freely let the algorithm probe the space; we had to constrain the problem to chewable pieces for this to work. This asked for a lot of work on the side of the experimenter to think hard about the questions we want to tackle and how they

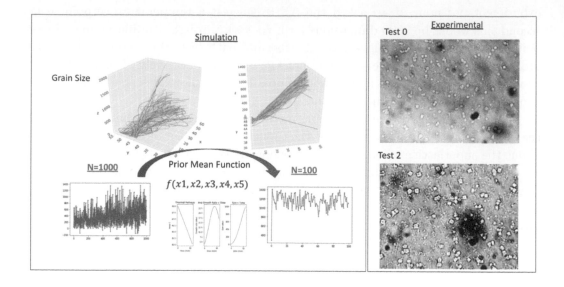

Figure 16.5 Prior mean function simulated and experimental result. (Left) Simulation of experimental thermal pathways without a prior mean and with a prior mean function which provided an initial model that served as a physics-informed trend in gpCAM.(Right) Experimental result optical microscope of initial thermal annealing N = 0 and N = 2.

want to parameterize those questions. This may not be desirable in many cases, and practitioners might attempt parameterizations that cant lead to the desired phase space or are too restrictive for future experiments. We came across this problem when we considered if instead of starting with monomers we start with monomers and a small number of nucleates. The information we collected for the monomer assembly pathway shares many aspects but differs dramatically in output size. It may be possible to map out experiments on a trajectory of complexity and work backward to the initial points, but it may be impossible to train a single Gaussian model to tackle all aspects and may inevitably require one to delve deeper into the mathematics of the problem which might be a bit overwhelming.

Repeatability is an extrinsic challenge for DNA nanotechnology, a simple change of the salt concentration will shift the melting temperature of the entire system, this was both a challenge and a strength of the autonomous approach. It means the results, from one experimental run to the next, needed to account for these shifts in building a model. Out of this challenge, we gain an experimenter that can handle probing similar spaces between runs that updates on the fly. For this to work we had to provide enough information for the experimenter so it could probe "similar" spaces between runs. We do this by providing a prior model but directing the model to assign a higher uncertainty to the model compared to actual findings, allowing for decisions to be made similar to the model but built off current real data points. These challenges will enable us to join multiple units together for complex self-assemblies with non-trivial thermal pathways.

In conclusion, we brought together an optical microscope, a camera, a machine learning algorithm (the Yolo object identifier), and a Gaussian-process-driven autonomous experimenter to explore the assembly of DNA nanotechnology. We described how to put together a closed-loop experimenter in the lab, and rationalized how to attack the problem of our choice, which in this case, is the optimization of the crystal size by optimization of the thermal pathway of assembly. In doing so we looked at the problem of parameterization, dimensionality, and interpretation of the experimenter's results in the context of nucleation and growth. Lastly, we looked at the incorporation of the model directly into the experimenter loop and how it can be leveraged to guide the experimenter with our domain knowledge.

CHAPTER 17

Aqueous Chemistry and Physics Constrained Autonomous-Closed-Loop Modeling for Quantifying Adsorption Processes as Applied to Metal-Mineral Interface Geochemistry

Elliot Chang

Seaborg Institute, Lawrence Livermore National Laboratory, Livermore, California, USA

Linda Beverly

Seaborg Institute, Lawrence Livermore National Laboratory, Livermore, California, USA

Haruko Wainwright

Department of Nuclear Science and Engineering, Massachusetts Institute of Technology, Cambridge, Massachusetts, USA

CONTENTS

DOI: 10.1201/9781003359593-17

THIS CHAPTER discusses the use of high-throughput, big data modeling approaches to investigate fluid-mineral interactions in subsurface geochemical processes. To this extent, working-data has been compiled in a FAIR formatted structure, using the Lawrence Livermore National Laboratory Surface Complexation/Ion Exchange (L-SCIE) database. Utilizing this consistently formatted set of data, the authors pose a test case using the mobility of metal in the subsurface at the soil-aqueous interfaces through adsorption processes. This chapter illustrates the application of Gaussian process (GP) modeling, informed by chemical aqueous speciation to develop predictive capabilities. A workflow is outlined based on autonomous optimization of the GP regressor, constrained by chemistry information—in this case mass conservation and statistical knowledge of high-variance regions. This approach demonstrates how chemical constraints can be incorporated into the generation of a data-driven adsorption model through an autonomous, closed-loop process.

17.1 INTRODUCTION

17.1.1 Metal Sorption Processes in the Natural Environment

The natural terrestrial environment—composed of complex interactions between abiotic and biotic components—supports a diverse body of various ecosystems. Such is the diversity of Nature, our planet is filled with unique climates and terrain; cold and treeless tundras are present in mountain ranges while warm tropical rainforests exist at equatorial bands with high annual rainfall. Within these variable ecosystems,

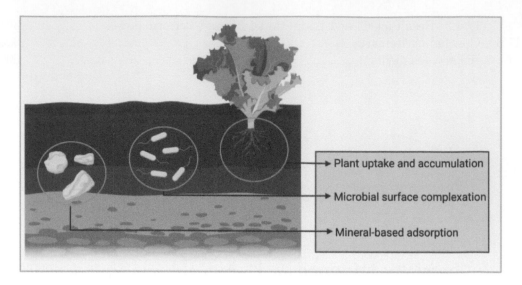

Figure 17.1 Schematic of subsurface geochemical processes that impact metal transport.

metals serve key roles to support various environmental functions. Naturally occurring metals, such as zinc, copper, and iron are essential for the support of healthy plant growth [141]. Humans have also exploited the use of various metals, such as rare earth elements (REEs), to deploy fertilizers in agricultural settings that improve crop yield and quality [391].

Upon the natural or anthropogenic release of metals into the environment through processes such as mineral weathering, mining activities, or agricultural practices, complex subsurface biogeochemical processes affect the availability, mobility, transport, and ultimate utilization of those metals. Plant organisms may absorb metals, accumulating them in roots [496] or utilizing them for photosynthetic purposes [595]. Microbes may assist in the transformation of redox-sensitive metals [557] and additionally participate in both absorption (internalization) and adsorption (exterior binding) of metals to influence overall mobility and subsurface transport. Minerals participate in surface complexation and ion-exchange of metals, directly affecting aqueous metal concentrations by becoming immobilized on solid mineral surfaces (Figure 17.1).

Among the many co-occurring subsurface biogeochemical processes, adsorption plays a particularly important role in metal mobility and transport. In this context, adsorption may be defined as the interaction between a metal sorbate and the reactive interface of a mineral sorbent, whereby the aqueous-phase sorbate is thermodynamically driven to electrostatically bind to the solid-phase surface. This process is affected by a variety of environmental conditions. Notably, pH not only affects the aqueous speciation and solubility of the metal ion, but it also impacts the protonation state of the solid phase. Because adsorption is largely driven by electrostatic interactions, the protonation state (positive, neutral, or negatively charged states) plays a key role in determining the effectiveness of the sorption process. Additionally, organic compounds produced from biological processes may inhibit the adsorption process by forming competing, strong aqueous metal complexes [548].

Thermodynamic models, known as surface complexation models (SCM), exist that describe the adsorption reaction from a numerical framework [342]. As metals adsorb onto the mineral surface, charge accumulates and must be accounted for in thermodynamic calculations; if a positive sorbate is complexing to a negatively charged substrate, as more of the sorbate is bound to the substrate, there is a reduction in surface charge, which in turn adversely affects the propensity for the sorbate to bind to the substrate. This phenomenon is commonly expressed through the formulation [212]

$$K_{effective} = K_{intrinsic}^{\frac{F\Psi}{RT}} \qquad (17.1)$$

where K represents the thermodynamic equilibrium constant, F is the Faraday constant $(\frac{C}{mol})$, R is the molar gas constant $(\frac{J}{Kmol})$, T is the absolute temperature (K), and Ψ is the surface potential (V).

17.2 THEORETICAL FRAMEWORKS

17.2.1 Virtual Instrument Conceptualization

This book explores the development and application of Autonomous Experimentation in a variety of unique disciplines. While experimentation typically has a connotation of laboratory activity and methods, the authors of this chapter employ a broader definition. Experimentation may be defined as performing a scientific procedure whereby inputs, such as geochemical conditions, metal concentrations, and abundance of minerals, are provided to generate an output (an experimental result). In this context, we define a virtual instrument that accepts experimental inputs and produces an output result. This virtual instrument will serve as the vehicle by which Autonomous Experimentation will be conducted. The aim of the automated procedure is to robustly optimize and fine-tune the instrument parameters such that the resultant outputs more accurately and precisely match literature data.

17.2.2 Instrument Part 1: Aqueous Speciation Calculator

Two modeling frameworks are coupled to build a virtual instrument: an aqueous speciation calculator and a Gaussian process model regressor. Aqueous speciation calculations are conducted by accepting a variety of geochemical inputs and transforming the inputs into a more detailed, valuable form of information. More specifically, the purpose of this step is to obtain key experimental information, such as the concentration of metal present, the amount of mineral available, and other geochemical factors, such as pH, ionic strength, and temperature, in order to simulate aqueous thermodynamic reactions to discern the speciation of metal in solution. Notably, this step allows the instrument to obtain and incorporate chemistry-informed knowledge in assisting the quantification of metal adsorption onto the mineral surface.

17.2.3 Instrument Part 2: Gaussian Process Modeling

Gaussian process models have been widely used in the environmental sciences mainly in the spatial context for interpolating point measurements or for optimizing sampling locations [235, 551, 497]. In addition, the Gaussian process has been used as a surrogate model for hydrological simulations [396]. However, to the authors knowledge, it has not been applied to geochemical systems, particularly for hybrid data-driven and physics-informed settings.

In this application, a Gaussian process (via gpCAM) is constructed as a regression of the equilibrium aqueous metal concentrations as a function of input variables, including water quality (e.g., pH, ionic strengths) and aqueous complexation species computed by PHREEQC (a geochemical modeling program). The residual is represented as a correlated multivariate Gaussian distribution in the multi-dimensional parameter space to interpolate the output values at unsampled locations. For the covariance functions, we explored several kernels and selected a radial-basis function.

Although there are multiple regression models available such as the random forest method [49], the uncertainty has to be quantified through bootstrapping or simulations. It is advantageous that Gaussian processes compute the uncertainty estimates without ensemble simulations. In addition, the concentration is expected to vary smoothly as a function of various aqueous chemistry conditions, making the Gaussian process method appropriate to use. The availability of multiple kernels additionally provides flexibility for fitting diverse sets of adsorption data as a function of geochemical conditions.

17.2.4 An Autonomous Experimentation Workflow

As both components of the virtual instrument have been described, we can now explore the workflow by which Autonomous Experimentation may be executed (Figure 17.2). Initially, consistently formatted experimental input data is obtained from the Lawrence Livermore National Laboratory Surface Complexation/Ion-Exchange (L-SCIE) database (step 1). Each entry obtained from L-SCIE database contains reference information, mineral type, temperature, electrolytes present in solution, pH, the concentration of the metal sorbate of interest, solid-phase sorbent concentration, site densities present on the sorbent, and gas phase fugacities. The following step involves inputting the experimental data into the virtual instrument (step 2). This step automatically assimilates the experimental input data into aqueous speciation models using Python and the PHREEQC geochemical modeling program. Upon the generation and execution of PHREEQC input files, relevant geochemical features (as will be discussed in Section 17.3.1) are extracted and utilized in step 3, the performance of a Gaussian process model to determine regression relationships between the PHREEQC-extracted values and equilibrium aqueous metal concentrations (the amount of metal present in solution excluding the sorbed fraction on the quartz mineral phase).

Steps 2 and 3 make up the two-component virtual instrument which will be looped in an autonomous workflow. To perform this autonomous process, step 4 extracts the GP output values and their associated variances (i.e., the posterior covariance). Step 5

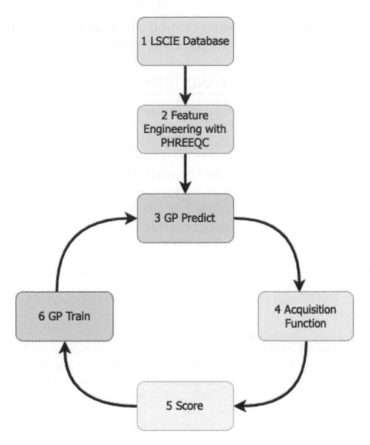

Figure 17.2 Autonomous experimentation loop for virtual instrument optimization.

is then triggered, calling upon a custom acquisition function (described in detail in Section 17.3.2) which determines the ranked scores of the data regions that should be sampled more to improve the virtual instrument performance. This acquisition function addresses high-variance regions of instrument outputs (such as from data paucity in a particular geochemical region) as well as penalizes unrealistic instrument outputs (as defined by the mass conservation law).

As step 5 ascertains the data region to sample more actively, step 6 then calls on the initial experimental input database extracted from L-SCIE to acquire more data to be inputted into the model as training. This step is considered the generation of surrogate data, as the "new" experimental data is in fact previously available information; in this manner, the addition of surrogate data may be thought of as a weighting scheme. This step also appends the "new" (surrogate) training data into the original experimental input database, effectively weighting data regions more heavily that are selected via the custom acquisition function. Upon the completion of step 6, the workflow reaches a closed-loop state, enabling an autonomous process to begin between steps 2 through 6. Each time the loop is executed, the hyperparameters associated with the virtual instrument are updated, allowing for each iteration of the autonomous workflow to enhance the performance of the virtual instrument. The optimization of the instrument includes chemistry-informed aqueous metal speciation data and physics-informed mass conservation constraints.

17.3   MODEL CONSTRUCTION AND TEST CASE

17.3.1   Chemistry-Informed Input Feature Development

U(VI)-quartz adsorption data extracted from the L-SCIE database consists of total U(VI) sorbate concentration, aqueous equilibrium U(VI) sorbate concentration, and associated metadata as defined by gas composition, quartz sorbent properties (concentration, surface area, reactive site density), background electrolyte concentrations, and pH conditions. The described metadata and total aqueous U(VI) sorbate concentration are used for each data entry to create PHREEQC simulations of solution chemistry conditions (Figure 17.3). The PHREEQC input file calculates the various

```
SELECTED_OUTPUT
-file C:\Users\chang58\Desktop\phreeqc-3.7.1-15876-x64\examples\python_trial_output.txt
-ionic_strength true
-user_punch true
-activities CO3-2 HCO3- UO2OH+ (UO2)2CO3(OH)3- UO2(OH)2
            UO2CO3 UO2+2 (UO2)3(OH)5+ (UO2)2(OH)2+2 UO2(CO3)2-2 UO2Cl+
            (UO2)4(OH)7+ UO2NO3+ (UO2)3(OH)4+2 (UO2)3(OH)7- (UO2)2OH+3
            UO2(CO3)3-4 UO2Cl2 (UO2)3O(OH)2(HCO3)+ UO2(OH)4-2 (UO2)3(CO3)6-6
            (UO2)11(CO3)6(OH)12-2 UO2(OH)3-
END

SOLUTION 0
units mol/L
pH 2.220
Na 0.010
Cl 0.010
F 0.000
U(+6) 0.00000021

GAS_PHASE 0
-fixed_pressure
-pressure 1.0
-volume 1.0
-temperature 25.0
 N2(g) 0.78973346
 O2(g) 0.19743337
 CO2(g) 0.00033623
 PHASES
 Fix_H+
 H+ = H+
 log_k 0.0


EQUILIBRIUM_PHASES 0
Fix_H+ -2.220 NaOH 10.0
 END
```

Figure 17.3   Automated generation of PHREEQC input file for a single data entry. Information is provided from L-SCIE database extraction for U(VI)-quartz adsorption.

aqueous species forms that are present for each data entry; this is specified by the "-activities CO3-2 HCO3- " line of code. Evidently, this step enables the GP regressor to receive geochemistry-informed information that addresses solution chemistry interactions (i.e., ionic strength impacts, ion-pairing, pH-dependent speciation, and precipitation effects).

In regards to the adsorption interactions between U(VI) sorbate and quartz sorbent, surface complexation of UO2+2 onto quartz is oftentimes described using a monodentate, inner-sphere reaction:

$$> SiOH + UO_2^{2+} + XH_2O \rightarrow SiOUO_2(OH)_X^{+1-X} + (1+X)H^+, \text{where } X = 03. \quad (17.2)$$

Additionally, the presence of aqueous carbonate species, such as HCO3- anions, may lead to the formation of uranyl-carbonate surface complexes:

$$> SiOH + UO_2^{2+} + YCO_3^{2-} \rightarrow SiOUO_2(CO_3)_Y^{+1-2Y} + 1H^+, \text{where } X = 03. \quad (17.3)$$

In an attempt to indirectly account for these surface complexation mechanisms, various U(VI) aqueous species and HCO3- concentrations are calculated in PHREEQC and provided as inputs into the GP regressor (adsorption model).

17.3.2 Gaussian Process Model Implementation

In our Gaussian process implementation, an anisotropic kernel function is used to create the covariance matrix [431]. We incorporated domain knowledge by customizing the acquisition function, which is passed through a mathematical optimization, [370] constraining the valid predictions by including mass conservation constraints. The chemistry constraints are applied via a scoring function that optimizes for out-of-chemistry bounds defined by predictions that exceed the total U(VI) concentration possible in the system (i.e., more adsorbed to the mineral than added to the system). The physics bounds are then coupled with prediction variance estimates to select areas that will help us gain the most information for the next round in the autonomously looped Gaussian process. Several functions are defined that are used to update the input to the Gaussian process until an acceptable stopping condition is achieved.

The input to the Gaussian process model is composed of PHREEQC output values in the form of ionic strength, HCO3- concentration, and U(VI) aqueous species concentrations, a calculated value from the experimental metadata in the form of mineral site density, and extracted experimental data in the form of equilibrium aqueous U(VI) concentration standard deviations. The standard deviations from the experimental data are used as the initial variance (measurement error).

The output of each iteration of the Gaussian process is a model that provides a posterior mean based on the learned covariance matrix. The hyperparameters, including the learned covariance matrix, are then used by the acquisition function we have provided to determine the "best" point to explore in the next iteration.

17.4 VIRTUAL INSTRUMENT PERFORMANCE AND ANALYTICS

17.4.1 Geochemical Feature Evaluation

Through the implementation of a closed-loop Gaussian process model, a virtual instrument was optimized by fine-tuning a suite of hyperparameters. As the Gaussian process model used in this example used an anisotropic radial basis function kernel, the aforementioned hyperparameters can be represented by length scales for each feature. Ultimately, the changes in the radial basis kernel function during each iteration of the Gaussian process model are proportional to the length scale as $\frac{1}{2(L)^2}$ where L refers to the length scale. Larger length scale hyperparameters quickly diminish a particular features correlation and thus impact the model prediction output.

In this test case of U(VI)-quartz adsorption, the autonomous Gaussian process loop was run 20 times, whereby each iteration allowed for new estimations of hyperparameters. The result of this autonomous loop generated length-scale hyperparameters to the radial basis kernel function, allowing for the assessment of the most impactful geochemical variables controlling the model predictions (Figure 17.4). Notably, the total U(VI) concentration present in the system (gray outer line) consistently played a major role in the Gaussian process model output; this feature was relevant in particular because of the physics-bounded constraint expressed in the acquisition function (predictions of U(VI) adsorption are heavily penalized if greater than the total U(VI) present in the system).

Quartz site density (mol of adsorption sites/m2) additionally played an important role in the best-performing Gaussian process model (yellow line); this information allows the scaling of U(VI) adsorption predictions based on the quantity of mineral present in the system—the more quartz is present, the more U(VI) can be removed from the solution. UO2NO3+ (light blue line) and ionic strength (orange line) were also present as relevant features in the best-performing Gaussian process model; this could be an artifact based on the large abundance of data points that contained nitric acid as an electrolyte in solution, and illustrates the continued work required to obtain more diverse datasets (such as solution conditions with varied electrolyte compositions and concentrations). The last notable feature was HCO3- concentration (dark blue line); this is indicative of secondary/ternary carbonate surface species of U(VI) that participate in quartz-based adsorption (refer to equation 3).

While a multitude of U(VI) aqueous species were added to the Gaussian process model as features, the autonomously looped model ultimately weighted five geochemical features above them: (1) total U(VI), (2) quartz site density, (3) UO2NO3+ concentration, (4) ionic strength, and (5) HCO3- concentration was chosen as the main relevant variables during the autonomous fitting process. This suggests that future studies may minimize computational intensity by first exploring standard key parameters involved in sorption modeling first - sorbate and sorbent concentrations, ionic strength, and liquid-gas exchange variables may all play an important role in achieving good initial predictions. For more complicated sorption systems, though, that involve multiple metals in solution alongside complexing anions, the chemistry-informed speciation information may play a more impactful role and may aid in better instrument predictions under specific geochemical conditions.

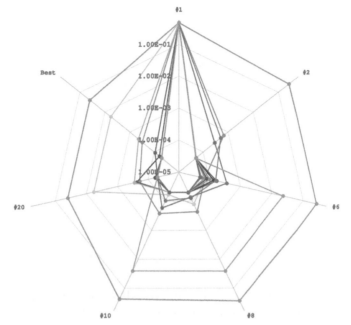

Figure 17.4 Geochemical feature weighting adjusted through a 20-iteration autonomous Gaussian process loop. Vertices indicate the iteration count. Values represent $\frac{1}{2(L)^2}$ units, where L refers to the length-scale hyperparameter for a radial basis kernel function used in the Gaussian process regressor. Colored lines indicate the geochemically relevant input features being assessed in the regressor.

17.4.2 Resultant Closed-Loop Model Fit and Performance

The best-performing Gaussian process virtual instrument that was determined from autonomous, closed-loop optimization was evaluated against an unused subset of the sorption dataset extracted from L-SCIE (i.e., a test dataset). The resultant virtual instrument generated values of equilibrium U(VI) aqueous concentrations consistent with experimentally generated data (Figure 17.5), with a Pearson correlation coefficient of 0.94. A coefficient greater than 0.90 is considered a satisfactory reproduction of the experimental data [599]. Through the implementation of a chemistry- and statistics-constrained acquisition function in a broader autonomous workflow, a Gaussian process adsorption model was successfully optimized and developed.

While most data points predicted in the closed-loop pipeline were selected based on their relatively high variance estimates, a small fraction was determined to be significantly influenced by the mass-conservation penalty as defined by the acquisition function (purple data points in Figure 17.6). Similar to the impacts of chemistry-informed geochemical feature engineering, physics-constrained search algorithms in the autonomous loop seemed to have a minimal impact on the model outputs.

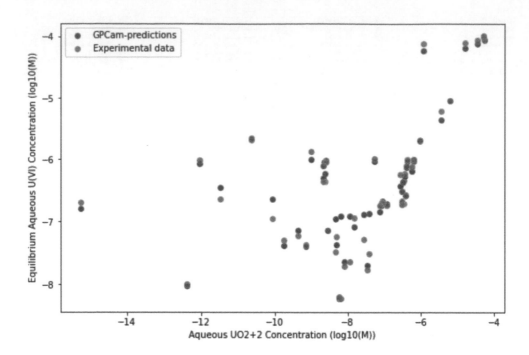

Figure 17.5 Best-performing Gaussian process adsorption model prediction results compared to experimentally generated data. This figure illustrates the strong performance of a fine-tuned Gaussian process model as supplemented by an autonomous optimization workflow.

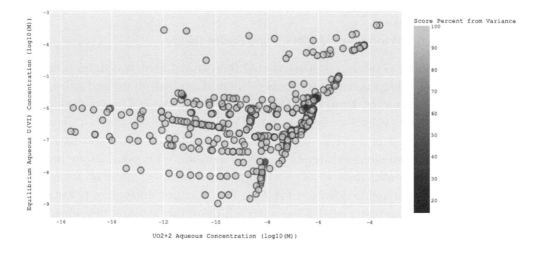

Figure 17.6 Gaussian process model prediction vs. input UO2+2 aqueous concentration plot. Darker colored data points indicate predictions where the acquisition function scoring was dominated by the mass conservation constraint penalty.

However, in more complex systems whereby geochemical conditions may not be so clearly correlated with adsorption behavior, the physics constraint provides a ground-truthing framework that allows for the generation of realistic outputs.

17.5 CONCLUDING REMARKS

This chapter illustrates the construction of an autonomous, closed-loop workflow to generate a fine-tuned adsorption model for metal exchange at the rock-aqueous interface. A broader definition of an "instrument" was employed to illustrate how the Gaussian process regressions could be coupled with geochemistry models. Aqueous speciation-based chemistry was used to engineer geochemically relevant explanatory variables, while a mass conservation law was applied to penalize unrealistic model outputs. Although the demonstration in this chapter is for a simple 1 metal: 1 mineral test case, the model built herein is certainly applicable to a broad range of geochemical systems that involve quantifying adsorption behavior found in Earths subsurface soils and sediments. The theoretical frameworks and test case implementation provided in this chapter will provide inspiration for scientists to creatively utilize Gaussian processes and Autonomous Experimentation of virtual instruments.

17.6 ACKNOWLEDGMENTS

This work was supported in part by the U.S. Department of Energys Earth and Environmental Systems Sciences Division of the Office of Science Biological and Environmental Research program under contract SCW1053. Additional funding was provided by the University Engagement program of LLNLs University Relations office under contract DE-AC52-07NA27344 to LLNL and DE-AC02- 05CH11231 to LBNL.

CHAPTER 18

Live Autonomous Beamline Experiments: Physics In the Loop

A. Gilad Kusne

Materials Measurement Science Division, National Institute of Standards and Technology, Gaithersburg, Maryland, USA

Austin McDannald

Materials Measurement Science Division, National Institute of Standards and Technology, Gaithersburg, Maryland, USA

Ichiro Takeuchi

Materials Science & Engineering Dept, University of Maryland, College Park, Maryland, USA

CONTENTS

18.1 INTRODUCTION

THE field of autonomous physical science (APS) aims to accelerate scientific understanding and discovery by helping scientists "fail smarter, learn faster, and spend fewer resources" in their studies [248]. In an APS, machine learning (ML) controls automated laboratory equipment, allowing for ML-driven experiment design,

DOI: 10.1201/9781003359593-18

execution, and analysis in a closed loop. Each subsequent experiment is selected to rapidly hone in on user goals. For instance, a user may want to maximize knowledge of the relationship between an experiment's input and output parameters, or the user may want to find an experiment with a target output. From each experiment, the APS learns and evolves to make better decisions.

APS holds great promise for materials science. Technology's voracious appetite for improved materials drives a constant search for ever-better materials. As researchers exhaust simple-to-synthesize materials, they are driven to explore ever more complex material systems. These complex materials involve a growing number of synthesis and processing parameters. However, with each new synthesis and processing parameter, the number of possible materials experiments grows exponentially. The relationship between how the material is made and its properties also increases in complexity. It is rapidly becoming infeasible for researchers to explore this complex, exponentially growing search space. The ML-driven closed-loop experiment cycle of APS promises to allow researchers to perform the minimum number of experiments necessary to explore the search space and identify improved technology-relevant materials.

APS success stories span a range of materials science challenges. APS-based materials process optimization was demonstrated with carbon nanotube growth—optimizing chemical vapor deposition parameters using in situ Raman spectroscopy data [352]. APS was used to optimize liquid chemical mixtures to tune the optical response of spin-coated materials [290]. APS-based Bayesian-inference-controlled neutron scattering measurements to hone in on the value of a physical parameter—the Neel temperature of magnetic materials [311] (see section on ANDiE below). The daunting challenge of solid-state materials discovery was also tackled (see section on CAMEO below), with a system capable of learning and exploiting the material synthesis-structure-property relationship toward material property optimization, resulting in the discovery of the new best-in-class phase change memory material [248].

Fundamental to the success of an APS system, is the data and ML pipeline (Figure 18.1). The pipeline begins with data collection from the experiment (either in the lab or in silico) followed by preprocessing the data to increase its utility for the experiment. The data is then analyzed, and the results are then extrapolated to predict the potential results of future experiments. These predictions are fed into an active learning algorithm—the ML field of optimal experiment design, to pick

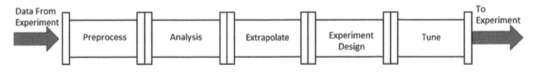

Figure 18.1 Machine learning pipeline for an autonomous physical science system. Data from an experiment is first preprocessed, then analyzed. Knowledge from performed experiments is then extrapolated to build predictions for future experiments. These predictions are then used to determine the best subsequent experiments to perform. The performance of the pipeline may also be tuned based on the given data and a performance measure, i.e., loss function.

the most information-rich experiments to perform next. Tuning the performance of these steps requires another addition to the pipeline - a loss function that quantifies performance.

Designing the machine learning pipeline requires the selection of multiple algorithms. A common first step is to identify easy-to-use, off-the-shelf machine learning tools that can be assembled into a preliminary ML pipeline. Many open-source software libraries exist for general ML [400] as well as tools targeted for various applications such as image processing [70, 541] and probabilistic analysis and prediction [58, 160, 1]. Based on closed-loop results of the APS system, the pipeline is re-engineered to improve performance.

A fundamental challenge exists in selecting each algorithm for the ML pipeline. Similar to materials science's large search space of potential experiments, ML algorithms often have large search spaces of potential solutions. As a result, the algorithms employ techniques to identify a "good" solution rather than the best solution based on some criteria. While these solutions may solve the posed challenge, they may not be useful to the APS system. For instance, algorithms can provide non-physical solutions. A live APS system that generates non-physically meaningful analysis can result in a dramatic waste of critical instrument time and resources, or drive experimental parameters beyond their practical bounds, potentially damaging expensive instruments. Additionally, there are potential computational gains that in searching a smaller solution space. Computational speed is critical to the value of an ML pipeline, as one cycle of the pipeline should take on the order of time of one experiment (in the case of small-number batch experiments), otherwise, computation time can lead to significant delays in experimentation.

Through the use of prior physical knowledge, one can narrow an algorithm's solution search space down to results that are practically useful and physically realizable in the lab. An array of methods exists for incorporating prior physical knowledge into ML, including constraint and probabilistic programming. While constraint programming tends to focus on building algorithmic bounds on the solutions, probabilistic programming defines probabilistic priors and employs uncertainty quantification and propagation to provide relevant solutions along with their probabilities and the model uncertainty.

An example challenge is shown in Figure 18.2. The APS system has collected X-ray diffraction patterns (XRD) for a set of samples spanning a phase region. The aim is to identify the XRD for the constituent phases from only this data. By collecting these diffraction patterns into a matrix Y, one can use matrix factorization to generate two matrices—X consisting of potential constituent XRD and β consisting of their abundances along with ϵ which collects additive noise values. However, in order to obtain a matrix factorization that corresponds to the constituent phases we must constrain both the factors and their abundances to be non-negative. The prior knowledge that β and X must contain values greater than or equal to 0 can be imposed using the Karush-Kuhn-Tucker conditions [247]. Additional prior knowledge can similarly be integrated into the algorithm including the fact that abundances should be values between 0 and 1 and for any sample should sum to 1; that XRD should smoothly vary as a function of inverse space (q); that samples similar in composition

$$
\begin{aligned}
&\begin{matrix} y_1^T \\ y_2^T \\ y_3^T \\ y_4^T \end{matrix}
= \begin{bmatrix} .2 & .8 \\ .4 & .6 \\ .6 & .4 \\ .8 & .2 \end{bmatrix} *
\begin{matrix} x_1^T \\ x_2^T \end{matrix}
\end{aligned}
$$

$$
\underbrace{\begin{bmatrix} \leftarrow & y_1^T & \rightarrow \\ \vdots & \ddots & \vdots \\ \leftarrow & y_N^T & \rightarrow \end{bmatrix}}_{Y}
= \underbrace{\begin{bmatrix} \beta_{11} & \cdots & \beta_{1L} \\ \vdots & \ddots & \vdots \\ \beta_{N1} & \cdots & \beta_{NL} \end{bmatrix}}_{\beta}
\underbrace{\begin{bmatrix} \leftarrow & x_1^T & \rightarrow \\ \vdots & \ddots & \vdots \\ \leftarrow & x_L^T & \rightarrow \end{bmatrix}}_{X}
+ \underbrace{\begin{bmatrix} \epsilon_{11} & \cdots & \epsilon_{1L} \\ \vdots & \ddots & \vdots \\ \epsilon_{N1} & \cdots & \epsilon_{NL} \end{bmatrix}}_{\epsilon}
$$

Potential Prior Physics-based Constraints

$x_{i,j} \geq 0$	Non-negative XRD
$\beta_{i,j} \geq 0$	Non-negative abundances
$y_i \sim N(\mu, K)$	XRD smoothness
$\beta_i \to \beta_j$ as $C_i \to C_j$	Composition-based similarity

Figure 18.2 A common challenge: Matrix factorization of high-dimensional data. Here we have a collection of X-ray diffraction patterns y_i. We'd like to identify the diffraction patterns of the constituent phases x_j as well as their abundances $\beta_{i,j}$. This challenge can be represented in terms of matrix factorization. Prior physics knowledge can then be introduced through constraint or probabilistic programming.

are likely composed of similar constituent phases and have similar abundances. Each of these prior physical rules can be integrated into the algorithm through constraint or probabilistic programming. With each new piece of physics knowledge, the solution space is narrowed to those more likely to be useful.

Many of these constraints already exist in off-the-shelf ML libraries, e.g., non-negative matrix factorization. Similarly, all ML algorithms have innate biases that either bound their solutions or provide preference to certain sets of solutions. For instance, the k-means clustering algorithm assumes that data clusters are spherical while spectral clustering assumes a graph-like connectivity structure between data points. Through the careful selection of off-the-shelf methods, one can build a pipeline that includes desired solution bounds and biases, thus reducing the solution space to those most useful. Alternatively, constraint and probabilistic programming frameworks allow a user to build bespoke ML algorithms that integrate their desired prior knowledge [58, 160, 85, 451, 1].

In this chapter, we will first discuss examples of using prior physical knowledge at different sections of the ML pipeline. We then discuss the ML pipeline of three autonomous systems and their use of physics-informed machine learning. For a broader discussion of recent autonomous systems, a good reference is [489]. The first two systems run live in control of beamline facilities - the Closed-loop autonomous material exploration and optimization (CAMEO) and the neutron diffraction agent (ANDiE). The third system we will discuss is the LEGO-based Low-cost Autonomous Scientist (LEGOLAS) - a kit used in courses to teach these skills to the next generation.

18.1.1 Preprocessing

Once data is collected, it must first be processed to ensure utility for the following steps in the ML pipeline. If the target signal in the data that we want to utilize is accompanied by other signals, we will want to remove these other signals to improve later data analysis. Prior knowledge of these signals can be used to aid in identifying the target signal and extracting it from the rest of the data. For instance, if the confounding secondary signals are background, knowing the behavior of the background can be used to differentiate the target signal from the background and remove it. Additional confounding signals can include unwanted, known diffraction peaks associated with the substrate or measurement artifacts. Using prior knowledge of these confounding signals, a signal processing step can be used to subtract the smoothly varying background and a peak detection algorithm can be used to identify and remove the unwanted substrate peak.

Another consideration is the representation of the data. For example, in some chemistry settings, it is common to see the components of a mixture represented as a ratio of A to B. This representation impacts ML analysis and results. The change in the ratio as you move from 1/4 to 1/1 is quite small compared to the change in the ratio as you move from 1/1 to 4/1 (the first move has a change in the ratio of 0.75 while the second has a change of 3.0). In describing mixtures, the space of ratios is asymmetric and non-Euclidean. Such a representation might be appropriate if the physics you intend to capture depends on the ratio, but care must be taken to ensure that the ML algorithms are compatible with such a non-Euclidean space. If instead, we represent those same mixtures as concentrations $(A/(A+B))$ this becomes: moving 0.20 to 0.50 has the same change as moving from 0.5 to 0.8 (both moves have a change of 0.3). The space of mixture concentrations is symmetric and Euclidean. Representing mixtures as concentrations might equivalently capture the physics while preserving the Euclidean space that many ML algorithms require.

18.1.2   Similarity/Dissimilarity

Many ML algorithms require precomputing the similarity or dissimilarity between each data observation. Here again, prior knowledge can be employed to define a dissimilarity (similarity) measure that quantifies the data features most relevant to the later data analysis and ignores those features that may confound later analysis. Without the use of a physics-informed measure, the resulting data may be so confounded by an extraneous signal that successful further analysis becomes highly unlikely. For example, if one is interested in computing the dissimilarity (similarity) between XRD and is most interested in the presence of peaks but not their intensities (as the primary indication of a phase are peak positions), one can use a measure that focuses on these features such as the cosine measure (or 1-cosine measure for similarity) which is scale invariant.

18.1.3   Analysis

For the ML data analysis step, the use of prior knowledge is again key. Here we can again point to the example of constituent phase XRD determination from a collection of XRD. Without the proper use of prior knowledge, the identified constituent XRDs and abundances may be non-physical and not useful for the APS system. For example, if simple matrix factorization is used, resulting in negative abundances, subsequent steps for extrapolation and experiment design will similarly be poisoned by negative abundances.

18.1.4 Extrapolation

When extrapolating from experiments with data to those without, it is important to incorporate prior knowledge of the relationship between these experiments. One common assumption is that if two experiments are performed with similar input parameters, their outputs are more likely to be similar. This statistical requirement can be imposed through a probabilistic graph constraint or the assumptions of a Gaussian process [570]. If using a Gaussian process, the user can impose certain assumptions of the behavior of the unknown function, including smoothness, variability in intensity, periodicity, etc.

18.1.5 Experiment Design

The selection of subsequent experiments can be greatly benefited by prior physical knowledge. As mentioned, the experiment search space tends to be vast and complex. With prior physical knowledge one can narrow in on relevant portions of the search space for the desired goals. For example, knowing that local and absolute optima may cluster together could direct subsequent experiments to search near prior local optima. Or alternatively, if one assumes that the experiment with the global optimum is likely to be far from local optima, one could use this knowledge to guide subsequent experiments a certain distance away from known optima. These prior knowledge-based modifications tend to be more easily made as the experiment design utility function is often a single function. Concrete examples are given in the following three APS examples.

18.1.6 Tuning the Pipeline

Once a set of ML algorithms are established, these algorithms often have hyperparameters that can be tuned to improve performance. Here the goal is to optimize cumulative performance—the ability for the APS system to more rapidly hone in those experiments that satisfy the user-defined goals. Prior knowledge of experiment behavior can be used to define how best to tune these various hyperparameters. For example, as data is collected, a prior model of the background signal may be refined, prior knowledge of extrapolation smoothness can be used to adjust bounds over extrapolation smoothness, and assumptions of the hidden function can guide subsequent tunning of the experiment design parameter associated with a step size.

18.2 EXAMPLES OF LIVE AUTONOMOUS PHYSICAL SCIENCE EXPERIMENTS

18.2.1 CAMEO

CAMEO [248] is an APS ML framework that combines two goals: learning the material synthesis-structure-property relationship and accelerating materials optimization and discovery. CAMEO was run live in control of a high-throughput X-ray diffraction system at the Stanford Synchrotron Radiation Lightsource. First a composition spread [180]—a wafer containing a set of material samples with varying composition within a defined material system—was loaded into the high-throughput XRD characterization system. CAMEO first directs XRD measurements for materials that are informative of the synthesis-structure relationship given by the phase map.

The ML pipeline built toward this goal combines both bespoke and off-the-shelf algorithms. For CAMEO's phase mapping operation:

Preprocessing. The XRD collected contains slowly varying background signals as a function of q and these background signals vary from material to material. Such variation in background signal can confound further analysis. Knowledge that the background signal is slowly varying relative to the XRD signal guided the selection of a preprocessing algorithm that identifies such slowly varying signals. Here a spline fit is used to identify the envelope of the signal and the the background is then subtracted.

Analysis. Analysis of the collected XRD then uses a bespoke algorithm for identifying potential phase regions and phase boundaries. The set of prior knowledge assumptions and how they are imposed is given in Figure 18.3. These assumptions include the Gibbs phase rule and the contiguous nature of phase regions, among others. Integrating the prior knowledge includes a collection of hard and soft constraints

Algorithm	Physical knowledge	Encoding Method
Data Analysis		
CAMEO Phase-mapping	Phase regions are contiguous and phase boundaries are continuous	1. Construct a graph of locations in the composition space with a Delaunay triangulation (which enforces no overlapping edges). Use a graph cut to assign the labels to the nodes. If two or more sets of vertices share the same phase region label but are not connected by vertex neighbors, differing labels are assigned to the disconnected sets. 2. The Markov Random Field smoothness constraint
	Materials of similar synthesis and processing parameters have similar properties	1.Markov Random Field smoothness constraint 2.Harmonic Energy Minimization for label propagation
	Abundances of phases is non-negative	Karush–Kuhn–Tucker conditions
	X-ray diffraction intensity is non-negative	Karush–Kuhn–Tucker conditions
	Soft Gibbs Phase Rule - Upper bound limit on number of constituent phases	Upper limit on number of endmember limits allowed in each phase region
	Identified endmembers should be physically realizable	Volume constraint on identified / predicted endmembers
Phase-mapping Prior	DFT phase map is predictive of bulk phase diagram. Structure is a good predictor of functional property and vice versa	Bayesian prior through similarity kernel For more information M1c Phase Mapping: Phase mapping prior.
Knowledge Propagation		
HEM	Phase regions are cohesive. Quantified likelihood for each sample belonging to each phase region due proximity in composition	Graph representation of composition space. Label propagation through graph. Labels uncertainty propagation.
Active Learning		
Risk Minimization	Each sample quantified for its potential impact on improving total phase map performance. Targets phase boundaries.	Minimize total phase region misclassification probability for the entire phase map.

Figure 18.3 Examples of algorithms considering physical knowledge.

as well as the use of a probabilistic graph representation for the composition-structure relationship.

Extrapolation. The next step in the ML pipeline is extrapolating knowledge gained for the materials that have been characterized for XRD to those that have not been characterized. Here again, a probabilistic graph representation of the materials is used to impose the prior knowledge that materials of similar composition are more likely to have similar properties. From this graph structure, the likelihood of each material belonging to each phase region is determined.

Experiment Design. Risk minimization - an off-the-shelf active learning algorithm for quantifying the utility of every potential subsequent measurement experiment is used. Risk minimization uses the estimates and uncertainties of the knowledge extrapolation step to identify the material most likely to reduce total error in classifying materials to the wrong phase regions.

The autonomous phase mapping ML pipeline was investigated for the impact of incorporating prior physical knowledge. Algorithms with different levels of prior physical knowledge were used for analysis, extrapolation, and decision-making. The more prior physics included, the greater the performance. Interestingly the portion of the ML pipeline with the largest impact on performance was the decision-making algorithm.

For the materials optimization and discovery, data preprocessing and analysis was unnecessary since the value obtained by the system was a scalar:

Extrapolation. Here a bespoke algorithm is used to impose the prior knowledge that a material's properties are tied to its lattice structure. A material property can have different composition dependence in one phase region than another, and these properties can have discontinuities at the phase boundaries. Thus, the property is modeled as a piecewise Gaussian process with potential discontinuities at phase boundaries and different composition-based trends within each phase region.

Experiment Design. A bespoke active learning utility function is used here. The phase region, most likely to contain an optimal material, is selected. The off-the-shelf Gaussian process upper confidence bounds algorithm is then modified to allow for greater importance for points either near phase boundaries or near the center of the phase region.

18.2.2 ANDiE

The autonomous neutron diffraction explorer (ANDiE) was used in live control over neutron diffraction experiments at both the National Institute of Standards and Technology's Center for Neutron Research and at the Oak Ridge National Laboratory's High Flux Isotope Reactor. The aim of ANDiE is to identify the temperature-structure behavior and relationship of a magnetic material (Figure 18.4). Of particular interest is identifying which of the known prototypical temperature-structure behavior the material should be classified under - first-order, Ising-type second-order, or Weiss-type second-order. Toward this goal, ANDiE also identified the critical parameter of the Neel temperature—the temperature at which magnetic order disappears, i.e., the magnetic structure dissipates. ANDiE was used to guide neutron diffraction

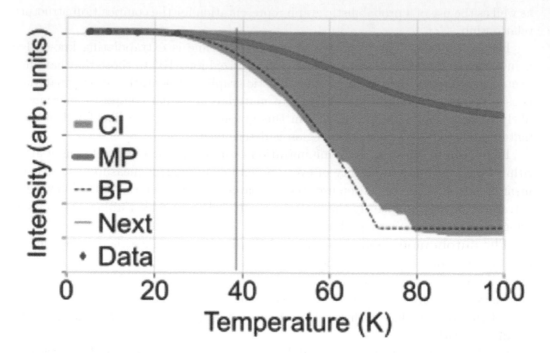

Figure 18.4 The temperature dependence of the magnetic component of the neutron diffraction intensity, for $Fe_{1.09}Te$ measured during an autonomous run using the WAND [352] instrument. The blue dots show the measured values with error bars. Additionally plotted: the confidence interval (CI) of the model, mean of the posterior (MP) distribution of the model, the model using the best (i.e., most likely) parameters (BP). The green vertical line shows the next temperature to perform the measurement, where the model uncertainty relative to the estimated measurement uncertainty exceeds a threshold.

angle to maximize knowledge of the magnetic and lattice structure as well as guide temperature of the experiments to maximize knowledge of the material's temperature-magnetic behavior. Here we focus on the latter autonomous ML pipeline.

Preprocess. In this step, the goal is to combine prior physical knowledge with neutron scattering data to compute a posterior probability for the magnetic peak intensity at a given experiment temperature. Prior knowledge includes: (1) input detector image data can be used to compute the statistics of the diffraction response (mean and distribution profile parameters) as a function of 2θ, (2) the target 2θ range is likely to contain 1 lattice peak and 1 magnetic peak. This knowledge is encoded using a Bayesian inference probabilistic programming framework. This framework allows one to define a physical model along with probabilistic priors, and then combine these with given data to compute a posterior probability for target parameters—in our case the magnetic peak intensity.

Analysis and Extrapolation. Magnetic peak intensity as a function of temperature is again input to a Bayesian inference model of the temperature-magnetic response. The output is a posterior over the model parameters. The statistics over model parameters are then used to extrapolate the given physical model to higher temperatures and provide a posterior for the Neel temperature. Analysis and extrapolation is guided by one of the three models—the Weiss model. The Weiss model has the flexibility to approximately represent the other three models. The Weiss model's more gradual slope toward the Neel temperature compared to the first-order model's discontinuity allows for more information to be gathered near the Neel temperature.

Decision Making. Here, we employ prior knowledge that the signal uncertainty has a lower bound of the measurement noise. The intensity of the signal is fundamentally a counting process and will therefore follow Poissonian counting statistics. From this, we know that the standard deviation is related to the square root of the measured intensity. Since the magnetic ordering can be hysteretic with temperature, we are constrained to only increase temperature. As we extrapolate from the current experiment temperature to higher temperatures, we can compare the model extrapolation uncertainty to the estimated measurement uncertainty. We can then choose the next experiment where the model uncertainty exceeds a threshold relative to the measurement uncertainty.

18.2.3 Low-Cost Autonomous Scientist: Educating the Next Generation Workforce

The Low-Cost Autonomous Scientist [456] allows students to explore the various aspects of APS with hands-on exercises. They can explore the physical skills of materials synthesis and characterization, the control systems challenges of managing automated equipment, the data science skills of data handling as well as analysis, extrapolation, and decision-making, and the statistics of uncertainty quantification and propagation.

Physics-informed exercises:

Exercise 1: Surrogate Model. In this exercise, the students are told that a Gaussian process with a radial basis function kernel will provide an adequate fit to the relationship. They are then asked to write an active learning loop that iteratively performs experiments with the goals of (1) maximizing knowledge of the relationship, or (2) finding the acid/base ratio that provides a target pH value.

The GP provides both analysis and extrapolation. For decision-making, toward the first challenge, the students use common exploration - setting the utility function to the output variance computed by the GP. For the second challenge, the common Upper Confidence Bounds algorithm is used.

Exercise 2: Model and Parameter Determination. For this exercise, students are provided a list of models that may describe the target relationship. They are asked to use Bayesian inference to (1) Determine the most likely model, and (2) For that model, determine the most likely parameter values. Here the prior knowledge is that the target model is one of a set of models.

For each iteration, analysis, and extrapolation is performed by fitting the provided models. Experiment design is performed by identifying the experiment that maximizes disagreement between the models. This is performed by quantifying the probabilistic entropy between the given models.

Exercise 3: Model and Parameter Discovery. For this exercise, the students are given the assumption of Occam's Razor—that low-complexity models are preferable to those more complex. They are asked to balance model fit error with model complexity.

Students use symbolic regression to fit the data, identifying the 5 models with the best pareto front score which balances fit error with model complexity. These models are then used for analysis and extrapolation. Students again use entropy across these models to determine the next experiment to perform.

18.3 CONCLUSION

As the field of APS continues to grow and tackle an ever-growing list of challenges, novel ML pipelines will be needed to handle the growing diversity of autonomous materials synthesis and characterization methods and the growing diversity of material systems, each described by a a different set of physical laws. By incorporating these physical laws, next-generation APS ML pipelines will accelerate materials exploration and discovery, bringing into reach the optimization of complex materials.

IV

A Guide through Autonomous Experimentation

CHAPTER 19

A Closed Loop of Diverse Disciplines

Marcus M. Noack

Applied Mathematics and Computational Research Division, Lawrence Berkeley National Laboratory, Berkeley, California, USA

Kevin G. Yager

Center for Functional Nanomaterials, Brookhaven National Laboratory, Upton, New York, USA

CONTENTS

19.1 INTRODUCTION

A UTONOMOUS experiment is a young field of research and it would be premature to define it in one narrow way or another. A single, correct definition might not exist since the term itself has different meanings in different communities. This was the main reason to offer the reader three separate introductions to autonomous experimentation instead of one, at the beginning of this book (Chapters 1, 2, 3). Autonomous experiments are rarely developed, planned, and executed by a single person or a small team, but instead by large, multi-disciplinary teams. Of course, inherent to the word "autonomous" is the ability to avoid relying on human input in the execution phase altogether. Figure 19.1 alludes to the reason why autonomous experimentation is such a diverse field of research; autonomy relies on all aspects of a loop to be robustly automated.

At the top of the loop, we have an instrument. It is either an experimental apparatus—an X-ray scattering beamline, a spectrometer, or perhaps an electron microscope—or a computer, acting as an operator to create outputs from inputs. Not only does the instrument have to be in good working order but interactions have to happen entirely electronically via electronic messages; no manual levers or knobs should be necessary for its operation. For this work, capable **instrument scientists** are needed to ensure that the process from receiving a command to communicating output raw data is robust and needs no human supervision. In a field where the idea of autonomous experimentation is very new, this first hurdle is often the greatest and the hardest to standardize. Unfortunately, this also means that there is no easy

DOI: 10.1201/9781003359593-19

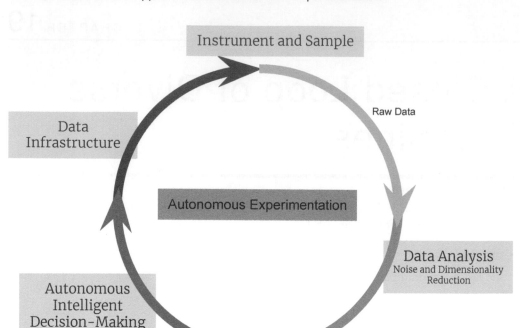

Figure 19.1 A simplified view of the autonomous loop. As data leaves the instrument, it has to be analyzed, de-noised, and dimensionality-reduced; this can include the selection of features. The reduced data are then fed into an autonomous decision-making engine, which suggests optimal new measurements to the instrument, closing the loop. The three components are linked by a robust communication infrastructure.

way to suggest a blanket strategy that turns an instrument autonomous-ready, which will depend on the type of instrument and the vendor. Many instruments come with proprietary software which makes low-level, hands-on practical work very difficult.

As raw data leaves the instrument, they are often comprised of high-dimensional structures—images, films, spectra—and contaminated with noise. Therefore, as a next step, the data have to be translated into a format that allows interpretation and decision-making—we will refer to this step broadly as **data analysis**—this step often includes a whole de-noising, feature selection, and dimensionality-reduction pipeline. Different communities have developed a wide variety of different workflows to accomplish this task. If the raw data are noisy spectra, for instance, they are often de-noised by various filters and then dimensionality-reduced by principle-component analysis or non-negative matrix factorization. Detector images originating from small-angle X-ray scattering are traditionally radially integrated and then peak heights and positions in the remaining one-dimensional signal are identified. One-dimensional signals can sometimes be clustered. More recently, neural networks have increasingly been used to derive low-dimensional embeddings of raw data.

After the analysis step, **intelligent autonomous decision-making** based on all collected observations can take place, and again, there is a variety of methods to choose from. If many similar experiments have been run in the past, we can train reinforcement learning algorithms to make optimal choices about experiment designs. If learning has to happen on the fly (or online), stochastic processes—especially Gaussian stochastic processes (see Chapter 4)—are often used to obtain a surrogate model together with associated uncertainties to create an acquisition functional. The acquisition function(al) assigns a value—in terms of a benefit—to every future measurement at the current state of knowledge. Constrained function optimization can then be used to find the most "valuable" next measurement, which is then communicated to the instrument, closing the loop.

These different building blocks of the autonomous loop have to be connected via a **robust data and communication infrastructure**. This becomes particularly important when some computations are run on remote computing architectures and therefore, steps have to be performed asynchronously.

It is this diversity of disciplines that makes autonomous experimentation so difficult to accomplish and serves as motivation to write this part of the book as a hands-on guide. The authors give examples of how to perform the various steps and present some pitfalls. The hope is that this part will serve as a guide for practitioners aiming to turn their experiment autonomous, starting with an instrument that can receive and automatically execute electronic messages.

CHAPTER 20

Analysis of Raw Data

Marcus M. Noack

Applied Mathematics and Computational Research Division, Lawrence Berkeley National Laboratory, Berkeley, California, USA

Kevin G. Yager

Center for Functional Nanomaterials, Brookhaven National Laboratory, Upton, New York, USA

CONTENTS

D ATA ANALYSIS is often the first step to be performed after raw data leave an instrument. The raw data are usually high-dimensional—dimensionality here refers to the number of outputs of one measurement—and noisy. Spectra, for instance, are often stored as some 1,000 real numbers. Decision-making, especially driven by uncertainty quantification, is infeasible on these large outputs which motivates the use of dimensionality-reduction techniques to extract a handful of characteristic values. In this chapter, we want to introduce a few tools that can be used for this task. This list is by no means complete; it is merely a compilation of tools that are used with some regularity across experimental facilities. We also include hand-crafted feature selection methods here; although they are not considered mainstream dimensionality-reduction tools, they can be very powerful in reducing the amount of data because the user can create them with their full domain expertise in mind.

DOI: 10.1201/9781003359593-20

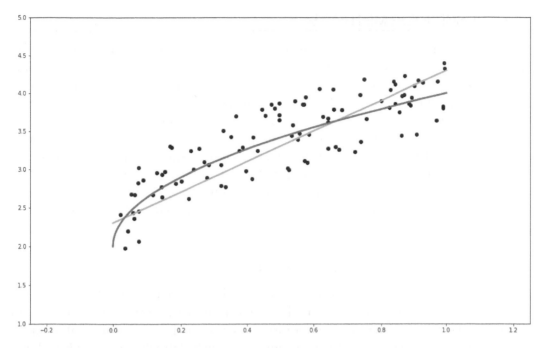

Figure 20.1 The basic idea behind linear and non-linear dimensionality reduction. Data is defined in a two-dimensional (embedding) space but can be well-approximated as just one-dimensional coordinates along an embedded line (linear, orange) or (non-linear, red) embedded curve. This could now easily be confused with regression; however, we are not looking for function values here but a way to represent data in low-dimensional spaces. In contrast to regression, the red curve could turn over itself, creating a closed curve.

Every practitioner who wants to perform dimensionality reduction has to make decisions on whether to select certain dimensions—called features by the machine-learning community—to project data onto a lower-dimensional space embedded in the high-dimensional space or to create an entirely new domain. In the case of embeddings, the lower-dimensional space can be assumed to be linear or non-linear. Of course, in the interest of generality, one could always assume non-linearity and discover linearity; however, non-linear dimensionality reduction is much more cumbersome, so it often proves useful to make a pre-selection. See Figure 20.1 for a visualization of the distinction between non-linear and linear embeddings. The third way to do dimensionality reduction operates without working with the original features directly but to create new descriptors. Examples of these kinds of algorithms are classification, clustering, and other operators whose output is not a subset of the original input space—we will loosely call these methods "descriptor methods"; they may not be considered mainstream dimensionality reduction methods. We include them here because of their practical purpose. The earlier-mentioned hand-crafted feature selection methods fall into this category.

20.1 DESCRIPTOR METHODS

Methods that do not assume some embedding of a lower-dimensional space in the original high-dimensional input space, but instead transform data points into a new domain we refer here to as descriptor methods. This is merely an umbrella term to be able to present some methods of data (dimensionality) reduction without being held back by definitions. These methods do find a lower-dimensional representation; it is just not a subset of the original input space. For instance, for a spectrum, one could just extract the average intensity and classify that quantity as A or B based on a threshold. The data amount has been reduced but this would not count formally as dimensionality reduction. These methods are often designed by the practitioner in order to include a wealth of domain knowledge; which makes these methods quite powerful and robust while often simple to implement.

20.1.1 Integrations and Point (Set) Evaluations

One of the simplest but also most robust ways of dimensionality reduction is through integrations, evaluations at points or slices, and peak-fitting procedures; robustness is one of the very important characteristics of the analysis step for autonomous experimentation—we would hate for the analysis step to fail when nobody is looking. The methods above are very commonly used for scattering data. Given a small-angle x-ray scattering detector image, for instance, we can radially integrate to obtain a one-dimensional function of intensity over radius (in Fourier space). We can subsequently find the intensity peak by simple optimization or a fitting procedure. The height or position of this peak is often used as a quantity of interest for autonomous experimentation. There are packages available online that will perform peak fitting, most notably the *SciAnalysis* package [251] and *MLExchange* [605].

20.1.2 Comparing to Theoretical Methods

Often, data coming from the instrument can be simulated, which allows us to perform dimensionality reduction by simply comparing the experimental result to the simulation result, reducing the dimensionality to one. However, simulation codes are often hard to come by, and running them can be costly. Grazing-incidence small-angle x-ray scattering data can, for instance, be modeled by the HipGisax tool [100, 467]. Simple distance metrics can then be used for comparisons.

20.1.3 Clustering

For many kinds of data structures, clustering can be an insightful tool for dimensionality reduction. As the name would suggest, clustering tries to assign data points to a given number of clusters. The labels of these clusters or distances thereof represent the dimensionality-reduced quantity. One of the most popular clustering algorithms is k-means clustering, which reduces the problem to a function optimization in which the average distance of data points to cluster centers is minimized. However, there are many more algorithms and methods for clustering to consider in practical applications. Python's *scikit-learn* has a great clustering library. Clustering works best for one-dimensional raw data, like spectra. Sometimes images can be clustered successfully as well.

20.1.4 Classification

Whenever data points are supposed to be assigned to different classes, based on their location in the input or some latent space, classification algorithms are the methods of choice. One could think of clustering as a kind of classification also, but classification is supervised (uses labeled data), while clustering is unsupervised. Classification algorithms can be divided into binary classification and multi-label classification. Most binary classification algorithms can, however, be used for multi-label classification by dividing all classes into two groups, one containing just one class, and one containing all the other classes. This can be done repeatedly to yield all classes.

20.1.5 Feature Selection

In the purest form of the phrase, feature selection means that an algorithm extracts user-defined features from the raw data. This gives the practitioner the chance to include the full, available set of domain knowledge. This is not an entirely new category and some of the already mentioned methods fall into this category as well. This method can be very powerful because it often only needs inexpensive calculations while offering powerful insights into raw data. This method is preferred when a wealth of domain knowledge would otherwise be difficult to include in a machine-learning-driven approach.

20.2 LINEAR DIMENSIONALITY REDUCTION

Linear dimensionality reduction methods are most often matrix factorization techniques. The idea behind all matrix factorization algorithms is that a data matrix $\mathbf{X} \in \mathbb{R}^{m \times n}$ can be approximated by $\tilde{\mathbf{X}} \in \mathbb{R}^{m \times l << n}$, where m is the number of data points (spectra, images) and n and l are the dimensions of the original data and the reduced data respectively.

20.2.1 Principal Component Analysis

Principal Component Analysis is one of the most common techniques for matrix factorization. PCA finds a new basis that allows the data to be represented in a lower-dimensional space with high accuracy. It can be interpreted as finding the eigenspace of the covariance matrix of the data or fitting a hyper-ellipsoid to the data. The result is the same: a lower-dimensional representation of high-dimensional data. See Figure 20.2 for a visual representation of PCA. PCA is the go-to technique for analyzing spectra.

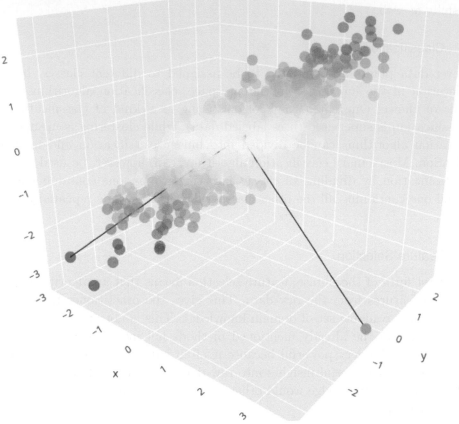

Figure 20.2 The principle of Principal Component Analysis (PCA). The algorithm defines a new orthogonal basis along the principal components of the data. In this new basis, every original data point has a lower-dimensional representation.

20.2.2 Non-Negative Matrix Factorization

Non-negative matrix factorization (NMF) assumes that the data matrix can be factorized as $\mathbf{X} = \mathbf{WH}$, where $\mathbf{W}$ and $\mathbf{H}$ are (hopefully) of much lower dimensionality than the matrix $\mathbf{X}$. NMF factorization is often preferred when data is naturally positive and will constrain the result to have the same property. NMF can be understood as identifying a set of positive components that can be added to give the measured signal.

20.3 NON-LINEAR DIMENSIONALITY REDUCTION

PCA and MNF are only able to find lower-dimensional approximations for data that are distributed linearly. If the structure of the data cannot be linearly approximated accurately, non-linear dimensionality reduction techniques are preferred. Often non-linear dimensionality-reduction can be understood as transforming the data non-linearly and then applying the former linear dimensionality-reduction technique to the data in that new space. Support Vector Machines (SVMs), for instance, are sometimes considered dimensionality reduction techniques, although they are typical classification algorithms; nevertheless, they can deal with non-linearity by transforming the dataset first and then acting in the transformed data. For large datasets, Uniform Manifold Approximation and Projection (UMAP), t-distributed stochastic neighbor embeddings, and neural networks are the preferred methods. For smaller datasets, kernel PCA and the Gaussian process latent variable model (GPlvm) [254, 524]—especially when uncertainties are needed—and its approximations are unrivaled.

20.3.1   T-Distributed Stochastic Neighbor Embedding

T-distributed stochastic neighbor embedding (t-SNE) is based on pairwise Gaussian joint distributions of points in the input domain p_{ij} and pairwise student-t-distributions q_{ij} of points in the output domain. The rest comes down to minimizing the sum of the Kullback-Leibler divergences between the resulting distributions. The simplest-to-use implementation is available as part of Python's *scikit-learn*. t-SNE does not scale well with dataset size and should therefore be preferred for small datasets.

20.3.2   Uniform Manifold Approximation and Projection

UMAP is based on constructing a weighted graph—a network of point-connecting edges based on fuzzy logic and non-constant Riemann metrics [313]. According to the authors of [313], UMAP "preserves more of the global structure with superior run-time performance" compared with t-SNE. We refer the reader to [313] for more information. UMAP might currently be the leading non-linear dimensionality-reduction algorithm for large datasets.

20.3.3   Autoencoders

Autoencoders are dimensionality-reduction algorithms based on neural networks. Their architecture is comprised of an encoder, a bottleneck, and a decoder. The encoder reduces the data from a high-dimensional to a low-dimensional space (the bottleneck). The decoder transforms the low-dimensional representation back to its original space, where it is compared to the input. Autoencoders are trained by varying the network architecture while minimizing the difference between the original input and the reconstructed set. When the autoencoder is trained, we can ignore the decoder and end up with a way to encode high-dimensional data in a lower-dimensional structure. Autoencoders are especially well known to encode images but can be used for most kinds of data.

20.3.4 Kernel PCA

Kernel PCA is essentially PCA in a non-linear feature space. Input points are transformed into latent point positions by applying the non-linear map $\mathbf{\Phi} : \mathbf{x} \rightarrow \mathbf{\Phi}(\mathbf{x})$ [473, 523]. Because we are never working in the feature space itself, kernel PCA does not yield the principal components of the new covariance matrix itself, but rather projections of the data onto those components. $\mathbf{\Phi}$ is never explicitly calculated and only accessed through an inner product, drawing parallels to other kernel methods.

20.3.5 gpLVM

The PCA algorithm we mentioned above has at least two probabilistic interpretations. In one of these interpretations—the so-called dual probabilistic PCA—we are optimizing a likelihood that, in essence, only depends on the inner product of the matrix of dimensionality-reduced points with itself $\mathbf{X}\mathbf{X}^T$ [254]

$$L = -\frac{DN}{2}ln(2\pi) - \frac{D}{2}ln(|\mathbf{X}\mathbf{X}^T + \sigma I|) - \frac{1}{2}trace((\mathbf{X}\mathbf{X}^T + \sigma I)^{-1}\mathbf{Y}\mathbf{Y}^T), \quad (20.1)$$

where $\mathbf{Y}$ is the data matrix, D is the dimensionality of the input space, and N is the number of input points. Optimizing 20.1 is a convex problem with a unique solution equivalent to PCA. As we learned earlier, PCA is only able to learn linear embeddings. However, the machinery of kernels can be used to expand PCA's capabilities to allow for non-linear embeddings to be learned. For that, we introduce the non-linear map $\mathbf{\Phi} : \mathbf{x} \rightarrow \mathbf{\Phi}(\mathbf{x})$, which leads to inner products $\langle \mathbf{\Phi}(\mathbf{X}), \mathbf{\Phi}(\mathbf{X})^T = \mathbf{K}$. In the case of gpLVM, we switch out the term $\mathbf{X}\mathbf{X}^T$ for the kernel matrix $\mathbf{K}$. gpLVM is a sophisticated technique that might lead to the most accurate non-linear embedding of any of the presented methods, but it does not scale well in its native form and is therefore recommended for small-to-moderate datasets.

CHAPTER 21

Autonomous Intelligent Decision Making

Marcus M. Noack

Applied Mathematics and Computational Research Division, Lawrence Berkeley National Laboratory, Berkeley, California, USA

Kevin G. Yager

Center for Functional Nanomaterials, Brookhaven National Laboratory, Upton, New York, USA

CONTENTS

A UTONOMOUS intelligent decision-making might just be the core of the autonomous loop and certainly represents the difference between automation and autonomy. While automation means no human interference is required for the experiment execution, autonomy implies a sense of artificial intelligence that is adapting to what is being learned during the data acquisition and as a result, is changing strategies throughout the experiment.

21.1   OVERVIEW OF METHOD CATEGORIES

Overall, we want to divide methods for autonomous intelligent decision-making into three sub-categories. First, classical methods use very robust and understandable local properties of the underlying function to make decisions. Often these methods are derivative-driven; in other words, data is collected where local gradients or curvatures are high. For instance, in one dimension, one can increase sampling frequency when the measured quantity started changing rapidly from one measurement to the next. Second, agent methods use past experiments to learn what best actions to perform in the current experiment. In this category, reinforcement learning is popular. And

DOI: 10.1201/9781003359593-21

third, greedy methods that update a model in every iteration, update their knowledge and use optimization to make decisions. This last category largely contains statistical methods that lead to uncertainty estimates that help the decision-making process. The Gaussian process is the most popular choice in this category. There are some other methods in this category, such as Voronoi surrogate modeling techniques [194]. Agent methods traditionally need training data from past experiments. Often, however, an experiment might be assumed to be unrelated to past ones in order to prevent bias. Even if only applied to the data from the current experiment, reinforcement learning is often based on neural networks, which need dense training data (a lot compared to the size of the space), an assumption often not satisfied in standard autonomous experimentation settings. We will focus our efforts on stochastic methods in this chapter. Since this is a guide, starting off with greedy stochastic methods is a great learning experience and leaves the door open to trying alternatives later. One more thing to consider is the timescales at which decision-making should occur. Stochastic methods can be costly requiring decision times of a small fraction of a second, which might mean other methods should be used. Agent methods can be very fast once trained.

21.2   A QUICK REVIEW OF GAUSSIAN PROCESSES

The idea behind Gaussian Processes (GPs) is to think of function values of a model to be jointly normally distributed. This is described in detail in Chapter 4. Learning a normal prior distribution and conditioning leads to a posterior probability distribution, which is understood as a distribution of the function values at unobserved places in the input domain. It is easy to see that this distribution is a very important and useful object for making decisions on where more data should be collected. The main difficulty in applying GPs for any task, but especially autonomous experimentation, is the wide variety of customizations for GPs and how the posterior is used for decision-making. In the next few sections, we want to give some practical tips when it comes to setting up a GP for an autonomous experiment.

21.3   SETTING UP GPS FOR AUTONOMOUS EXPERIMENTATION

In most situations we come across when attempting an autonomous experiment, we have a clear understanding of what parameters can be changed, which directly translates into the parameter space we want to explore. That space can be made up of motor positions, temperatures, atmospheric pressure, fluid ratios, and so on. The number of these parameters is at the same time the dimensionality of the parameter space. The question we should ask ourselves is what function value we want to model over the input space. This so-called quantity of interest (QoI) is often extracted from the raw data in every step before the decision-making happens. Often the quantity of interest is a scalar quantity but can also be vector-valued. In that case, multi-task GPs should be used for decision-making. We will assume that we are after a scalar quantity for now. The extension to multiple QoIs is straightforward with the exception of the kernel design, which needs some more thought in the multi-task case.

In what follows, we want to set up a GP for autonomous, intelligent decision-making step by step. Note that what follows is just pseudocode, and will change based on the particular API that is used.

Let's assume a simple example in which we want to intelligently decide some motor positions (x, y) of an x-ray scattering beamline. The input dimension is two, and we are after the Bragg-peak position as our quantity of interest. This translates into

```
1  #set up the space
2  bounds = array([[0,1],[0,1]])
3
4  #create initial points randomly drawn from the input domain
5  x_data = random(bounds)
6
7  #ask the instrument for the QoI at initial points
8  y_data, variances = call_instrument(x_data)
9
10 my_gp = GP(2,x_data,y_data,variances) #setting up the GP
```

Listing 21.1: Setting up a Gaussian Process.

The call_instrument function is assumed to eat parameters (points), move the motors, collect raw data, analyze the raw data, and return the QoI; needless to say, a lot of work goes into setting up this function. We now have a GP that can be trained by a variety of methods.

```
1  my_gp.train(method = ...)
```

Listing 21.2: Training the GP.

Now we assume the intuitive situation in which areas in the domain with high uncertainties are indeed good places for data acquisition. All GP APIs will have a method to query the posterior at a certain point in the parameter space. How we use the point-wise output of a GP to inform decision-making is encoded in the acquisition function, which can be arbitrarily complicated. Again, in this case, we make it easy on ourselves by using the posterior variance.

```
1  var = my_gp.posterior_variance(x_pred)
```

Listing 21.3: Querying the posterior variance.

But of course, what we are looking for is not the posterior variance (uncertainty) at a point but the point of maximum uncertainty. It is common, but often suboptimal, to calculate the posterior variance on a grid and then select the grid point with the highest variance. A much more accurate and at the same time often faster approach is to use function optimizers. The *gpCAM* library, for instance, implements

```
1  best_point = my_gp.ask(acquisiton_function = 'variance')
```

Listing 21.4: A simple setup of intelligent decision making inside a loop.

to help users make the optimization easier. Of course, at this point, all we have done is ask for one new point given the initial data. Organizing the steps above in a loop, and appending the newly collected data to the dataset, results in a working autonomous loop.

```
1  #set up the space
2  bounds = array([[0,1],[0,1]])
3
4  #create initial points randomly drawn from the input domain
5  x_data = random(bounds)
6
7  #ask the instrument for the QoI at initial points
8  y_data, variances = call_instrument(x_data)
9
10 while break condition not triggered:
11     my_gp = GP(2,x_data_y_data,variances) #setting up the GP
12     my_gp.train(method = ...)
13     best_point = my_gp.ask(acquisiton_function = 'variance')
14     best_point_return, variance = call_instrument(best_point)
15     x_data = append(x_data,best_point)
16     y_data = append(y_data,best_point_return)
17     variances = append(variances, variance)
```

Listing 21.5: A simple setup of intelligent decision-making inside a loop.

This code is more symbolic than realistic and many APIs will have methods that execute this loop for the user. The "call instrument" function usually contains all the complexities of a real experiment, from the motor control and raw-data collection to analysis and dimensionality reduction. This basic loop has several important building blocks that should be customized for optimal performance of the experiment.

21.4   ACQUISITION FUNCTIONS

For anything beyond very standard situations, more sophisticated acquisition functions than the posterior variance have to be used. We learned earlier in this book that we are actually misusing the term "acquisition function" and *acquisition functional* is actually the correct term because we are speaking of a function eating the posterior (a function itself) and spitting out a scalar. This is the only requirement for the acquisition functional and the customization opportunities are therefore endless. We can, for instance, formulate acquisition functionals that focus on high probabilities that the model function is within a user-defined interval. The function

$$f_a(\mathbf{x}) = p(-\infty \le f(\mathbf{x}) \le b) = \frac{1}{\sigma\sqrt{2\pi}} \int_{-\infty}^{b} e^{\frac{f(\mathbf{x})-m(\mathbf{x})^2}{2\sigma^2}} df = 0.5\left(1 + \text{erf}\left(\frac{b - m(\mathbf{x})}{\sigma\sqrt{2}}\right)\right),$$

$$(21.1)$$

for instance, focuses on regions where the probability that the function value is between $-\infty$ and b. Of course, at the beginning of every experiment, the probability will be quite high everywhere, which leads to a nice transition of exploration to exploitation without any user interference. Many other acquisition function(al)s are known and a complete list is out of the scope of this chapter. The most important ones beyond those already mentioned here include mutual information, lower and upper confidence bound, knowledge gradient, and expected improvement [151].

21.5   KERNELS

Kernels are one of the most important building blocks of GPs—that is why GPs are also categorized as kernel methods. In the GP framework, kernels are considered covariance functions because the predicted covariance of two function values $f(x_1)$ and $f(x_2)$ is approximated by calling the kernel $k(x_1, x_2)$. In other words, kernels give us the power to say something about the similarity of any two function values without having to evaluate them. Stationary kernels do so by simply looking at the distance between two points, disregarding the respective positions of the points. Non-stationary kernels depend on the positions explicitly. Kernels affect everything from the flexibility and accuracy of the function approximation and uncertainty quantification to domain awareness of the predictions. It is shocking, given this importance, that more than nine out of ten studies using GPs are conducted using the squared-exponential kernel [407]. While we can't even get close to a comprehensive description and explanation of kernel designs here, we can offer three suggestions. (1) The exponential kernel is rarely the appropriate choice, with the exception of clearly non-differential and very simple model functions. Those happen mostly for imaging applications. Models arising from physics are almost always differentiable at least once. (2) The squared-exponential kernel is too smooth because it constrains the model to infinite-order differentiability which is unrealistic in most circumstances. (3) Strongly consider some kind of non-stationary kernel; they model uncertainty quantification much better than stationary kernels. For autonomous experimentation, stationary kernels lead to a sophisticated space-filling operation. A note on a common misconception about Gaussian processes. The "Gaussian" in the name has nothing to do with the shape of the kernel. The name comes from the normal distribution over functions, the kernel can be any symmetric positive semi-definite function.

21.6   NOISE

The accurate treatment of noise is one of the strengths of Bayesian methods such as the GP. However; that is only true if noise is either accurately estimated during the measurement process or the GPs have enough flexibility to estimate the noise. One of the most important rules for using GPs is to not set the noise to zero. For autonomous experimentation, setting the noise to any constant across the domain can render any intelligent decision-making less efficient than even random point choice. Either let the GP find its own parametric noise model—as simple as a constant value across the domain—or communicate measured noise to the GP whenever available (See Figure 21.1).

21.7   SUMMARY

The beauty of GP-driven autonomous experimentation is that it is very interpretable and ranges in complexity from very simple and ready in minutes to very complex with customized kernel functions, noise models, prior mean functions, and acquisition functionals. This flexibility and unlimited scalability of complexity render GPs one of the best choices to make autonomous intelligent decisions during an experiment.

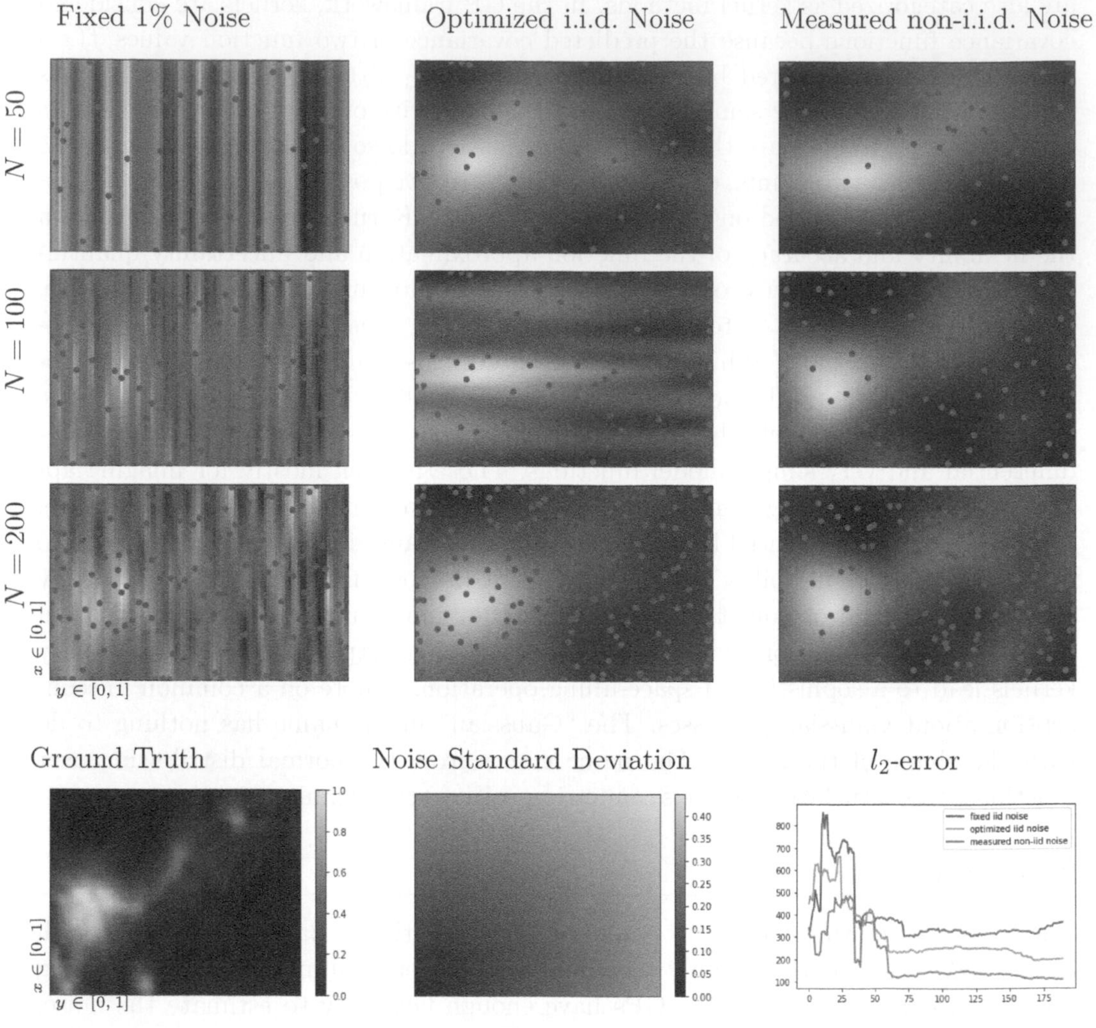

Figure 21.1 The impact of noise on an autonomous experiment. From the left: fixed 1% noise, constant (over the domain) but optimized noise, measured noise. For each case, we see the progression of an autonomous experiment at snapshots $N = 50, 100, 200$. The bottom row shows the ground truth, the synthetic noise model, and the error during the autonomous experiment. The key takeaway from this figure is that noise has a significant influence on the success of an autonomous experiment. Especially fixed noise can mislead the hyperparameter training and render function approximation and uncertainty quantification utterly useless, misleading decision-making and leading to worse-than-random point suggestions. Figure courtesy of [364].

CHAPTER 22

Data Infrastructure

Marcus M. Noack

Applied Mathematics and Computational Research Division, Lawrence Berkeley National Laboratory, Berkeley, California, USA

Kevin G. Yager

Center for Functional Nanomaterials, Brookhaven National Laboratory, Upton, New York, USA

CONTENTS

D ATA infrastructure as a term has wide-ranging interpretations across communities. What we want to focus on, in this chapter, is ways to communicate information between the different components of an autonomous loop in a robust way (Figure 22.1).

22.1   INTEGRATED VS. DISTRIBUTED

To create a functional data infrastructure, we can basically follow two trains of thought. First, we can embed analysis, intelligent decision-making, and instrument control into one integrated AI/ML module. When that module hits a bug, it will have to be fixed and the entire pipeline has to be restarted. On the upside, it can easily be migrated to other instruments once developed. Second, control, analysis and decision-making can all run separately and only be connected by a simple communications protocol. This idea is displayed in Figure 22.1. In this case, the three components run independently, waiting for input. When the input is received, the module performs its designated task and the result is sent to the next module. When one of the modules hits a bug, the other components keep running, the bug can be fixed and only the affected module requires a restart. If a module has to be switched out during an experiment or in between experiments, the others will be unaffected. This structure also supports best practices for software maintainability and abstraction. Furthermore, it is very easy to run any of the components on a remote cluster.

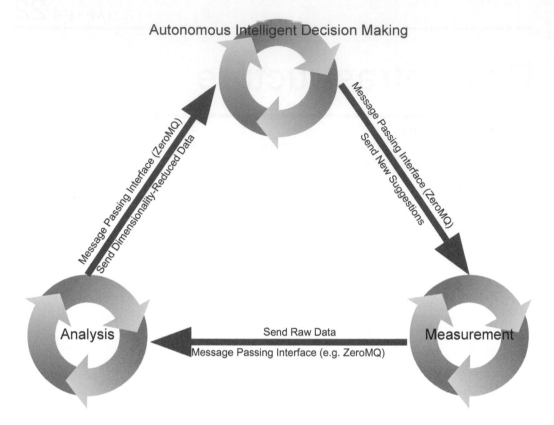

Figure 22.1 Communication infrastructure used at beamlines—most notably the Complex-Materials-Scattering beamline at Brookhaven National Laboratory. The framework consists of three separate building blocks that only interact via messages passed through a message-passing interface (ZMQ). All three modules run the entire time, waiting for messages to be passed to them. When a message is received, work is performed and the output is sent to the next member in the workflow. This framework of independent asynchronous agents acting on received messages has a few advantages compared to more centralized options. Bugs can be fixed and the code restarted in one algorithm without even stopping the others. When results aren't as expected looking at the past messages passed can help identify where things went wrong. Changes to code setups can be accomplished while the experiment is running. In the event that one module crashes all others keep running.

This workload distribution is a change that can also be done on-the-fly without any impact on the other components or the experiment process. The changes only affect the communications protocol. We focus here on the distributed data infrastructure. For more practical information and success stories, we refer the reader to Chapter 8 and [517].

22.2 HANDS-ON DESCRIPTION OF DISTRIBUTED EXECUTION

In this section, we want to provide actual codes/ pseudo codes to help practitioners set up the distributed execution of an autonomous experiment. We only describe this particular option because integrated options need setups on a case-by-case basis, dependent on the software used.

22.2.1 Decision Making

The following code builds upon algorithm 21.4, in particular the "call_instrument" function.

```
1  import pickle
2  import time
3  import zlib
4  import numpy as np
5
6
7  port = "5555"
8  context = zmq.Context()
9  socket = context.socket(zmq.PAIR)
10 socket.bind("tcp://*:%s" % port)
11 ###another computer has to connect to the same port at my ip
12
13 def send_zipped_pickle(obj, socket, flags=0, protocol=-1):
14     """pack and compress an object with pickle and zlib."""
15     pobj = pickle.dumps(obj, protocol)
16     zobj = zlib.compress(pobj)
17     return socket.send(zobj, flags=flags)
18
19 def recv_zipped_pickle(socket, flags=0):
20     """reconstruct a Python object sent with zipped_pickle"""
21     zobj = socket.recv(flags)
22     pobj = zlib.decompress(zobj)
23     return pickle.loads(pobj)
24
25
26 ####the client has to send a starting message here
27 msg = recv_zipped_pickle(socket)
28 print(msg)
29
30 def call_instrument(data):
31     send_zipped_pickle(data, socket)
32     data = recv_zipped_pickle(socket)
33     return data
```

Listing 22.1: Function call_instrument(data) using zmq

This code binds to a socket and sends suggestions to the instrument and then listens for messages back. When a message is received, the next decision is made. When any of the other modules—analysis or instrument control—breaks this code will just continue to wait.

22.2.2 Instrument Control

The instrument can use codes similar to the ones shown above. The only change is that the instrument has to listen to messages coming from the decision-making module and send messages where the data analysis is executed. The instrument does not need any other information than the instructions to perform the next measurement; however, it might be useful to let the instrument choose alternative ways to select measurement parameters in case the decision-making is taking too long to respond, or crashes. This will protect the instrument from idling and wasting valuable experiment time.

22.2.3 Analysis

The analysis code follows the same pattern as the decision-making and instrument control codes. The only changes are the addresses of the sockets to listen to and send information to. The analysis code receives messages from the instrument and sends analyzed data to the decision-maker. The analysis incorporates de-noising and dimensionality reduction as described earlier in this guide.

Bibliography

[1] Martín Abadi, Ashish Agarwal, Paul Barham, Eugene Brevdo, Zhifeng Chen, Craig Citro, Greg S. Corrado, Andy Davis, Jeffrey Dean, Matthieu Devin, Sanjay Ghemawat, Ian Goodfellow, Andrew Harp, Geoffrey Irving, Michael Isard, Yangqing Jia, Rafal Jozefowicz, Lukasz Kaiser, Manjunath Kudlur, Josh Levenberg, Dandelion Mané, Rajat Monga, Sherry Moore, Derek Murray, Chris Olah, Mike Schuster, Jonathon Shlens, Benoit Steiner, Ilya Sutskever, Kunal Talwar, Paul Tucker, Vincent Vanhoucke, Vijay Vasudevan, Fernanda Viégas, Oriol Vinyals, Pete Warden, Martin Wattenberg, Martin Wicke, Yuan Yu, and Xiaoqiang Zheng. TensorFlow: Large-scale machine learning on heterogeneous systems, 2015. Software available from tensorflow.org.

[2] Asmaa Abbas, Mohammed M. Abdelsamea, and Mohamed Medhat Gaber. Classification of covid-19 in chest x-ray images using detrac deep convolutional neural network. *Applied Intelligence*, 51(2):854–864, 2021.

[3] Moloud Abdar, Farhad Pourpanah, Sadiq Hussain, Dana Rezazadegan, Li Liu, Mohammad Ghavamzadeh, Paul Fieguth, Xiaochun Cao, Abbas Khosravi, U. Rajendra Acharya, et al. A review of uncertainty quantification in deep learning: Techniques, applications and challenges. *Information Fusion*, 76:243–297, 2021.

[4] Oludare Isaac Abiodun, Aman Jantan, Abiodun Esther Omolara, Kemi Victoria Dada, Nachaat AbdElatif Mohamed, and Humaira Arshad. State-of-the-art in artificial neural network applications: A survey. *Heliyon*, 4(11):e00938, 2018.

[5] Milad Abolhasani and Eugenia Kumacheva. The rise of self-driving labs in chemical and materials sciences. *Nature Synthesis*, 2:483–492, 2023.

[6] Mark A. Abramson, Charles Audet, James W. Chrissis, and Jennifer G. Walston. Mesh adaptive direct search algorithms for mixed variable optimization. *Optimization Letters*, 3(1):35–47, 2009.

[7] Yaser S Abu-Mostafa, Malik Magdon-Ismail, and Hsuan-Tien Lin. *Learning from data*, volume 4. AMLBook New York, 2012.

[8] David H. Ackley, Geoffrey E. Hinton, and Terrence J. Sejnowski. A learning algorithm for Boltzmann machines. *Cognitive Science*, 9:147–169, 1985.

[9] Ajay Agrawal, Joshua Gans, and Avi Goldfarb. *Prediction Machines: The Simple Economics of Artificial Intelligence*. Harvard Business Press, 2018.

[10] D. Aisa, S. Aisa, E. Babucci, F. Barocchi, A. Cunsolo, A. De Francesco, F. Formisano, T. Gahl, E. Guarini, A. Laloni, H. Mutka, A. Orecchini, C. Petrillo, W.-C. Pilgrim, A. Piluso, F. Sacchetti, J.-B. Suck, and G. Venturi. Brisp: A new thermal-neutron spectrometer for small-angle studies of disordered matter. *Journal of Non-Crystalline Solids*, 352(42):5130–5135, 2006. Proceedings of the 5th International Discussion Meeting on Relaxations in Complex Systems.

[11] D. Aisa, E. Babucci, F. Barocchi, A. Cunsolo, F. D'Anca, A. De Francesco, F. Formisano, T. Gahl, E. Guarini, S. Jahn, A. Laloni, H. Mutka, A. Orecchini, C. Petrillo, F. Sacchetti, J.-B. Suck, and G. Venturi. The development of the BRISP spectrometer at the Institut Laue-Langevin. *Nuclear Instruments and Methods in Physics Research Section A: Accelerators, Spectrometers, Detectors and Associated Equipment*, 544(3):620–642, 2005.

[12] Shivaji Alaparthi and Manit Mishra. Bidirectional encoder representations from transformers (bert): A sentiment analysis odyssey. *arXiv preprint arXiv:2007.01127*, 2020.

[13] Francis J. Alexander, James Ang, Jenna A Bilbrey, Jan Balewski, Tiernan Casey, Ryan Chard, Jong Choi, Sutanay Choudhury, Bert Debusschere, Anthony M. DeGennaro, Nikoli Dryden, J. Austin Ellis, Ian Foster, Cristina Garcia Cardona, Sayan Ghosh, Peter Harrington, Yunzhi Huang, Shantenu Jha, Travis Johnston, Ai Kagawa, Ramakrishnan Kannan, Neeraj Kumar, Zhengchun Liu, Naoya Maruyama, Satoshi Matsuoka, Erin McCarthy, Jamaludin Mohd-Yusof, Peter Nugent, Yosuke Oyama, Thomas Proffen, David Pugmire, Sivasankaran Rajamanickam, Vinay Ramakrishniah, Malachi Schram, Sudip K. Seal, Ganesh Sivaraman, Christine Sweeney, Li Tan, Rajeev Thakur, Brian Van Essen, Logan Ward, Paul Welch, Michael Wolf, Sotiris S. Xantheas, Kevin G. Yager, Shinjae Yoo, and Byung-Jun Yoon. Co-design center for exascale machine learning technologies (exalearn). *The International Journal of High Performance Computing Applications*, 35(6):598–616, 2021.

[14] Francis J. Alexander, James Ang, Jenna A. Bilbrey, Jan Balewski, Tiernan Casey, Ryan Chard, Jong Choi, Sutanay Choudhury, Bert Debusschere, Anthony M. DeGennaro, Nikoli Dryden, J. Austin Ellis, Ian Foster, Cristina Garcia Cardona, Sayan Ghosh, Peter Harrington, Yunzhi Huang, Shantenu Jha, Travis Johnston, Ai Kagawa, Ramakrishnan Kannan, Neeraj Kumar, Zhengchun Liu, Naoya Maruyama, Satoshi Matsuoka, Erin McCarthy, Jamaludin Mohd-Yusof, Peter Nugent, Yosuke Oyama, Thomas Proffen, David Pugmire, Sivasankaran Rajamanickam, Vinay Ramakrishniah, Malachi Schram, Sudip K. Seal, Ganesh Sivaraman, Christine Sweeney, Li Tan, Rajeev Thakur, Brian Van Essen, Logan Ward, Paul Welch, Michael Wolf, Sotiris S. Xantheas, Kevin G. Yager, Shinjae Yoo, and Byung-Jun Yoon. Co-design center for exascale machine learning technologies (exalearn). *The International Journal of High Performance Computing Applications*, 35(6):598–616, 2021.

[15] Daniel Allan, Thomas Caswell, Stuart Campbell, and Maksim Rakitin. Bluesky's ahead: A multi-facility collaboration for an a la Carte software project for data acquisition and management. *Synchrotron Radiation News*, 32(3):19–22, 2019.

[16] Benjamin Alldritt, Fedor Urtev, Niko Oinonen, Markus Aapro, Juho Kannala, Peter Liljeroth, and Adam S. Foster. Automated tip functionalization via machine learning in scanning probe microscopy. *Computer Physics Communications*, 273:108258, 2022.

[17] Jeffery Allen, Ethan Young, Pietro Bortolotti, Ryan King, and Garrett Barter. Blade planform design optimization to enhance turbine wake control. *Wind Energy*, 25(5):811–830, 2022.

[18] J. Als-Nielsen and D. McMorrow. *Elements of Modern X-Ray Physics*. John Wiley & Sons, Inc. New York, NY, USA, 2001.

[19] Mauricio A. Alvarez, Lorenzo Rosasco, Neil D. Lawrence, et al. Kernels for vector-valued functions: A review. *Foundations and Trends® in Machine Learning*, 4(3):195–266, 2012.

[20] Sebastian Ament, Maximilian Amsler, Duncan R. Sutherland, Ming-Chiang Chang, Dan Guevarra, Aine B. Connolly, John M. Gregoire, Michael O. Thompson, Carla P. Gomes, and R. Bruce van Dover. Autonomous materials synthesis via hierarchical active learning of nonequilibrium phase diagrams. *Science Advances*, 7(51):eabg4930, 2021.

[21] Eric J. Amis, Xiao-Dong Xiang, and Ji-Cheng Zhao. Combinatorial materials science: What's new since edison? *MRS Bulletin*, 27(4):295300, 2002.

[22] Hadis Anahideh, Jay Rosenberger, and Victoria Chen. High-dimensional black-box optimization under uncertainty. *Computers & Operations Research*, 137:105444, 2022.

[23] Yasmin Ansari, Mariangela Manti, Egidio Falotico, Yoan Mollard, Matteo Cianchetti, and Cecilia Laschi. Towards the development of a soft manipulator as an assistive robot for personal care of elderly people. *International Journal of Advanced Robotic Systems*, 14(2):1729881416687132, 2017.

[24] Araujo, Silva, Medeiros, Parkinson, Hexemer, Carneiro, and Ushizima. Reverse image search for scientific data within and beyond the visible spectrum. *Expert Systems with Applications*, 109:35–48, 2018.

[25] Araujo, Silva, Resende, Ushizima, Medeiros, Carneiro, and Bianchi. Deep learning for cell image segmentation and ranking. *Computerized Medical Imaging and Graphics*, 72:13–21, 2019.

[26] Aleksandr Y. Aravkin, Bradley M. Bell, James V. Burke, and Gianluigi Pillonetto. The connection between bayesian estimation of a gaussian random field and rkhs. *IEEE Transactions on Neural Networks and Learning Systems*, 26(7):1518–1524, 2014.

[27] Carrie Arnold. Cloud labs: where robots do the research. *Nature*, 606(7914):612–613, 2022.

[28] Brookhaven Science Associates. National Synchrotron Light Source II (NSLS-II) website.

[29] Charles Audet and J. E. Dennis. Mesh adaptive direct search algorithms for constrained optimization. *SIAM Journal on Optimization*, 17(1):188–217, 2006.

[30] Charles Audet, Gilles Savard, and Walid Zghal. A mesh adaptive direct search algorithm for multiobjective optimization. *European Journal of Operational Research*, 204(3):545–556, 2010.

[31] Ido Azuri, Irit Rosenhek-Goldian, Neta Regev-Rudzki, Georg Fantner, and Sidney R. Cohen. The role of convolutional neural networks in scanning probe microscopy: a review. *Beilstein Journal of Nanotechnology*, 12(1):878–901, 2021.

[32] Aly Badran, Dula Parkinson, Daniela Ushizima, David Marshall, and Emmanuel Maillet. Validation of deep learning segmentation of ct images of fiber-reinforced composites. *Journal of Composites Science*, 6(2):60, 2022. https://doi.org/10.3390/jcs6020060

[33] Vijay Badrinarayanan, Alex Kendall, and Roberto Cipolla. Segnet: A deep convolutional encoder-decoder architecture for image segmentation. *IEEE Transactions on Pattern Analysis and Machine Intelligence*, 39(12):2481–2495, 2017.

[34] Suwon Bae, Marcus M. Noack, and Kevin G. Yager. Surface enrichment dictates block copolymer orientation. *Nanoscale*, 15:6901–6912, 2023.

[35] Suwon Bae and Kevin G Yager. Chain redistribution stabilizes coexistence phases in block copolymer blends. *ACS nano*, 16(10):17107–17115, 2022.

[36] Jacob Baelum, Sharon Borglin, Romy Chakraborty, Julian L. Fortney, Regina Lamendella, Olivia U. Mason, Manfred Auer, Marcin Zemla, Markus Bill, Mark E. Conrad, et al. Deep-sea bacteria enriched by oil and dispersant from the deepwater horizon spill. *Environmental Microbiology*, 14(9):2405–2416, 2012.

[37] Chandrajit Bajaj, Yi Wang, and Yunhao Yang. Reinforcement learning of self enhancing camera image and signal processing, 653, 2022. Springer, Singapore. https://doi.org/10.1007/978-981-99-0981-0_22

[38] Prasanna V. Balachandran, Dezhen Xue, James Theiler, John Hogden, and Turab Lookman. Adaptive strategies for materials design using uncertainties. *Scientific Reports*, 6(1):19660, 2016.

[39] Prasanna Balaprakash, Michael Salim, Thomas D. Uram, Venkat Vishwanath, and Stefan M. Wild. Deephyper: Asynchronous hyperparameter search for deep neural networks. In *2018 IEEE 25th International Conference on High Performance Computing (HiPC)*, pages 42–51, 2018.

[40] P. Baldi, P. Sadowski, and D. Whiteson. Searching for exotic particles in high-energy physics with deep learning. *Nature Communications*, 5(1):4308, 2014.

[41] Leon Balents. Spin liquids in frustrated magnets. *Nature*, 464(7286):199–208, 2010.

[42] Sudipto Banerjee, Alan E. Gelfand, Andrew O. Finley, and Huiyan Sang. Gaussian predictive process models for large spatial data sets. *Journal of the Royal Statistical Society: Series B (Statistical Methodology)*, 70(4):825–848, 2008.

[43] Lars Banko, Phillip M. Maffettone, Dennis Naujoks, Daniel Olds, and Alfred Ludwig. Deep learning for visualization and novelty detection in large x-ray diffraction datasets. *npj Computational Materials*, 7(1):104, 2021.

[44] Andi Barbour, Stuart Campbell, Thomas Caswell, Masafumi Fukuto, Marcus Hanwell, Andrew Kiss, Tatiana Konstantinova, Ricarda Laasch, Phillip Maffettone, Bruce Ravel, and Daniel Olds. Advancing discovery with artificial intelligence and machine learning at nsls-ii. *Synchrotron Radiation News*, 35:44–50, 2022.

[45] J. Bardeen. Tunnelling from a many-particle point of view. *Phys. Rev. Lett.*, 6:57–59, 1961.

[46] R. Barer, A. R. H. Cole, and H. W. Thompson. Infra-red spectroscopy with the reflecting microscope in physics, chemistry and biology. *Nature*, 163(4136):198–201, 1949.

[47] Gabriel Barth-Maron, Matthew W. Hoffman, David Budden, Will Dabney, Dan Horgan, Dhruva Tb, Alistair Muldal, Nicolas Heess, and Timothy Lillicrap. Distributed distributional deterministic policy gradients. *arXiv preprint arXiv:1804.08617*, 2018.

[48] Rohit Batra, Le Song, and Rampi Ramprasad. Emerging materials intelligence ecosystems propelled by machine learning. *Nature Reviews Materials*, 6(8):655–678, 2021.

[49] Bahareh Beigzadeh, Mehdi Bahrami, Mohammad Javad Amiri, and Mohammad Reza Mahmoudi. A new approach in adsorption modeling using random forest regression, Bayesian multiple linear regression, and multiple linear regression: 2,4-D adsorption by a green adsorbent. *Water Science and Technology*, 82(8):1586–1602, 2020.

[50] Richard E Bellman. *Adaptive Control Processes: A Guided Tour*. Princeton University Press, 2015.

[51] Samy Bengio, Oriol Vinyals, Navdeep Jaitly, and Noam Shazeer. Scheduled sampling for sequence prediction with recurrent neural networks. *Advances in Neural Information Processing Systems*, 28, 2015.

[52] Liane G. Benning, V. R. Phoenix, Nathan Yee, and K. O. Konhauser. The dynamics of cyanobacterial silicification: an infrared micro-spectroscopic investigation. *Geochimica et Cosmochimica Acta*, 68(4):743–757, 2004.

[53] Liane G. Benning, V. R. Phoenix, Nathan Yee, and M. J. Tobin. Molecular characterization of cyanobacterial silicification using synchrotron infrared micro-spectroscopy. *Geochimica et Cosmochimica Acta*, 68(4):729–741, 2004.

[54] James Bergstra, Brent Komer, Chris Eliasmith, Dan Yamins, and David D. Cox. Hyperopt: a python library for model selection and hyperparameter optimization. *Computational Science &amp: Discovery*, 8(1):014008, 2015.

[55] Alain Berlinet and Christine Thomas-Agnan. *Reproducing Kernel Hilbert Spaces in Probability and Statistics*. Springer Science & Business Media, 2011.

[56] E. Bernard. Introduction to machine learning. Champaign. Wolfram Media, 2021.

[57] Dimitri P. Bertsekas. Value and policy iterations in optimal control and adaptive dynamic programming. *IEEE Transactions on Neural Networks and Learning Systems*, 28(3):500–509, 2015.

[58] Eli Bingham, Jonathan P. Chen, Martin Jankowiak, Fritz Obermeyer, Neeraj Pradhan, Theofanis Karaletsos, Rohit Singh, Paul Szerlip, Paul Horsfall, and Noah D. Goodman. Pyro: Deep universal probabilistic programming. *The Journal of Machine Learning Research*, 20(1):973–978, 2019.

[59] G. Binnig, H. Rohrer, Ch. Gerber, and E. Weibel. Surface studies by scanning tunneling microscopy. *Phys. Rev. Lett.*, 49:57–61, 1982.

[60] Christopher M Bishop. *Pattern Recognition and Machine Learning*. Springer, 2006.

[61] Elkan R. Blout and Robert C. Mellors. Infrared spectra of tissues. *Science*, 110(2849):137–138, 1949.

[62] Bluesky. Project Website: `https://nsls-ii.github.io/bluesky/`.

[63] M. Boehm, S. Martynov, B. Roessli, G. Petrakovskii, and J. Kulda. Spin-wave spectrum of copper metaborate in the commensurate phase 10k¡t¡21k. *Journal of Magnetism and Magnetic Materials*, 250:313–318, 2002.

[64] M. Boehm, P. Steffens, J. Kulda, M. Klicpera, S. Roux, P. Courtois, P. Svoboda, J. Saroun, and V. Sechovsky. Thalesthree axis low energy spectroscopy for highly correlated electron systems. *Neutron News*, 26(3):18–21, 2015.

[65] Rishi Bommasani, Drew A. Hudson, Ehsan Adeli, Russ Altman, Simran Arora, Sydney von Arx, Michael S. Bernstein, Jeannette Bohg, Antoine Bosselut, Emma Brunskill, Erik Brynjolfsson, Shyamal Buch, Dallas Card, Rodrigo Castellon, Niladri Chatterji, Annie Chen, Kathleen Creel, Jared Quincy Davis, Dora Demszky, Chris Donahue, Moussa Doumbouya, Esin Durmus, Stefano Ermon, John Etchemendy, Kawin Ethayarajh, Li Fei-Fei, Chelsea Finn, Trevor Gale, Lauren Gillespie, Karan Goel, Noah Goodman, Shelby Grossman, Neel Guha, Tatsunori Hashimoto, Peter Henderson, John Hewitt, Daniel E. Ho, Jenny Hong, Kyle Hsu, Jing Huang, Thomas Icard, Saahil Jain, Dan Jurafsky, Pratyusha Kalluri, Siddharth Karamcheti, Geoff Keeling, Fereshte Khani, Omar Khattab, Pang Wei Koh, Mark Krass, Ranjay Krishna, Rohith Kuditipudi, Ananya Kumar, Faisal Ladhak, Mina Lee, Tony Lee, Jure Leskovec, Isabelle Levent, Xiang Lisa Li, Xuechen Li, Tengyu Ma, Ali Malik, Christopher D. Manning, Suvir Mirchandani, Eric Mitchell, Zanele Munyikwa, Suraj Nair, Avanika Narayan, Deepak Narayanan, Ben Newman, Allen Nie, Juan Carlos Niebles, Hamed Nilforoshan, Julian Nyarko, Giray Ogut, Laurel Orr, Isabel Papadimitriou, Joon Sung Park, Chris Piech, Eva Portelance, Christopher Potts, Aditi Raghunathan, Rob Reich, Hongyu Ren, Frieda Rong, Yusuf Roohani, Camilo Ruiz, Jack Ryan, Christopher Ré, Dorsa Sadigh, Shiori Sagawa, Keshav Santhanam, Andy Shih, Krishnan Srinivasan, Alex Tamkin, Rohan Taori, Armin W. Thomas, Florian Tramèr, Rose E. Wang, William Wang, Bohan Wu, Jiajun Wu, Yuhuai Wu, Sang Michael Xie, Michihiro Yasunaga, Jiaxuan You, Matei Zaharia, Michael Zhang, Tianyi Zhang, Xikun Zhang, Yuhui Zhang, Lucia Zheng, Kaitlyn Zhou, and Percy Liang. On the opportunities and risks of foundation models, 2021.

[66] A. J. Booker, J. E. Dennis, P. D. Frank, D. B. Serafini, V. Torczon, and M. W. Trosset. A rigorous framework for optimization of expensive functions by surrogates. *Structural Optimization*, 17(1):1–13, 1999.

[67] Hanen Borchani, Gherardo Varando, Concha Bielza, and Pedro Larrañaga. A survey on multi-output regression. *Wiley Interdisciplinary Reviews: Data Mining and Knowledge Discovery*, 5(5):216–233, 2015.

[68] Christopher K. H. Borg, Eric S. Muckley, Clara Nyby, James E. Saal, Logan Ward, Apurva Mehta, and Bryce Meredig. Quantifying the performance of machine learning models in materials discovery. *Digital Discovery*, 2:327–338, 2023.

[69] G. E. P Box and K. B. Wilson. On the experimental attainment of optimum conditions. *Journal of the Royal Statistical Society. Series B (Methodological)*, 13(1):1–45, 1951.

[70] Gary Bradski and Adrian Kaehler. *Learning OpenCV: Computer vision with the OpenCV library*. "O'Reilly Media, Inc.", 2008.

[71] Leo Breiman. Statistical modeling: The two cultures (with comments and a rejoinder by the author). *Statistical Science*, 16(3):199–231, 2001.

[72] Greg Brockman, Vicki Cheung, Ludwig Pettersson, Jonas Schneider, John Schulman, Jie Tang, and Wojciech Zaremba. Openai gym. *arXiv preprint arXiv:1606.01540*, 2016.

[73] Tom B. Brown, Benjamin Mann, Nick Ryder, Melanie Subbiah, Jared Kaplan, Prafulla Dhariwal, Arvind Neelakantan, Pranav Shyam, Girish Sastry, Amanda Askell, Sandhini Agarwal, Ariel Herbert-Voss, Gretchen Krueger, Tom Henighan, Rewon Child, Aditya Ramesh, Daniel M. Ziegler, Jeffrey Wu, Clemens Winter, Christopher Hesse, Mark Chen, Eric Sigler, Mateusz Litwin, Scott Gray, Benjamin Chess, Jack Clark, Christopher Berner, Sam McCandlish, Alec Radford, Ilya Sutskever, and Dario Amodei. Language models are few-shot learners, 2020.

[74] Lukas Brunke, Melissa Greeff, Adam W. Hall, Zhaocong Yuan, Siqi Zhou, Jacopo Panerati, and Angela P. Schoellig. Safe learning in robotics: From learning-based control to safe reinforcement learning. *Annual Review of Control, Robotics, and Autonomous Systems*, 5(1):411–444, 2022.

[75] Christian Buck, Jannis Bulian, Massimiliano Ciaramita, Wojciech Gajewski, Andrea Gesmundo, Neil Houlsby, and Wei Wang. Ask the right questions: Active question reformulation with reinforcement learning. *arXiv preprint arXiv:1705.07830*, 2017.

[76] Míriam Bellver Bueno, Xavier Giró-i Nieto, Ferran Marqués, and Jordi Torres. Hierarchical object detection with deep reinforcement learning. *Deep Learning for Image Processing Applications*, 31(164):3, 2017.

[77] Yuri Burda, Harrison Edwards, Amos Storkey, and Oleg Klimov. Exploration by random network distillation. *arXiv preprint arXiv:1810.12894*, 2018.

[78] Benjamin Burger, Phillip M. Maffettone, Vladimir V. Gusev, Catherine M. Aitchison, Yang Bai, Xiaoyan Wang, Xiaobo Li, Ben M. Alston, Buyi Li, Rob Clowes, et al. A mobile robotic chemist. *Nature*, 583(7815):237–241, 2020.

[79] Paul-Christian Bürkner, Ilja Kröker, Sergey Oladyshkin, and Wolfgang Nowak. The sparse polynomial chaos expansion: a fully bayesian approach with joint priors on the coefficients and global selection of terms. *arXiv preprint arXiv:2204.06043*, 2022.

[80] Keith T. Butler, Manh Duc Le, Jeyan Thiyagalingam, and Toby G. Perring. Interpretable, calibrated neural networks for analysis and understanding of inelastic neutron scattering data. *Journal of Physics: Condensed Matter*, 33(19):194006, 2021.

[81] Juan C. Caicedo and Svetlana Lazebnik. Active object localization with deep reinforcement learning. In *Proceedings of the IEEE International Conference on Computer Vision*, pages 2488–2496, 2015.

[82] Stuart Campbell, Daniel B. Allan, Andi Barbour, Daniel Olds, Maksim Rakitin, Reid Smith, and Stuart B. Wilkins. Outlook for artificial intelligence and machine learning at the nsls-ii. *Machine Learning: Science and Technology*, 2:013001, 2020.

[83] Stuart I. Campbell, Daniel B. Allan, Andi M. Barbour, Daniel Olds, Maksim S. Rakitin, Reid Smith, and Stuart B. Wilkins. Outlook for artificial intelligence and machine learning at the nsls-ii. *Machine Learning: Science and Technology*, 2(1):013001, 2021.

[84] Bing Cao, Lawrence A. Adutwum, Anton O. Oliynyk, Erik J. Luber, Brian C. Olsen, Arthur Mar, and Jillian M. Buriak. How to optimize materials and devices via design of experiments and machine learning: Demonstration using organic photovoltaics. *ACS Nano*, 12(8):7434–7444, 2018. PMID: 30027732.

[85] Bob Carpenter, Andrew Gelman, Matthew D. Hoffman, Daniel Lee, Ben Goodrich, Michael Betancourt, Marcus Brubaker, Jiqiang Guo, Peter Li, and Allen Riddell. Stan: A probabilistic programming language. *Journal of Statistical Software*, 76(1):1–32, 2017.

[86] J. Carson Meredith, Alamgir Karim, and Eric J. Amis. Combinatorial methods for investigations in polymer materials science. *MRS Bulletin*, 27(4):330335, 2002.

[87] Arturo Casadevall and Ferric C. Fang. Reforming science: methodological and cultural reforms, 80:891–896, 2012.

[88] Rafael Celestre, Oleg Chubar, Thomas Roth, Manuel Sanchez del Rio, and Raymond Barrett. Recent developments in x-ray lens modelling with SRW. In Kawal Sawhney and Oleg Chubar, editors, *Advances in Computational Methods for X-Ray Optics V*. SPIE, 2020.

[89] CERN. Mad-methodological accelerator design.

[90] Alexander N. Chaika. High resolution stm imaging. In Challa S. S. R. Kumar, editor, *Surface Science Tools for Nanomaterials Characterization*, pages 561–619. Springer, 2015.

[91] Emory M. Chan, Chenxu Xu, Alvin W. Mao, Gang Han, Jonathan S. Owen, Bruce E. Cohen, and Delia J. Milliron. Reproducible, high-throughput synthesis of colloidal nanocrystals for optimization in multidimensional parameter space. *Nano Letters*, 10(5):1874–1885, 2010. PMID: 20387807.

[92] I-Ju Chen, Markus Aapro, Abraham Kipnis, Alexander Ilin, Peter Liljeroth, and Adam S. Foster. Precise atom manipulation through deep reinforcement learning. *Nature Communications*, 13(1):7499, 2022.

[93] W. Chen, V. Madhavan, T. Jamneala, and M. F. Crommie. Scanning tunneling microscopy observation of an electronic superlattice at the surface of clean gold. *Physical Review Letters*, 80:1469–1472, 1998.

[94] Yushi Chen, Hanlu Jiang, Chunyang Li, Xiuping Jia, and Pedram Ghamisi. Deep feature extraction and classification of hyperspectral images based on convolutional neural networks. *IEEE Transactions on Geoscience and Remote Sensing*, 54(10):6232–6251, 2016.

[95] Yu-Chen Karen Chen-Wiegart, Iradwikanari Waluyo, Andrew Kiss, Stuart Campbell, Lin Yang, Eric Dooryhee, Jason R. Trelewicz, Yiyang Li, Bruce Gates, Mark Rivers, and Kevin G. Yager. Multimodal synchrotron approach: Research needs and scientific vision. *Synchrotron Radiation News*, 33(1):44–47, 2020.

[96] Yanyan Cheng, Wei Xia, Matteo Detto, and Christine A. Shoemaker. A framework to calibrate ecosystem demography models within earth system models using parallel surrogate global optimization. *Water Resources Research*, 59(1):e2022WR032945, 2023.

[97] Siwar Chibani and François-Xavier Coudert. Machine learning approaches for the prediction of materials properties. *APL Materials*, 8(8):080701, 2020.

[98] Soo-Yeon Cho, Youhan Lee, Sangwon Lee, Hohyung Kang, Jaehoon Kim, Junghoon Choi, Jin Ryu, Heeeun Joo, Hee-Tae Jung, and Jihan Kim. Finding hidden signals in chemical sensors using deep learning. *Analytical Chemistry*, 92(9):6529–6537, 2020.

[99] Li Chonghe, Guo Jin, Qin Pei, Chen Ruiliang, and Chen Nianyi. Some regularities of melting points of ab-type intermetallic compounds. *Journal of Physics and Chemistry of Solids*, 57(12):1797–1802, 1996.

[100] Slim T. Chourou, Abhinav Sarje, Xiaoye S. Li, Elaine R. Chan, and Alexander Hexemer. Hipgisaxs: a high-performance computing code for simulating grazing-incidence x-ray scattering data. *Journal of Applied Crystallography*, 46(6):1781–1795, 2013.

[101] Paul Christiano, Jan Leike, Tom B. Brown, Miljan Martic, Shane Legg, and Dario Amodei. Deep reinforcement learning from human preferences, 30, 2023.

[102] Julie Christodoulou, Lisa E. Friedersdorf, Linda Sapochak, and James A. Warren. The second decade of the materials genome initiative. *JOM*, 73(12):3681–3683, 2021.

[103] John Coates. Interpretation of infrared spectra, a practical approach. *Encyclopedia of Analytical Chemistry: Applications, Theory and Instrumentation*, 2006.

[104] Samuel Cohen, Rendani Mbuvha, Tshilidzi Marwala, and Marc Deisenroth. Healing products of gaussian process experts. In *International Conference on Machine Learning*, pages 2068–2077. PMLR, 2020.

[105] D. A. Cohn, Z. Ghahramani, and M. I. Jordan. Active learning with statistical models. 4, 1996.

[106] M. J. Cooper and R. Nathans. The resolution function in neutron diffractometry. I. The resolution function of a neutron diffractometer and its application to phonon measurements. *Acta Crystallographica*, 23(3):357–367, 1967.

[107] Noel Cressie. The origins of kriging. *Mathematical Geology*, 22(3):239–252, 1990.

[108] Noel Cressie and Gardar Johannesson. Fixed rank kriging for very large spatial data sets. *Journal of the Royal Statistical Society: Series B (Statistical Methodology)*, 70(1):209–226, 2008.

[109] Geoff Currie, K. Elizabeth Hawk, Eric Rohren, Alanna Vial, and Ran Klein. Machine learning and deep learning in medical imaging: Intelligent imaging. *Journal of Medical Imaging and Radiation Sciences*, 50(4):477–487, 2019.

[110] Felipe Leno Da Silva, Ruben Glatt, and Anna Helena Reali Costa. MOO-MDP: An object-oriented representation for cooperative multiagent reinforcement learning. *IEEE Transactions on Cybernetics*, 49:567–579 2017.

[111] Chengyu Dai and Sharon C. Glotzer. Efficient phase diagram sampling by active learning. *The Journal of Physical Chemistry B*, 124(7):1275–1284, 2020.

[112] Rituparna Datta and Rommel G. Regis. A surrogate-assisted evolution strategy for constrained multi-objective optimization. *Expert Systems with Applications*, 57:270–284, 2016.

[113] T. Davies and Tom Fearn. Back to basics: the principles of principal component analysis. *Spectroscopy Europe*, 16(6):20, 2004.

[114] Eddie Davis and Marianthi Ierapetritou. A kriging based method for the solution of mixed-integer nonlinear programs containing black-box functions. *Journal of Global Optimization*, 43(2):191–205, 2009.

[115] A. De Francesco, E. Guarini, U. Bafile, F. Formisano, and L. Scaccia. Bayesian approach to the analysis of neutron Brillouin scattering data on liquid metals. *Physical Reviews E*, 94:023305, 2016.

[116] Alessio De Francesco, Luisa Scaccia, R. Bruce Lennox, Eleonora Guarini, Ubaldo Bafile, Peter Falus, and Marco Maccarini. Model-free description of polymer-coated gold nanoparticle dynamics in aqueous solutions obtained by Bayesian analysis of neutron spin echo data. *Physical Reviews E*, 99:052504, 2019.

[117] Santiago González de la Hoz, Carles Acosta-Silva, Javier Aparisi Pozo, Jose del Peso, Álvaro Fernández Casani, José Flix Molina, Esteban Fullana Torregrosa, Carlos García Montoro, Julio Lozano Bahilo, Almudena Montiel,

Andrés Pacheco Pages, Javier Sánchez Martínez, José Salt Cairols, and Aresh Vedaee. Computing activities at the spanish tier-1 and tier-2s for the ATLAS experiment towards the LHC run3 and high-luminosity periods. *EPJ Web of Conferences*, 245:07027, 2020.

[118] Juan J. de Pablo, Nicholas E. Jackson, Michael A. Webb, Long-Qing Chen, Joel E. Moore, Dane Morgan, Ryan Jacobs, Tresa Pollock, Darrell G. Schlom, Eric S. Toberer, et al. New frontiers for the materials genome initiative. *npj Computational Materials*, 5(1):41, 2019.

[119] Alexandre Fioravante de Siqueira, Daniela Mayumi Ushizima, and Stéfan van der Walt. A reusable pipeline for large-scale fiber segmentation on unidirectional fiber beds using fully convolutional neural networks, 9, 2021.

[120] James J. De Yoreo. A holistic view of nucleation and self-assembly. *MRS Bulletin*, 42(7):525–536, 2017.

[121] K. Deb, A. Pratap, S. Agarwal, and T. Meyarivan. A fast and elitist multiobjective genetic algorithm: NSGA-II. *IEEE Transactions on Evolutionary Computation*, 6(2):182–197, 2002.

[122] Kalyanmoy Deb and Himanshu Jain. An evolutionary many-objective optimization algorithm using reference-point-based nondominated sorting approach, Part I: Solving problems with box constraints. *IEEE Transactions on Evolutionary Computation*, 18(4):577–601, 2014.

[123] Allison McCarn Deiana, Nhan Tran, Joshua Agar, Michaela Blott, Giuseppe Di Guglielmo, Javier Duarte, Philip Harris, Scott Hauck, Mia Liu, Mark S. Neubauer, et al. Applications and techniques for fast machine learning in science. *Frontiers in Big Data*, 5:787421, 2022.

[124] Frederik Michel Dekking, Cornelis Kraaikamp, Hendrik Paul Lopuhaä, and Ludolf Erwin Meester. *A Modern Introduction to Probability and Statistics: Understanding Why and How*, volume 488. Springer, 2005.

[125] Volker L. Deringer, Albert P. Bartók, Noam Bernstein, David M. Wilkins, Michele Ceriotti, and Gábor Csányi. Gaussian process regression for materials and molecules. *Chemical Reviews*, 121(16):10073–10141, 2021. PMID: 34398616.

[126] Gregory S. Doerk, Aaron Stein, Suwon Bae, Marcus M. Noack, Masafumi Fukuto, and Kevin G. Yager. Autonomous discovery of emergent morphologies in directed self-assembly of block copolymer blends. *Science Advances*, 9(2):eadd3687, 2023.

[127] Gregory S. Doerk, Aaron Stein, Suwon Bae, Marcus M. Noack, Masafumi Fukuto, and Kevin G. Yager. Autonomous discovery of emergent morphologies in directed self-assembly of block copolymer blends. *Science Advances*, 9(2):eadd3687, 2023.

[128] Gregory S. Doerk and Kevin G. Yager. Beyond native block copolymer morphologies. *Molecular Systems Design & Engineering*, 2:518–538, 2017.

[129] G. S. Doerk, A. Stein, Bae S., M. M. Noack, M. Fukuto, and K. G. Yager. Autonomous discovery of emergent morphologies in directed self-assembly of block copolymer blends. *Science Advances*, 9:eadd3687, 2023.

[130] J. Donatelli, M. Haranczyk, A. Hexemer, H. Krishnan, X. Li, L. Lin, F. Maia, S. Marchesini, D. Parkinson, T. Perciano, et al. Camera: The center for advanced mathematics for energy research applications. *Synchrotron Radiation News*, 28(2):4–9, 2015.

[131] David L. Donoho and Carrie Grimes. Hessian eigenmaps: Locally linear embedding techniques for high-dimensional data. *Proceedings of the National Academy of Sciences*, 100(10):5591–5596, 2003.

[132] Alexey Dosovitskiy, Lucas Beyer, Alexander Kolesnikov, Dirk Weissenborn, Xiaohua Zhai, Thomas Unterthiner, Mostafa Dehghani, Matthias Minderer, Georg Heigold, Sylvain Gelly, et al. An image is worth 16x16 words: Transformers for image recognition at scale. In *Proceedings of the IEEE/CVF Conference on Computer Vision and Pattern Recognition*, pages 8798–8808, 2021.

[133] Paul Dumas, Ganesh D. Sockalingum, and Josep Sule-Suso. Adding synchrotron radiation to infrared microspectroscopy: what's new in biomedical applications? *Trends in Biotechnology*, 25(1):40–44, 2007.

[134] Vasilios Duros, Jonathan Grizou, Abhishek Sharma, S. Hessam, M. Mehr, Andrius Bubliauskas, Przemysław Frei, Haralampos N. Miras, and Leroy Cronin. Intuition-enabled machine learning beats the competition when joint human-robot teams perform inorganic chemical experiments. *Journal of Chemical Information and Modeling*, 59(6):2664–2671, 2019.

[135] G. Eckold and O. Sobolev. Analytical approach to the 4d-resolution function of three axes neutron spectrometers with focussing monochromators and analysers. *Nuclear Instruments and Methods in Physics Research Section A: Accelerators, Spectrometers, Detectors and Associated Equipment*, 752:54–64, 2014.

[136] EcoFAB. Advancing health and environmental science through standardized laboratory microbial ecosystems. `http://eco-fab.org/meetings-and-annual-reports/`, 2018.

[137] Bradley Efron. Bayes' theorem in the 21st century. *Science*, 340(6137):1177–1178, 2013.

[138] K. C. Elbert, T. Thi Vo, N. M. Krook, W. Zygmunt, J. Park, K. G. Yager, R. J. Composto, S. C. Glotzer, and C. B. Murray. Dendrimer Ligand Directed Nanoplate Assembly. *ACS Nano*, 13:14241–14251, 2019.

[139] Jeffrey L Elman. Finding structure in time. *Cognitive Science*, 14(2):179–211, 1990.

[140] D. Eriksson, M. Pearce, J.R. Gardner, R. Turner, and M. Poloczek. Scalable global optimization via local Bayesian optimization. In *33rd Conference on Neural Information Processing Systems (NeurIPS 2019)*, 2019.

[141] N. K. Fageria, V. C. Baligar, and R. B. Clark. Micronutrients in crop production. In Donald L. Sparks, editor, *Advances in Agronomy*, volume 77 of *Advances in Agronomy*, pages 185–268. Academic Press, 2002.

[142] Noah Fahlgren, Maximilian Feldman, Malia A. Gehan, Melinda S. Wilson, Christine Shyu, Douglas W. Bryant, Steven T. Hill, Colton J. McEntee, Sankalpi N. Warnasooriya, Indrajit Kumar, Tracy Ficor, Stephanie Turnipseed, Kerrigan B. Gilbert, Thomas P. Brutnell, James C. Carrington, Todd C. Mockler, and Ivan Baxter. A versatile phenotyping system and analytics platform reveals diverse temporal responses to water availability in setaria. *Molecular Plant*, 8(10):1520–1535, 2015.

[143] Hatem Fakhruldeen, Gabriella Pizzuto, Jakub Glawucki, and Andrew Ian Cooper. Archemist: Autonomous robotic chemistry system architecture. *IEEE International Conference on Robotics and Automation*, 2022.

[144] Steff Farley, Jo E. A. Hodgkinson, Oliver M. Gordon, Joanna Turner, Andrea Soltoggio, Philip J. Moriarty, and Eugenie Hunsicker. Improving the segmentation of scanning probe microscope images using convolutional neural networks. *Machine Learning: Science and Technology*, 2(1):015015, 2020.

[145] Matthias Feurer and Frank Hutter. Hyperparameter optimization. In *Automated Machine Learning*, pages 3–33. Springer, Cham, 2019.

[146] Andrew O. Finley, Huiyan Sang, Sudipto Banerjee, and Alan E. Gelfand. Improving the performance of predictive process modeling for large datasets. *Computational Statistics & Data Analysis*, 53(8):2873–2884, 2009.

[147] B. Fåk and B. Dorner. Phonon line shapes and excitation energies. *Physica B: Condensed Matter*, 234-236:1107–1108, 1997. Proceedings of the First European Conference on Neutron Scattering.

[148] Daniel Foreman-Mackey, Eric Agol, Sivaram Ambikasaran, and Ruth Angus. Fast and scalable gaussian process modeling with applications to astronomical time series. *The Astronomical Journal*, 154(6):220, 2017.

[149] Alessio De Francesco, Alessandro Cunsolo, and Luisa Scaccia. Bayesian approach for x-ray and neutron scattering spectroscopy. In Alessandro Cunsolo, Margareth K. K. D. Franco, and Fabiano Yokaichiya, editors, *Inelastic X-Ray Scattering and X-Ray Powder Diffraction Applications*, chapter 2. IntechOpen, Rijeka, 2020.

[150] Alessio De Francesco, Luisa Scaccia, Martin Bohem, and Alessandro Cunsolo. Bayesian inference as a tool to optimize spectral acquisition in scattering experiments. In Niansheng Tang, editor, *Bayesian Inference*, chapter 5. IntechOpen, Rijeka, 2022.

[151] Peter I. Frazier. A tutorial on bayesian optimization. *arXiv preprint arXiv:1807.02811*, 2018.

[152] Jerome H. Friedman. Multivariate Adaptive Regression Splines. *The Annals of Statistics*, 19(1):1–67, 1991.

[153] Naohiro Fujinuma, Brian DeCost, Jason Hattrick-Simpers, and Samuel E. Lofland. Why big data and compute are not necessarily the path to big materials science. *Communications Materials*, 3(1):59, 2022.

[154] Ryosuke Furuta, Naoto Inoue, and Toshihiko Yamasaki. Pixelrl: Fully convolutional network with reinforcement learning for image processing, 22(7):1704–1719, 2019.

[155] Iason Gabriel. Artificial intelligence, values, and alignment. *Minds and Machines*, 30(3):411–437, 2020.

[156] Pei Gao, Antti Honkela, Magnus Rattray, and Neil D. Lawrence. Gaussian process modelling of latent chemical species: applications to inferring transcription factor activities. *Bioinformatics*, 24(16):i70–i75, 2008.

[157] Yinghua Gao, Naiqi Li, Ning Ding, Yiming Li, Tao Dai, and Shu-Tao Xia. Generalized local aggregation for large scale gaussian process regression. In *2020 International Joint Conference on Neural Networks (IJCNN)*, pages 1–8. IEEE, 2020.

[158] Jacob R. Gardner, Geoff Pleiss, David Bindel, Kilian Q. Weinberger, and Andrew Gordon Wilson. GPyTorch: Blackbox matrix-matrix Gaussian process inference with GPU acceleration. In *Proceedings of the 32nd International Conference on Neural Information Processing Systems*, pages 7587–7597, Red Hook, NY, USA, 2018. Curran Associates Inc.

[159] M. Garg and K. Kern. Attosecond coherent manipulation of electrons in tunneling microscopy. *Science*, 367(6476):411–415, 2020.

[160] Hong Ge, Kai Xu, and Zoubin Ghahramani. Turing: a language for flexible probabilistic inference. In *International Conference on Artificial Intelligence and Statistics*, pages 1682–1690. PMLR, 2018.

[161] Nathan Gesmundo, Kevin Dykstra, James Douthwaite, Babak Mahjour, Ron Ferguson, Spencer Dreher, Berengere Sauvagnat, Josep Sauri, and Tim Cernak. Miniaturization of popular reactions from the medicinal chemists' toolbox for ultrahigh-throughput experimentation. 2022.

[162] Pedram Ghamisi, Yushi Chen, and Xiao Xiang Zhu. A self-improving convolution neural network for the classification of hyperspectral data. *IEEE Geoscience and Remote Sensing Letters*, 13(10):1537–1541, 2016.

[163] H. K. Gibbs and J. M. Salmon. Mapping the world's degraded lands. *Applied Geography*, 57:12–21, 2015.

[164] W. R. Gilks, S. Richardson, and D. Spiegelhalter. *Markov Chain Monte Carlo in Practice*. Chapman and Hall/CRC, 1995.

[165] David Ginsbourger, Xavier Bay, Olivier Roustant, and Laurent Carraro. Argumentwise invariant kernels for the approximation of invariant functions. *Annales de la Faculté de Sciences de Toulouse*, Tome 21(numéro 3):p. 501–527, 2012. https://afst.centre-mersenne.org/item/10.5802/afst.1343.pdf

[166] David Ginsbourger, Nicolas Durrande, and Olivier Roustant. Kernels and designs for modelling invariant functions: From group invariance to additivity. In *mODa 10–Advances in Model-Oriented Design and Analysis*, pages 107–115. Springer, 2013.

[167] Madelyn Glymour, Judea Pearl, and Nicholas P. Jewell. *Causal Inference in Statistics: A Primer*. John Wiley & Sons, 2016.

[168] Tushar Goel, Raphael T. Haftka, Wei Shyy, and Nestor V. Queipo. Ensemble of surrogates. *Structural and Multidisciplinary Optimization*, 33(3):199–216, 2007.

[169] Daniel Golovin, Benjamin Solnik, Subhodeep Moitra, Greg Kochanski, John Karro, and D. Sculley. Google vizier: A service for black-box optimization. In *Proceedings of the 23rd ACM SIGKDD International Conference on Knowledge Discovery and Data Mining*, KDD '17, page 14871495, New York, NY, USA, 2017. Association for Computing Machinery.

[170] Ian Goodfellow, Yoshua Bengio, Aaron Courville, and Yoshua Bengio. *Deep Learning*, volume 1. MIT Press, Cambridge, 2016.

[171] Ian Goodfellow, Jean Pouget-Abadie, Mehdi Mirza, Bing Xu, David Warde-Farley, Sherjil Ozair, Aaron Courville, and Yoshua Bengio. Generative adversarial nets. In *Advances in Neural Information Processing Systems*, pages 2672–2680, 2014.

[172] Abhijith M. Gopakumar, Prasanna V. Balachandran, Dezhen Xue, James E. Gubernatis, and Turab Lookman. Multi-objective optimization for materials discovery via adaptive design. *Scientific Reports*, 8(1):3738, 2018.

[173] Prashun Gorai, Vladan Stevanović, and Eric S. Toberer. Computationally guided discovery of thermoelectric materials. *Nature Reviews Materials*, 2(9):1–16, 2017.

[174] O. Gordon, P. D'Hondt, L. Knijff, S. E. Freeney, F. Junqueira, P. Moriarty, and I. Swart. Scanning tunneling state recognition with multi-class neural network ensembles. *Review of Scientific Instruments*, 90(10):103704, 2019.

[175] Oliver M. Gordon, Jo E. A. Hodgkinson, Steff M. Farley, Eugénie L. Hunsicker, and Philip J. Moriarty. Automated searching and identification of self-organized nanostructures. *Nano Letters*, 20(10):7688–7693, 2020. PMID: 32866019.

[176] Oliver M. Gordon, Filipe L. Q. Junqueira, and Philip J. Moriarty. Embedding human heuristics in machine-learning-enabled probe microscopy. *Machine Learning: Science and Technology*, 1(1):015001, 2020.

[177] I. Goumiri. Gaussian process regression algorithm (muygps) science applications. Poster presentation, CoDA 2023, Santa Fe, NM, USA, March 2023.

[178] gpCAM. Project Website: https://gpcam.lbl.gov/home.

[179] Jarosław M. Granda, Liva Donina, Vincenza Dragone, De-Liang Long, and Leroy Cronin. Controlling an organic synthesis robot with machine learning to search for new reactivity. *Nature*, 559(7714):377–381, 2018.

[180] Martin L. Green, C. L. Choi, J. R. Hattrick-Simpers, A. M. Joshi, I. Takeuchi, S. C. Barron, E. Campo, T. Chiang, S. Empedocles, J. M. Gregoire, et al. Fulfilling the promise of the materials genome initiative with high-throughput experimental methodologies. *Applied Physics Reviews*, 4(1):011105, 2017.

[181] Léo Grinsztajn, Edouard Oyallon, and Gaël Varoquaux. Why do tree-based models still outperform deep learning on tabular data? *arXiv preprint arXiv:2207.08815*, 2022.

[182] Kai Guo, Zhenze Yang, Chi-Hua Yu, and Markus J. Buehler. Artificial intelligence and machine learning in design of mechanical materials. *Materials Horizons*, 8(4):1153–1172, 2021.

[183] Moritz Hardt, Ben Recht, and Yoram Singer. Train faster, generalize better: Stability of stochastic gradient descent. In *International Conference on Machine Learning*, pages 1225–1234. PMLR, 2016.

[184] Gus L. W. Hart, Tim Mueller, Cormac Toher, and Stefano Curtarolo. Machine learning for alloys. *Nature Reviews Materials*, 6(8):730–755, 2021.

[185] Terry C. Hazen, Eric A. Dubinsky, Todd Z. DeSantis, Gary L. Andersen, Yvette M. Piceno, Navjeet Singh, Janet K. Jansson, Alexander Probst, Sharon E. Borglin, Julian L. Fortney, et al. Deep-sea oil plume enriches indigenous oil-degrading bacteria. *Science*, 330(6001):204–208, 2010.

[186] Ji He, Jianshu Chen, Xiaodong He, Jianfeng Gao, Lihong Li, Li Deng, and Mari Ostendorf. Deep reinforcement learning with a natural language action space. *arXiv preprint arXiv:1511.04636*, 2015.

[187] Gary W. Heiman. *Understanding Research Methods and Statistics: An Integrated Introduction for Psychology.* Houghton, Mifflin and Company, 2001.

[188] Todd Hester, Michael Quinlan, and Peter Stone. Generalized model learning for reinforcement learning on a humanoid robot. In *Robotics and Automation (ICRA), 2010 IEEE International Conference on*, pages 2369–2374. IEEE, 2010.

[189] Todd Hester and Peter Stone. Generalized model learning for reinforcement learning in factored domains. In *Proceedings of The 8th International Conference on Autonomous Agents and Multiagent Systems-Volume 2*, pages 717–724. International Foundation for Autonomous Agents and Multiagent Systems, 2009.

[190] John Hill, Stuart Campbell, Gabriella Carini, Yu-Chen Karen Chen-Wiegart, Yong Chu, Andrei Fluerasu, Masafumi Fukuto, Mourad Idir, Jean Jakoncic, Ignace Jarrige, Peter Siddons, Toshi Tanabe, and Kevin G. Yager. Future trends in synchrotron science at nsls-ii. *Journal of Physics: Condensed Matter*, 32(37):374008, 2020.

[191] Sepp Hochreiter and Jürgen Schmidhuber. Long short-term memory. *Neural Computation*, 9(8):1735–1780, 1997.

[192] Sebastian Höfer, Kostas Bekris, Ankur Handa, Juan Camilo Gamboa, Melissa Mozifian, Florian Golemo, Chris Atkeson, Dieter Fox, Ken Goldberg, John Leonard, et al. Sim2real in robotics and automation: Applications and challenges. *IEEE Transactions on Automation Science and Engineering*, 18(2):398–400, 2021.

[193] Thomas Hofmann, Bernhard Schölkopf, and Alexander J. Smola. Kernel methods in machine learning. *The Annals of Statistics*, 36(3):1171–1220, 2008.

[194] Elizabeth A. Holman, Yuan-Sheng Fang, Liang Chen, Michael DeWeese, Hoi-Ying N. Holman, and Paul W. Sternberg. Autonomous adaptive data acquisition for scanning hyperspectral imaging. *Communications Biology*, 3(1):1–7, 2020.

[195] Hoi-Ying N. Holman, Hans A. Bechtel, Zhao Hao, and Michael C. Martin. Synchrotron ir spectromicroscopy: chemistry of living cells, 82, 8757–8765, 2010.

[196] Hoi-Ying N. Holman, Karl Nieman, Darwin L. Sorensen, Charles D. Miller, Michael C. Martin, Thomas Borch, Wayne R. McKinney, and Ronald C. Sims. Catalysis of pah biodegradation by humic acid shown in synchrotron infrared studies. *Environmental Science & Technology*, 36(6):1276–1280, 2002.

[197] Hoi-Ying N. Holman, Dale L. Perry, Michael C. Martin, Geraldine M. Lamble, Wayne R. McKinney, and Jennie C. Hunter-Cevera. Real-time characterization of biogeochemical reduction of cr (vi) on basalt surfaces by sr-ftir imaging. *Geomicrobiology Journal*, 16(4):307–324, 1999.

[198] Hoi-Ying N. Holman, Eleanor Wozei, Zhang Lin, Luis R. Comolli, David A. Ball, Sharon Borglin, Matthew W. Fields, Terry C. Hazen, and Kenneth H. Downing. Real-time molecular monitoring of chemical environment in obligate anaerobes during oxygen adaptive response. *Proceedings of the National Academy of Sciences*, 106(31):12599–12604, 2009.

[199] Alexander T. Holmes. Simple bayesian method for improved analysis of quasi-two-dimensional scattering data. *Physical Review B*, 90:024514, 2014.

[200] Benedikt Hoock, Santiago Rigamonti, and Claudia Draxl. Advancing descriptor search in materials science: feature engineering and selection strategies. *New Journal of Physics*, 24(11):113049, 2022.

[201] Dan Horgan, John Quan, David Budden, Gabriel Barth-Maron, Matteo Hessel, Hado van Hasselt, and David Silver. Distributed prioritized experience replay. *CoRR*, abs/1803.00933, 2018.

[202] Kurt Hornik, Maxwell Stinchcombe, and Halbert White. Multilayer feedforward networks are universal approximators. *Neural Networks*, 2:359–366, 1989.

[203] Jianjun Hu, Stanislav Stefanov, Yuqi Song, Sadman Sadeed Omee, Steph-Yves Louis, Edirisuriya M.D. Siriwardane, Yong Zhao, and Lai Wei. Materialsatlas. org: a materials informatics web app platform for materials discovery and survey of state-of-the-art. *npj Computational Materials*, 8(1):65, 2022.

[204] Ying Huang, David Perlmutter, Andrea Fei-Huei Su, Jerome Quenum, Pavel Shevchenko, Dilworth Y. Parkinson, Iryna V. Zenyuk, and Daniela Ushizima. Detecting lithium plating dynamics in a solid-state battery with operando x-ray computed tomography using machine learning. *npj Computational Materials*, 9(1):93, 2023.

[205] Evan Hubinger, Chris van Merwijk, Vladimir Mikulik, Joar Skalse, and Scott Garrabrant. Risks from learned optimization in advanced machine learning systems, 2019. https://intelligence.org/learned-optimization/

[206] Kazunori Iwata. Extending the peak bandwidth of parameters for softmax selection in reinforcement learning. *IEEE Transactions on Neural Networks and Learning Systems*, 28(8):1865–1877, 2017.

[207] Lauren K. Jabusch, Peter W. Kim, Dawn Chiniquy, Zhiying Zhao, Bing Wang, Benjamin Bowen, Ashley J. Kang, Yasuo Yoshikuni, Adam M. Deutschbauer, Anup K. Singh, and Trent R. Northen. Microfabrication of a chamber for high-resolution, in situ imaging of the whole root for plantmicrobe interactions. *International Journal of Molecular Sciences*, 22(15):7880, 2021.

[208] Ryan Jacobs, Tam Mayeshiba, Ben Afflerbach, Luke Miles, Max Williams, Matthew Turner, Raphael Finkel, and Dane Morgan. The materials simulation toolkit for machine learning (mast-ml): An automated open source toolkit

to accelerate data-driven materials research. *Computational Materials Science*, 176:109544, 2020.

[209] Alon Jacovi. Trends in explainable ai (xai) literature. *arXiv preprint arXiv:2301.05433*, 2023.

[210] Andrew Jaegle, Felix Gimeno, Andy Brock, Oriol Vinyals, Andrew Zisserman, and Joao Carreira. Perceiver: General perception with iterative attention. In *International Conference on Machine Learning*, pages 4651–4664. PMLR, 2021.

[211] Anil K. Jain, Jianchang Mao, and K.M. Mohiuddin. Artificial neural networks: a tutorial. *Computer*, 29(3):31–44, 1996.

[212] E. Jenne. *Adsorption of Metals by Geomedia: Variables, Mechanisms, and Model Applications*. Elsevier Science, 1998.

[213] Hao Jiang, Zhanchi Wang, Yusong Jin, Xiaotong Chen, Peijin Li, Yinghao Gan, Sen Lin, and Xiaoping Chen. Hierarchical control of soft manipulators towards unstructured interactions. *The International Journal of Robotics Research*, 40(1):411–434, 2021.

[214] Wei-Cheng Jiang, Vignesh Narayanan, and Jr-Shin Li. Model learning and knowledge sharing for cooperative multiagent systems in stochastic environment. *IEEE Transactions on Cybernetics*, 51:5717–5727, 2020.

[215] Yangwei Jiang and Shi-Jie Chen. Rldock method for predicting rna-small molecule binding modes. *Methods*, 197:97–105, 2022.

[216] Donald R. Jones. A taxonomy of global optimization methods based on response surfaces. *Journal of Global Optimization*, 21(4):345–383, 2001.

[217] Donald R. Jones, Matthias Schonlau, and William J. Welch. Efficient global optimization of expensive black-box functions. *Journal of Global Optimization*, 13(4):455–492, 1998.

[218] Prateek Joshi. *Artificial Intelligence with Python: A Comprehensive Guide to Building Intelligent Apps for Python Beginners and Developers*. Packt Publishing Ltd, 2017.

[219] Frédéric Joucken, John L. Davenport, Zhehao Ge, Eberth A. Quezada-Lopez, Takashi Taniguchi, Kenji Watanabe, Jairo Velasco, Jérôme Lagoute, and Robert A. Kaindl. Denoising scanning tunneling microscopy images with machine learning, *Physical Review Materials*, 6, 2022.

[220] James Joyce. Bayes' theorem. The Stanford Encyclopedia of Philosophy (Fall 2021 Edition), Edward N. Zalta (ed.), https://plato.stanford.edu/archives/fall2021/entries/bayes-theorem/. 2003.

[221] Sham M. Kakade. A natural policy gradient. *Advances in Neural Information Processing Systems*, 14, (NIPS 2001).

[222] Sergei V. Kalinin, Ondrej Dyck, Stephen Jesse, and Maxim Ziatdinov. Exploring order parameters and dynamic processes in disordered systems via variational autoencoders. *Science Advances*, 7(17):eabd5084, 2021.

[223] Sergei V. Kalinin, Mani Valleti, Rama K. Vasudevan, and Maxim Ziatdinov. Exploration of lattice hamiltonians for functional and structural discovery via gaussian process-based explorationexploitation. *Journal of Applied Physics*, 128(16):164304, 2020.

[224] Sergei V. Kalinin, Maxim Ziatdinov, Jacob Hinkle, Stephen Jesse, Ayana Ghosh, Kyle P. Kelley, Andrew R. Lupini, Bobby G. Sumpter, and Rama K. Vasudevan. Automated and autonomous experiments in electron and scanning probe microscopy. *ACS Nano*, 15(8):12604–12627, 2021. PMID: 34269558.

[225] Shivaram Kalyanakrishnan and Peter Stone. Batch reinforcement learning in a complex domain. In *Proceedings of the Sixth International Joint Conference on Autonomous Agents and Multiagent Systems*, page 94. ACM, 2007.

[226] Motonobu Kanagawa, Philipp Hennig, Dino Sejdinovic, and Bharath K. Sriperumbudur. Gaussian processes and kernel methods: A review on connections and equivalences. *arXiv preprint arXiv:1807.02582*, 2018.

[227] George Em Karniadakis, Ioannis G. Kevrekidis, Lu Lu, Paris Perdikaris, Sifan Wang, and Liu Yang. Physics-informed machine learning. *Nature Reviews Physics*, 3(6):422–440, 2021.

[228] Yukari Katsura, Masaya Kumagai, Takushi Kodani, Mitsunori Kaneshige, Yuki Ando, Sakiko Gunji, Yoji Imai, Hideyasu Ouchi, Kazuki Tobita, Kaoru Kimura, et al. Data-driven analysis of electron relaxation times in pbte-type thermoelectric materials. *Science and Technology of Advanced Materials*, 20(1):511–520, 2019.

[229] Matthias Katzfuss and Joseph Guinness. A general framework for vecchia approximations of gaussian processes. *Statistical Science*, 36(1):124–141, 2021.

[230] Ke, Brewster, Yu, Yang, Ushizima, and Sauter. A convolutional neural network-based screening tool for x-ray serial crystallography. *Journal of Synchrotron Radiation*, 25:655–670, 2018.

[231] Kyle P. Kelley, Maxim Ziatdinov, Liam Collins, Michael A. Susner, Rama K. Vasudevan, Nina Balke, Sergei V. Kalinin, and Stephen Jesse. Fast scanning probe microscopy via machine learning: Non-rectangular scans with compressed sensing and gaussian process optimization. *Small*, 16(37):2002878, 2020.

[232] Yaser Keneshloo, Tian Shi, Naren Ramakrishnan, and Chandan K. Reddy. Deep reinforcement learning for sequence-to-sequence models. *IEEE Transactions on Neural Networks and Learning Systems*, 31(7):2469–2489, 2019.

[233] Ross D. King, Kenneth E. Whelan, Ffion M. Jones, Philip G. K. Reiser, Christopher H. Bryant, Stephen H. Muggleton, Douglas B. Kell, and Stephen G. Oliver. Functional genomic hypothesis generation and experimentation by a robot scientist. *Nature*, 427(6971):247–252, 2004.

[234] Diederik P. Kingma and Jimmy Ba. Adam: A method for stochastic optimization. December 2014.

[235] Peter K. Kitanidis and Efstratios G. Vomvoris. A geostatistical approach to the inverse problem in groundwater modeling (steady state) and one-dimensional simulations. *Water Resources Research*, 19(3):677–690, 1983.

[236] Tim Kittel, Finn Müller-Hansen, Rebekka Koch, Jobst Heitzig, Guillaume Deffuant, Jean-Denis Mathias, and Jürgen Kurths. From lakes and glades to viability algorithms: automatic classification of system states according to the topology of sustainable management. *The European Physical Journal Special Topics*, 230:3133–3152, 2021.

[237] Lucas J. Koerner, Thomas A. Caswell, Daniel B. Allan, and Stuart I. Campbell. A python instrument control and data acquisition suite for reproducible research. *IEEE Transactions on Instrumentation and Measurement*, 69(4):1698–1707, 2020.

[238] Teuvo Kohonen and Timo Honkela. Kohonen network. *Scholarpedia*, 2(1):1568, 2007.

[239] Tatiana Konstantinova, Phillip M. Maffettone, Bruce Ravel, Stuart I. Campbell, Andi M. Barbour, and Daniel Olds. Machine learning enabling high-throughput and remote operations at large-scale user facilities. *Digital Discovery*, 1(4):413–426, 2022.

[240] Tatiana Konstantinova, Lutz Wiegart, Maksim Rakitin, Anthony M. DeGennaro, and Andi M. Barbour. Noise reduction in x-ray photon correlation spectroscopy with convolutional neural networks encoder–decoder models. *Scientific Reports*, 11(1):14756, 2021.

[241] Tatiana Konstantinova, Lutz Wiegart, Maksim Rakitin, Anthony M. DeGennaro, and Andi M. Barbour. Noise reduction in x-ray photon correlation spectroscopy with convolutional neural networks encoder–decoder models. *Scientific Reports*, 11(1):14756, 2021.

[242] Tatiana Konstantinova, Lutz Wiegart, Maksim Rakitin, Anthony M. DeGennaro, and Andi M. Barbour. Machine learning for analysis of speckle dynamics: quantification and outlier detection. *Physical Review Research*, 4:033228, 2022.

[243] B. Koslowski, C. Dietrich, A. Tschetschetkin, and P. Ziemann. Evaluation of scanning tunneling spectroscopy data: Approaching a quantitative determination of the electronic density of states. *Physical Review B*, 75:035421, 2007.

[244] R. Krishnan, S. Jagannathan, and V. A. Samaranayake. Direct error driven learning for deep neural networks with applications to bigdata. *Procedia Computer Science*, 144:89–95, 2018.

[245] Anders Krogh. What are artificial neural networks? *Nature Biotechnology*, 26(2):195–197, 2008.

[246] A. Krull, P. Hirsch, C. Rother, A. Schiffrin, and C. Krull. Artificial-intelligence-driven scanning probe microscopy. *Communications Physics*, 3(1):54, 2020.

[247] Harold W. Kuhn and Albert W. Tucker. Nonlinear programming. In *Traces and Emergence of Nonlinear Programming*, pages 247–258. Springer, 2013.

[248] A. Gilad Kusne, Heshan Yu, Changming Wu, Huairuo Zhang, Jason Hattrick-Simpers, Brian DeCost, Suchismita Sarker, Corey Oses, Cormac Toher, Stefano Curtarolo, et al. On-the-fly closed-loop materials discovery via bayesian active learning. *Nature Communications*, 11(1):5966, 2020.

[249] Aaron Gilad Kusne, Tieren Gao, Apurva Mehta, Liqin Ke, Manh Cuong Nguyen, Kai-Ming Ho, Vladimir Antropov, Cai-Zhuang Wang, Matthew J. Kramer, Christian Long, and Ichiro Takeuchi. On-the-fly machine-learning for high-throughput experiments: search for rare-earth-free permanent magnets. *Scientific Reports*, 4(1):6367, 2014.

[250] D. Kusnetzky. *Virtualization: A Manager's Guide*. O'Reilly, 2011.

[251] Brookhaven National Laboratory. Scianalysis. https://github.com/CFN-softbio/SciAnalysis, 2015.

[252] Lauro Langosco, Jack Koch, Lee Sharkey, Jacob Pfau, Laurent Orseau, and David Krueger. Goal misgeneralization in deep reinforcement learning, *Proceedings of Machine Learning Research*, 162:12004–12019, 2021.

[253] Jakob Lass, Magnus Egede Bøggild, Per Hedegard, and Kim Lefmann. Multinomial, poisson and gaussian statistics in count data analysis. *Journal of Neutron Research*, 23:69–92, 2021.

[254] Neil Lawrence and Aapo Hyvärinen. Probabilistic non-linear principal component analysis with gaussian process latent variable models. *Journal of Machine Learning Research*, 6(11), 1783–1816, 2005.

[255] S. Le Digabel and S. Wild. A taxonomy of constraints in simulation-based optimization. arXiv:1505.07881, 2015.

[256] Sébastien Le Digabel. Algorithm 909: NOMAD: Nonlinear optimization with the MADS algorithm. *ACM Transactions on Mathematical Software*, 37(4), 2011.

[257] Yann LeCun, Yoshua Bengio, and Geoffrey Hinton. Deep learning. *Nature*, 521(7553):436, 2015.

[258] Dong-Hyun Lee, Saizheng Zhang, Asja Fischer, and Yoshua Bengio. Difference target propagation. In *Joint European Conference on Machine Learning and Knowledge Discovery in Databases*, pages 498–515. Springer, 2015.

[259] H. Lee, R. Gramacy, C. Linkletter, and G. Gray. Optimization subject to hidden constraints via statistical emulation. Technical Report UCSC-SOE-10-10, University of California, Santa Cruz, 2010.

[260] Drew A. Leins, Steven B. Haase, Mohammed Eslami, Joshua Schrier, and Jared T. Freeman. Collaborative methods to enhance reproducibility and accelerate discovery. *Digital Discovery*, 2:12–27, 2023.

[261] Matteo Leonetti, Luca Iocchi, and Peter Stone. A synthesis of automated planning and reinforcement learning for efficient, robust decision-making. *Artificial Intelligence*, 241:103–130, 2016.

[262] Erika Levenson, Philippe Lerch, and Michael C. Martin. Infrared imaging: Synchrotrons vs. arrays, resolution vs. speed. *Infrared Physics & Technology*, 49(1-2):45–52, 2006.

[263] Erika Levenson, Philippe Lerch, and Michael C. Martin. Spatial resolution limits for synchrotron-based spectromicroscopy in the mid-and near-infrared. *Journal of Synchrotron Radiation*, 15(4):323–328, 2008.

[264] Sergey Levine, Chelsea Finn, Trevor Darrell, and Pieter Abbeel. End-to-end training of deep visuomotor policies. *The Journal of Machine Learning Research*, 17(1):1334–1373, 2016.

[265] Frank L. Lewis, Draguna Vrabie, and Kyriakos G. Vamvoudakis. Reinforcement learning and feedback control: Using natural decision methods to design optimal adaptive controllers. *IEEE Control Systems Magazine*, 32(6):76–105, 2012.

[266] Bin Li, Steven S. Kaye, Conor Riley, Doron Greenberg, Daniel Galang, and Mark S. Bailey. Hydrogen storage materials discovery via high throughput ball milling and gas sorption. *ACS Combinatorial Science*, 14(6):352–358, 2012. PMID: 22616741.

[267] Jiwei Li, Will Monroe, Alan Ritter, Michel Galley, Jianfeng Gao, and Dan Jurafsky. Deep reinforcement learning for dialogue generation, *Proceedings of the 2016 Conference on Empirical Methods in Natural Language Processing*, 1192–1202, 2016.

[268] Jiwei Li, Will Monroe, Tianlin Shi, Sébastien Jean, Alan Ritter, and Dan Jurafsky. Adversarial learning for neural dialogue generation, *Proceedings of the 2017 Conference on Empirical Methods in Natural Language Processing*, 2157–2169, 2017.

[269] Shuang Li, Shuai Xiao, Shixiang Zhu, Nan Du, Yao Xie, and Le Song. Learning temporal point processes via reinforcement learning. *Advances in Neural Information Processing Systems*, 31, 2018.

[270] Zhi Li, Philip W Nega, Mansoor Ani Najeeb Nellikkal, Chaochao Dun, Matthias Zeller, Jeffrey J. Urban, Wissam A. Saidi, Joshua Schrier, Alexander J. Norquist, and Emory M. Chan. Dimensional control over metal halide perovskite crystallization guided by active learning. *Chemistry of Materials*, 34(2):756–767, 2022.

[271] Richard Liaw, Eric Liang, Robert Nishihara, Philipp Moritz, Joseph E. Gonzalez, and Ion Stoica. Tune: A research platform for distributed model selection and training. *arXiv preprint arXiv:1807.05118*, 2018.

[272] Angelica Liguori, Luciano Caroprese, Marco Minici, Bruno Veloso, Francesco Spinnato, Mirco Nanni, Giuseppe Manco, and Joao Gama. Modeling events and interactions through temporal processes–a survey. *arXiv preprint arXiv:2303.06067*, 2023.

[273] Lizhen Lin, Niu Mu, Pokman Cheung, and David Dunson. Extrinsic gaussian processes for regression and classification on manifolds. *Bayesian Analysis*, 14(3):887–906, 2019.

[274] Long-Ji Lin. Self-improving reactive agents based on reinforcement learning, planning and teaching. *Machine Learning*, 8(3-4):293–321, 1992.

[275] Zhouhan Lin, Minwei Feng, Cicero Nogueira dos Santos, Mo Yu, Bing Xiang, Bowen Zhou, and Yoshua Bengio. A structured self-attentive sentence embedding. *arXiv preprint arXiv:1703.03130*, 2017.

[276] Chen Ling. A review of the recent progress in battery informatics. *npj Computational Materials*, 8(1):33, 2022.

[277] Daochang Liu and Tingting Jiang. Deep reinforcement learning for surgical gesture segmentation and classification, *Medical Image Computing and Computer Assisted Intervention – MICCAI*, 2018.

[278] Naijia Liu, Tianxing Ma, Chaoqun Liao, Guannan Liu, Rodrigo Miguel Ojeda Mota, Jingbei Liu, Sungwoo Sohn, Sebastian Kube, Shaofan Zhao, Jonathan P. Singer, et al. Combinatorial measurement of critical cooling rates in aluminum-base metallic glass forming alloys. *Scientific Reports*, 11(1):1–9, 2021.

[279] Shuai Liu, Charles N. Melton, Singanallur Venkatakrishnan, Ronald J. Pandolfi, Guillaume Freychet, Dinesh Kumar, Haoran Tang, Alexander Hexemer, and Daniela M. Ushizima. Convolutional neural networks for grazing incidence x-ray scattering patterns: thin film structure identification. *MRS Communications*, pages 1–7, 2019.

[280] Yongtao Liu, Kyle P. Kelley, Rama K. Vasudevan, Hiroshi Funakubo, Maxim A. Ziatdinov, and Sergei V. Kalinin. Experimental discovery of structure–property relationships in ferroelectric materials via active learning. *Nature Machine Intelligence*, 4(4):341–350, 2022.

[281] Yue Liu, Tianlu Zhao, Wangwei Ju, and Siqi Shi. Materials discovery and design using machine learning. *Journal of Materiomics*, 3(3):159–177, 2017.

[282] Yun Liu, Oladapo Christopher Esan, Zhefei Pan, and Liang An. Machine learning for advanced energy materials. *Energy and AI*, 3:100049, 2021.

[283] Olimpia Lombardi, Federico Holik, and Leonardo Vanni. What is shannon information? *Synthese*, 193:1983–2012, 2016.

[284] T. Lookman, P. V. Balachandran, D. Xue, G. Pilania, T. Shearman, J. Theiler, J. E. Gubernatis, J. Hogden, K. Barros, E. BenNaim, and F. J. Alexander. *A Perspective on Materials Informatics: State-of-the-Art and Challenges*, pages 3–12. Springer International Publishing, Cham, 2016.

[285] Turab Lookman, Francis J. Alexander, and Alan R. Bishop. Perspective: Codesign for materials science: An optimal learning approach. *APL Materials*, 4(5):053501, 2016.

[286] Turab Lookman, Prasanna V. Balachandran, Dezhen Xue, John Hogden, and James Theiler. Statistical inference and adaptive design for materials discovery. *Current Opinion in Solid State and Materials Science*, 21(3):121–128, 2017. Materials Informatics: Insights, Infrastructure, and Methods.

[287] Alfred Ludwig. Discovery of new materials using combinatorial synthesis and high-throughput characterization of thin-film materials libraries combined with computational methods. *npj Computational Materials*, 5(1):70, 2019.

[288] Minh-Thang Luong, Hieu Pham, and Christopher D. Manning. Effective approaches to attention-based neural machine translation. *arXiv preprint arXiv:1508.04025*, 2015.

[289] Steven A. Macenka and Michael P. Chrisp. Airborne Visible/Infrared Imaging Spectrometer (Aviris) Spectrometer Design And Performance. In Gregg Vane, editor, *Imaging Spectroscopy II*, volume 0834, pages 32–43. International Society for Optics and Photonics, SPIE, 1987.

[290] Benjamin P. MacLeod, Fraser G. L. Parlane, Thomas D. Morrissey, Florian Häse, Loïc M. Roch, Kevan E. Dettelbach, Raphaell Moreira, Lars P. E. Yunker, Michael B. Rooney, Joseph R. Deeth, et al. Self-driving laboratory for accelerated discovery of thin-film materials. *Science Advances*, 6(20):eaaz8867, 2020.

[291] Phillip Maffettone, Daniel Allan, Stuart I. Campbell, Matthew R. Carbone, Thomas Caswell, Brian L. DeCost, Dmitri Gavrilov, Marcus Hanwell, Howie Joress, Joshua Lynch, Bruce Ravel, Stuart Wilkins, Jakub Wlodek, and Daniel Olds. Self-driving multimodal studies at user facilities. In *AI for Accelerated Materials Design NeurIPS 2022 Workshop*, 2022.

[292] Phillip M. Maffettone, Daniel B. Allan, Stuart I. Campbell, Matthew R. Carbone, Thomas A. Caswell, Brian L. DeCost, Dmitri Gavrilov, Marcus D. Hanwell, Howie Joress, Joshua Lynch, et al. Self-driving multimodal studies at user facilities. *arXiv preprint arXiv:2301.09177*, 2023.

[293] Phillip M. Maffettone, Lars Banko, Peng Cui, Yury Lysogorskiy, Marc A. Little, Daniel Olds, Alfred Ludwig, and Andrew I. Cooper. Crystallography companion agent for high-throughput materials discovery. *Nature Computational Science*, 1(4):290–297, 2021.

[294] Phillip M. Maffettone, Stuart Campbell, Marcus D. Hanwell, Stuart Wilkins, and Daniel Olds. Delivering real-time multi-modal materials analysis with enterprise beamlines. *Cell Reports Physical Science*, 3(11):101–112, 2022.

[295] Phillip M. Maffettone, Aidan C. Daly, and Daniel Olds. Constrained non-negative matrix factorization enabling real-time insights of in situ and high-throughput experiments. *Applied Physics Reviews*, 8(4):041410, 2021.

[296] Phillip M. Maffettone, Pascal Friederich, Sterling G. Baird, Ben Blaiszik, Keith A. Brown, Stuart I. Campbell, Orion A. Cohen, Tantum Collins, Rebecca L. Davis, Ian T. Foster, Navid Haghmoradi, Mark Hereld, Howie Joress, Nicole Jung, Ha-Kyung Kwon, Gabriella Pizzuto, Jacob Rintamaki, Casper Steinmann, Luca Torresi, and Shijing Sun. What is missing in autonomous discovery: Open challenges for the community, 2023.

[297] Phillip M. Maffettone, Joshua K. Lynch, Thomas A. Caswell, Clara E. Cook, Stuart I. Campbell, and Daniel Olds. Gaming the beamlines—employing reinforcement learning to maximize scientific outcomes at large-scale user facilities. *Machine Learning: Science and Technology*, 2, 2021.

[298] Phillip M. Maffettone, Joshua K. Lynch, Thomas A. Caswell, Clara E. Cook, Stuart I. Campbell, and Daniel Olds. Gaming the beamlines—employing reinforcement learning to maximize scientific outcomes at large-scale user facilities. *Machine Learning: Science and Technology*, 2(2):025025, 2021.

[299] P. M. Maffettone, L. Banko, P. Cui, Y. Lysogorskiy, M. A. Little, D. Olds, A. Ludwig, and A. I. Cooper. Crystallography companion agent for high-throughput materials discovery. *Nature Computational Science*, 1:290–297, 2021.

[300] Wilhelm F. Maier, Klaus Stoewe, and Simone Sieg. Combinatorial and high-throughput materials science. *Angewandte Chemie International Edition*, 46(32):6016–6067, 2007.

[301] Jonathan H. Manton, Pierre-Olivier Amblard, et al. A primer on reproducing kernel hilbert spaces. *Foundations and Trends® in Signal Processing*, 8(1–2):1–126, 2015.

[302] Sergei Manzhos and Manabu Ihara. On the optimization of hyperparameters in gaussian process regression. *arXiv preprint arXiv:2112.01374*, 2021.

[303] Stefano Marelli and Bruno Sudret. Uqlab user manual–polynomial chaos expansions. *Chair of Risk, Safety & Uncertainty Quantification, ETH Zürich, 0.9-104 edition*, pages 97–110, 2015.

[304] Nadia Martinez, Hadis Anahideh, Jay M. Rosenberger, Diana Martinez, Victoria C. P. Chen, and Bo Ping Wang. Global optimization of non-convex piecewise linear regression splines. *Journal of Global Optimization*, 68(3):563–586, 2017.

[305] Olivia U. Mason, Terry C. Hazen, Sharon Borglin, Patrick S. G. Chain, Eric A. Dubinsky, Julian L. Fortney, James Han, Hoi-Ying N. Holman, Jenni Hultman, Regina Lamendella, et al. Metagenome, metatranscriptome and single-cell sequencing reveal microbial response to deepwater horizon oil spill. *The ISME Journal*, 6(9):1715–1727, 2012.

[306] Amir massoud Farahmand, Azad Shademan, Martin Jagersand, and Csaba Szepesvári. Model-based and model-free reinforcement learning for visual servoing. In *Robotics and Automation, 2009. ICRA'09. IEEE International Conference on*, pages 2917–2924. IEEE, 2009.

[307] G. Matheron. Principles of geostatistics. *Economic Geology*, 58:1246–1266, 1963.

[308] Andrew R. McCluskey, Joshaniel F. K. Cooper, Tom Arnold, and Tim Snow. A general approach to maximise information density in neutron reflectometry analysis. *Machine Learning: Science and Technology*, 1(3):035002, 2020.

[309] Ian McCue, Joshua Stuckner, Mitsu Murayama, and Michael J. Demkowicz. Gaining new insights into nanoporous gold by mining and analysis of published images. *Scientific Reports*, 8(1):1–11, 2018.

[310] Warren McCulloch and Walter Pitts. A logical calculus of ideas immanent in nervous activity. *Bulletin of Mathematical Biophysics*, 5:127–147, 1943.

[311] Austin McDannald, Matthias Frontzek, Andrei T. Savici, Mathieu Doucet, Efrain E. Rodriguez, Kate Meuse, Jessica Opsahl-Ong, Daniel Samarov, Ichiro Takeuchi, William Ratcliff, et al. On-the-fly autonomous control of neutron diffraction via physics-informed bayesian active learning. *Applied Physics Reviews*, 9(2):021408, 2022.

[312] Austin McDannald, Matthias Frontzek, Andrei T. Savici, Mathieu Doucet, Efrain E. Rodriguez, Kate Meuse, Jessica Opsahl-Ong, Daniel Samarov, Ichiro Takeuchi, William Ratcliff, and A. Gilad Kusne. On-the-fly autonomous control of neutron diffraction via physics-informed bayesian active learning. *Applied Physics Reviews*, 9(2):021408, 2022.

[313] Leland McInnes, John Healy, and James Melville. Umap: Uniform manifold approximation and projection for dimension reduction. *arXiv preprint arXiv:1802.03426*, 2018.

[314] Charles N. Melton, Marcus M. Noack, Taisuke Ohta, Thomas E. Beechem, Jeremy Robinson, Xiaotian Zhang, Aaron Bostwick, Chris Jozwiak, Roland J. Koch, Petrus H. Zwart, Alexander Hexemer, and Eli Rotenberg. K-means-driven gaussian process data collection for angle-resolved photoemission spectroscopy. *Machine Learning: Science and Technology*, 1(4):045015, oct 2020.

[315] Charles Micchelli and Massimiliano Pontil. Kernels for multi–task learning. *Advances in Neural Information Processing Systems*, 17, (NIPS 2004).

[316] Aaron Noam Michelson. *The Design of Complex Material Aided by DNA Nanotechnology*. PhD thesis, Columbia University, 2022.

[317] Andreia Valentina Miclea, Romulus Terebes, and Serban Meza. One dimensional convolutional neural networks and local binary patterns for hyperspectral image classification. In *2020 IEEE International Conference on Automation, Quality and Testing, Robotics (AQTR)*, pages 1–6, May 2020.

[318] S. Miramontes, T. Pierges, L. Grinberg, and D. M. Ushizima. Accelerating quantitative microscopy with u-net-based cell counting. *IEEE International Symposium on Biomedical Imaging (ISBI)*, April 2021.

[319] S. Miramontes, A. M. H. Piergies, L. T. Grinberg, and D. Ushizima. The role of biomarkers in cell counting with u-net cnn. *Alzheimer's & Dementia*, 17:e051135, 2021.

[320] Tom M. Mitchell. *Machine Learning*. McGraw-Hill, New York, NY, 2010.

[321] Volodymyr Mnih, Adria Puigdomenech Badia, Mehdi Mirza, Alex Graves, Timothy Lillicrap, Tim Harley, David Silver, and Koray Kavukcuoglu. Asynchronous methods for deep reinforcement learning. In *International Conference on Machine Learning*, pages 1928–1937. PMLR, 2016.

[322] Volodymyr Mnih, Koray Kavukcuoglu, David Silver, Andrei A. Rusu, Joel Veness, Marc G. Bellemare, Alex Graves, Martin Riedmiller, Andreas K. Fidjeland, Georg Ostrovski, et al. Human-level control through deep reinforcement learning. *Nature*, 518(7540):529–533, 2015.

[323] Robert Munro Monarch. *Human-in-the-Loop Machine Learning: Active Learning and Annotation for Human-Centered AI*. Simon and Schuster, 2021.

[324] D. Monserrat, A. Vispa, L. C. Pardo, R. Tolchenov, S. Mukhopadhyay, and F. Fernandez-Alonso. Fabada goes mantid to answer an old question: How many lines are there? *Journal of Physics: Conference Series*, 663(1):012009, 2015.

[325] T. W. Morris, M. Rakitin, A. Giles, J. Lynch, A. L. Walter, B. Nash, D. Abell, P. Moeller, I. Pogorelov, and N. Goldring. On-the-fly optimization of synchrotron beamlines using machine learning. In *Optical System Alignment, Tolerancing, and Verification XIV*, volume 12222. SPIE, 2022.

[326] Martin Mourigal, Mechthild Enderle, Axel Klöpperpieper, Jean-Sébastien Caux, Anne Stunault, and Henrik M. Rønnow. Fractional spinon excitations in the quantum heisenberg antiferromagnetic chain. *Nature Physics*, 9(7):435–441, 2013.

[327] Juliane Müller. MISO: mixed-integer surrogate optimization framework. *Optimization and Engineering*, 17(1):177–203, 2016.

[328] Juliane Müller. SOCEMO: Surrogate optimization of computationally expensive multiobjective problems. *INFORMS Journal on Computing*, 29(4):581–596, 2017.

[329] Juliane Müller and Marcus Day. Surrogate optimization of computationally expensive black-box problems with hidden constraints. *INFORMS Journal on Computing*, 31(4):689–702, 2019.

[330] Juliane Müller, Wim Lavrijsen, Costin Iancu, and Wibe de Jong. Accelerating noisy VQE optimization with Gaussian processes. In *2022 IEEE International Conference on Quantum Computing and Engineering (QCE)*, pages 215–225, 2022.

[331] Juliane Müller, Jangho Park, Reetik Sahu, Charuleka Varadharajan, Bhavna Arora, Boris Faybishenko, and Deborah Agarwal. Surrogate optimization of deep neural networks for groundwater predictions. *Journal of Global Optimization*, 81(1):203–231, 2021.

[332] Juliane Müller and Christine A. Shoemaker. Influence of ensemble surrogate models and sampling strategy on the solution quality of algorithms for computationally expensive black-box global optimization problems. *Journal of Global Optimization*, 60(2):123–144, 2014.

[333] Juliane Müller, Christine A. Shoemaker, and Robert Piché. SO-MI: A surrogate model algorithm for computationally expensive nonlinear mixed-integer black-box global optimization problems. *Computers & Operations Research*, 40(5):1383–1400, 2013.

[334] Juliane Müller, Christine A. Shoemaker, and Robert Piché. SO-I: a surrogate model algorithm for expensive nonlinear integer programming problems including global optimization applications. *Journal of Global Optimization*, 59(4):865–889, 2014.

[335] Juliane Müller and Joshua D. Woodbury. GOSAC: global optimization with surrogate approximation of constraints. *Journal of Global Optimization*, 69(1):117–136, 2017.

[336] Kevin P. Murphy. *Probabilistic Machine Learning: An Introduction.* MIT Press, 2022.

[337] A. Muyskens, B. Priest, I. Goumirii, and M. Schneider. MuyGPs: Scalable gaussian process hyperparameter estimation using local cross-validation. arXiv:2104.14581, 4 2021.

[338] Débora N. Diniz, Mariana T. Rezende, Andrea G. C. Bianchi, Claudia M. Carneiro, Eduardo J. S. Luz, Gladston J. P. Moreira, Daniela M. Ushizima, Fátima N. S. de Medeiros, and Marcone J. F. Souza. A deep learning ensemble method to assist cytopathologists in pap test image classification. *Journal of Imaging*, 7(7):111, 2021.

[339] Gyoung S. Na and Hyunju Chang. A public database of thermoelectric materials and system-identified material representation for data-driven discovery. *npj Computational Materials*, 8(1):214, 2022.

[340] Anusha Nagabandi, Kurt Konolige, Sergey Levine, and Vikash Kumar. Deep dynamics models for learning dexterous manipulation. In *Conference on Robot Learning*, pages 1101–1112. PMLR, 2020.

[341] Anusha Nagabandi, Guangzhao Yang, Thomas Asmar, Ravi Pandya, Gregory Kahn, Sergey Levine, and Ronald S Fearing. Learning image-conditioned dynamics models for control of underactuated legged millirobots. In *2018 IEEE/RSJ International Conference on Intelligent Robots and Systems (IROS)*, pages 4606–4613. IEEE, 2018.

[342] Sreejesh Nair, Lotfollah Karimzadeh, and Broder J. Merkel. Surface complexation modeling of uranium(vi) sorption on quartz in the presence and absence of alkaline earth metals. *Environmental Earth Sciences*, 71(4):1737–1745, Feb 2014.

[343] Anis Najar and Mohamed Chetouani. Reinforcement learning with human advice: a survey. *Frontiers in Robotics and AI*, 8:584075, 2021.

[344] V. Narayanan and S. Jagannathan. Event-triggered distributed control of nonlinear interconnected systems using online reinforcement learning with exploration. *IEEE Transactions on Cybernetics*, 48(9):2510–2519, 2018.

[345] Kumpati S. Narendra and Anuradha M. Annaswamy. *Stable Adaptive Systems.* Courier Corporation, 2012.

[346] Michael J. Nasse, Michael J. Walsh, Eric C. Mattson, Ruben Reininger, André Kajdacsy-Balla, Virgilia Macias, Rohit Bhargava, and Carol J. Hirschmugl. High-resolution fourier-transform infrared chemical imaging with multiple synchrotron beams. *Nature Methods*, 8(5):413–416, 2011.

[347] United Nations. Protect, restore and promote sustainable use of terrestrial ecosystems, sustainably manage forests, combat desertification, and halt and reverse land degradation and halt biodiversity loss. `https://unstats.un.org/sdgs/report/2019/goal-15/`, 2019.

[348] E. Nazaretski, D. S. Coburn, W. Xu, J. Ma, H. Xu, R. Smith, X. Huang, Y. Yang, L. Huang, M. Idir, A. Kiss, and Y. S. Chu. A new Kirkpatrick–Baez-based scanning microscope for the Submicron Resolution X-ray Spectroscopy (SRX) beamline at NSLS-II. *Journal of Synchrotron Radiation*, 29(5):1284–1291, 2022.

[349] Hai Nguyen and Hung La. Review of deep reinforcement learning for robot manipulation. In *2019 Third IEEE International Conference on Robotic Computing (IRC)*, pages 590–595, 2019.

[350] P. Nikolaev, D. Hooper, F. Webber, R. Rao, K. Decker, M. Krein, J. Poleski, R. Barto, and B. Maruyama. Autonomy in materials research: a case study in carbon nanotube growth. *npj Computational Materials*, 2:16031, 2016.

[351] Pavel Nikolaev, Daylond Hooper, Nestor Perea-López, Mauricio Terrones, and Benji Maruyama. Discovery of wall-selective carbon nanotube growth conditions via automated experimentation. *ACS Nano*, 8(10):10214–10222, 2014. PMID: 25299482.

[352] Pavel Nikolaev, Daylond Hooper, Frederick Webber, Rahul Rao, Kevin Decker, Michael Krein, Jason Poleski, Rick Barto, and Benji Maruyama. Autonomy in materials research: a case study in carbon nanotube growth. *npj Computational Materials*, 2(1):1–6, 2016.

[353] Jan Nitzbon, Jobst Heitzig, and Ulrich Parlitz. Sustainability, collapse and oscillations in a simple world-earth model. *Environmental Research Letters*, 12(7):074020, 2017.

[354] Tong Niu and Mohit Bansal. Polite dialogue generation without parallel data. *Transactions of the Association for Computational Linguistics*, 6:373–389, 2018.

[355] M. Noack and J. Sethian. Advanced stationary and nonstationary kernel designs for domain-aware Gaussian processes. *Communications in Applied Mathematics and Computational Science*, 17(1):131–156, 2022.

[356] M. M. Noack, G. S. Doerk, R. Li, J. K. Streit, R. A. Vaia, K. G. Yager, and M. Fukuto. Autonomous materials discovery driven by gaussian process regression with inhomogeneous measurement noise and anisotropic kernels. *Scientific Reports*, 10(1):17663, 2020.

[357] Marcus Noack, James Sethian, Kevin Yager, Masafumi Fukuto, Petrus Zwart, Daniela Ushizima, Harinarayan Krishnan, Steven Lee, Ronald Pandolfi, Ian Humphrey, David Perryman, Pablo Enfedaque, Alexander Hexemer, Suchismita

Sarker, Tobias Weber, Apurva Mehta, Ruipeng Li, Gregory Doerk, and Martin Boehm. gpCAM, Feb. 2022.

[358] Marcus M. Noack, Gregory S. Doerk, Ruipeng Li, Masafumi Fukuto, and Kevin G. Yager. Advances in Kriging-based autonomous x-ray scattering experiments. *Scientific Reports*, 10(1):1325, 2020.

[359] Marcus M. Noack, Gregory S. Doerk, Ruipeng Li, Jason K. Streit, Richard A. Vaia, Kevin G. Yager, and Masafumi Fukuto. Autonomous materials discovery driven by Gaussian process regression with inhomogeneous measurement noise and anisotropic kernels. *Scientific Reports*, 10(1):17663, 2020.

[360] Marcus M. Noack, et al. Autonomous materials discovery driven by gaussian process regression with inhomogeneous measurement noise and anisotropic kernels. *Scientific Reports*, 10:17663, 2020.

[361] Marcus M. Noack and Simon W. Funke. Hybrid genetic deflated newton method for global optimisation. *Journal of Computational and Applied Mathematics*, 325:97–112, 2017.

[362] Marcus M. Noack, Harinarayan Krishnan, Mark D. Risser, and Kristofer G. Reyes. Exact gaussian processes for massive datasets via non-stationary sparsity-discovering kernels. *Scientific Reports*, 13(1):3155, 2023.

[363] Marcus M. Noack, Harinarayan Krishnan, Mark D. Risser, and Kristofer G. Reyes. Exact Gaussian processes for massive datasets via non-stationary sparsity-discovering kernels. *Scientific Reports*, 13(1):3155, 2023.

[364] Marcus M. Noack and Kristofer G. Reyes. Mathematical nuances of gaussian process-driven autonomous experimentation. *MRS Bulletin*, pages 1–11, 2023.

[365] Marcus M. Noack and James A. Sethian. Advanced stationary and nonstationary kernel designs for domain-aware gaussian processes. *Communications in Applied Mathematics and Computational Science*, 17(1):131–156, 2021.

[366] Marcus M. Noack and James A. Sethian. Advanced stationary and nonstationary kernel designs for domain-aware gaussian processes. *Communications in Applied Mathematics and Computational Science*, 17(1):131–156, 2022.

[367] Marcus M. Noack, Kevin G. Yager, Masafumi Fukuto, Gregory S. Doerk, Ruipeng Li, and James A. Sethian. A kriging-based approach to autonomous experimentation with applications to x-ray scattering. *Scientific Reports*, 9(1):11809, 2019.

[368] Marcus M. Noack, Kevin G. Yager, Masafumi Fukuto, Gregory S. Doerk, Ruipeng Li, and James A. Sethian. A Kriging-Based Approach to Autonomous Experimentation with Applications to X-Ray Scattering. *Scientific Reports*, 9(1):11809, 2019.

[369] Marcus M. Noack, Petrus H. Zwart, Daniela M. Ushizima, Masafumi Fukuto, Kevin G. Yager, Katherine C. Elbert, Christopher B. Murray, Aaron Stein, Gregory S. Doerk, Esther H. R. Tsai, Ruipeng Li, Guillaume Freychet, Mikhail Zhernenkov, Hoi-Ying N. Holman, Steven Lee, Liang Chen, Eli Rotenberg, Tobias Weber, Yannick Le Goc, Martin Boehm, Paul Steffens, Paolo Mutti, and James A. Sethian. Gaussian processes for autonomous data acquisition at large-scale synchrotron and neutron facilities. *Nature Reviews Physics*, 3(10):685–697, 2021.

[370] Marcus M. Noack, Petrus H. Zwart, Daniela M. Ushizima, Masafumi Fukuto, Kevin G. Yager, Katherine C. Elbert, Christopher B. Murray, Aaron Stein, Gregory S. Doerk, Esther H. R. Tsai, Ruipeng Li, Guillaume Freychet, Mikhail Zhernenkov, Hoi-Ying N. Holman, Steven Lee, Liang Chen, Eli Rotenberg, Tobias Weber, Yannick Le Goc, Martin Boehm, Paul Steffens, Paolo Mutti, and James A. Sethian. Gaussian processes for autonomous data acquisition at large-scale synchrotron and neutron facilities. *Nature Reviews Physics*, 3(10):685–697, 2021.

[371] Marcus Michael Noack, David Perryman, Harinarayan Krishnan, and Petrus H. Zwart. High-performance hybrid-global-deflated-local optimization with applications to active learning. In *2021 3rd Annual Workshop on Extreme-scale Experiment-in-the-Loop Computing (XLOOP)*, pages 24–29. IEEE, 2021.

[372] M. M. Noack. gpcam - autonomous data acquisition, uncertainty quantification and hpc optimization.

[373] M. M. Noack, G. S. Doerk, R. Li, M. Fukuto, and K. G. Yager. Advances in Kriging-Based Autonomous X-Ray Scattering Experiments. *Scientific Reports*, 10:1325, 2020.

[374] M. M. Noack, G. S. Doerk, R. Li, J. K. Streit, R. A. Vaia, K .G. Yager, and M. Fukuto. Autonomous materials discovery driven by Gaussian process regression with inhomogeneous measurement noise and anisotropic kernels. *Scientific Reports*, 10:17663, 2020.

[375] M. M. Noack et al. Gaussian processes for autonomous data acquisition at large-scale x-ray and neutron scattering facilities. accepted, *Nature Reviews Physics*, 2021.

[376] Arild Nøkland. Direct feedback alignment provides learning in deep neural networks. In *Advances in Neural Information Processing Systems*, pages 1037–1045, 2016.

[377] Vlastimil Novak, Peter F. Andeer, Benjamin P. Bowen, Yezhang Ding, Kateryna Zhalnina, Connor Tomaka, Amber N. Golini, Suzanne M. Kosina, and Trent R. Northen. Reproducible growth of brachypodium distachyon in fabricated ecosystems (ecofab 2.0) reveals that nitrogen form and starvation modulate root exudation. *bioRxiv*, 2023.

[378] S. R. Nowak, N. Tiwale, G. S. Doerk, C.-Y. Nam, C. T. Black, and K. G. Yager. Responsive blends of block copolymers stabilize the hexagonally perforated lamellae morphology. *Soft Matter*, 19:2594–2604, 2023.

[379] Marcel Nunez and Dionisios G. Vlachos. Multiscale modeling combined with active learning for microstructure optimization of bifunctional catalysts. *Industrial & Engineering Chemistry Research*, 58(15):6146–6154, 2018.

[380] United States Department of Energy. Doe update on cyber incident related to solar winds compromise.

[381] Alexey G. Okunev, Mikhail Yu. Mashukov, Anna V. Nartova, and Andrey V. Matveev. Nanoparticle recognition on scanning probe microscopy images using computer vision and deep learning. *Nanomaterials*, 10(7):1285, 2020.

[382] OpenAI, Christopher Berner, Greg Brockman, Brooke Chan, Vicki Cheung, Przemysław Dybiak, Christy Dennison, David Farhi, Quirin Fischer, Shariq Hashme, Chris Hesse, Rafal Józefowicz, Scott Gray, Catherine Olsson, Jakub Pachocki, Michael Petrov, Henrique P. d. O. Pinto, Jonathan Raiman, Tim Salimans, Jeremy Schlatter, Jonas Schneider, Szymon Sidor, Ilya Sutskever, Jie Tang, Filip Wolski, and Susan Zhang. Dota 2 with large scale deep reinforcement learning, 2019.

[383] OpenAI. AI research and deployment company. `https://openai.com/`, 2023.

[384] Jonas Oppenlaender. The creativity of text-to-image generation. In *Proceedings of the 25th International Academic Mindtrek Conference*, Academic Mindtrek '22, page 192202, New York, NY, USA, 2022. Association for Computing Machinery.

[385] Gizem Ortac and Giyasettin Ozcan. Comparative study of hyperspectral image classification by multidimensional convolutional neural network approaches to improve accuracy. *Expert Systems with Applications*, 182:115280, 2021.

[386] Brenden R Ortiz, Jesse M. Adamczyk, Kiarash Gordiz, Tara Braden, and Eric S. Toberer. Towards the high-throughput synthesis of bulk materials: thermoelectric pbte–pbse–snte–snse alloys. *Molecular Systems Design & Engineering*, 4(2):407–420, 2019.

[387] Long Ouyang, Jeffrey Wu, Xu Jiang, Diogo Almeida, Carroll Wainwright, Pamela Mishkin, Chong Zhang, Sandhini Agarwal, Katarina Slama, Alex Ray, et al. Training language models to follow instructions with human feedback. *Advances in Neural Information Processing Systems*, 35:27730–27744, 2022.

[388] Ling Pan, Qingpeng Cai, and Longbo Huang. Softmax deep double deterministic policy gradients. *Advances in Neural Information Processing Systems*, 33:11767–11777, 2020.

[389] Piyush Pandita, Ilias Bilionis, and Jitesh Panchal. Deriving information acquisition criteria for sequentially inferring the expected value of a black-box function. In *International Design Engineering Technical Conferences and Computers and Information in Engineering Conference*, volume 51760, page V02BT03A057. American Society of Mechanical Engineers, 2018.

[390] Piyush Pandita, Ilias Bilionis, and Jitesh Panchal. Bayesian optimal design of experiments for inferring the statistical expectation of expensive black-box functions. *Journal of Mechanical Design*, 141(10), 2019.

[391] Xin Pang, Decheng Li, and An Peng. Application of rare-earth elements in the agriculture of china and its environmental behavior in soil. *Environmental Science and Pollution Research*, 9(2):143–148, 2002.

[392] Kishore Papineni, Salim Roukos, Todd Ward, and Wei-Jing Zhu. Bleu: a method for automatic evaluation of machine translation. In *Proceedings of the 40th Annual Meeting of the Association for Computational Linguistics*, pages 311–318, 2002.

[393] Mario Teixeira Parente, Georg Brandl, Christian Franz, Uwe Stuhr, Marina Ganeva, and Astrid Schneidewind. Ai-assisted neutron spectroscopy using active learning with log-gaussian processes, 2023.

[394] Emilio Parisotto and Ruslan Salakhutdinov. Neural map: Structured memory for deep reinforcement learning. *arXiv preprint arXiv:1702.08360*, 2017.

[395] Adam Paszke, Sam Gross, Francisco Massa, Adam Lerer, James Bradbury, Gregory Chanan, Trevor Killeen, Zeming Lin, Natalia Gimelshein, Luca Antiga, Alban Desmaison, Andreas Kopf, Edward Yang, Zachary DeVito, Martin Raison, Alykhan Tejani, Sasank Chilamkurthy, Benoit Steiner, Lu Fang, Junjie Bai, and Soumith Chintala. Pytorch: An imperative style, high-performance deep learning library. In H. Wallach, H. Larochelle, A. Beygelzimer, F. d'Alché-Buc, E. Fox, and R. Garnett, editors, *Advances in Neural Information Processing Systems*, volume 32. Curran Associates, Inc., 2019.

[396] George Shu Heng Pau, Yingqi Zhang, Stefan Finsterle, Haruko Wainwright, and Jens Birkholzer. Reduced order modeling in itough2. *Computers & Geosciences*, 65:118–126, 2014. TOUGH Symposium 2012.

[397] Joel A. Paulson, Edward A. Buehler, and Ali Mesbah. Arbitrary polynomial chaos for uncertainty propagation of correlated random variables in dynamic systems. *IFAC-PapersOnLine*, 50(1):3548–3553, 2017.

[398] Emory M. Payne, Daniel A. Holland-Moritz, Shuwen Sun, and Robert T. Kennedy. High-throughput screening by droplet microfluidics: Perspective into key challenges and future prospects. *Lab on a Chip*, 20(13):2247–2262, 2020.

[399] Ryan Pederson, Bhupalee Kalita, and Kieron Burke. Machine learning and density functional theory. *Nature Reviews Physics*, 4(6):357–358, 2022.

[400] F. Pedregosa, G. Varoquaux, A. Gramfort, V. Michel, B. Thirion, O. Grisel, M. Blondel, P. Prettenhofer, R. Weiss, V. Dubourg, J. Vanderplas, A. Passos, D. Cournapeau, M. Brucher, M. Perrot, and E. Duchesnay. Scikit-learn: Machine learning in Python. *Journal of Machine Learning Research*, 12:2825–2830, 2011.

[401] Ian M. Pendleton, Gary Cattabriga, Zhi Li, Mansoor Ani Najeeb, Sorelle A. Friedler, Alexander J. Norquist, Emory M. Chan, and Joshua Schrier. Experiment specification, capture and laboratory automation technology (escalate): a software pipeline for automated chemical experimentation and data management. *MRS Communications*, 9(3):846–859, 2019.

[402] Jan Peters, Katharina Mulling, and Yasemin Altun. Relative entropy policy search. In *Proceedings of the AAAI Conference on Artificial Intelligence*, volume 24, pages 1607–1612, 2010.

[403] Jan Peters and Stefan Schaal. Policy gradient methods for robotics. In *2006 IEEE/RSJ International Conference on Intelligent Robots and Systems*, pages 2219–2225. IEEE, 2006.

[404] S. Petit. Spinwave - an easy-to-use and versatile software to calculate spin waves in any abitrary magnetic lattice.

[405] Mathijs Pieters and Marco A. Wiering. Q-learning with experience replay in a dynamic environment. In *SSCI*, pages 1–8, 2016.

[406] Ghanshyam Pilania, Chenchen Wang, Xun Jiang, Sanguthevar Rajasekaran, and Ramamurthy Ramprasad. Accelerating materials property predictions using machine learning. *Scientific Reports*, 3(1):2810, 2013.

[407] Karl Ezra Pilario, Mahmood Shafiee, Yi Cao, Liyun Lao, and Shuang-Hua Yang. A review of kernel methods for feature extraction in nonlinear process monitoring. *Processes*, 8(1):24, 2020.

[408] Athanasios S. Polydoros and Lazaros Nalpantidis. Survey of model-based reinforcement learning: Applications on robotics. *Journal of Intelligent & Robotic Systems*, 86(2):153–173, 2017.

[409] M. Popovici. On the resolution of slow-neutron spectrometers. IV. The triple-axis spectrometer resolution function, spatial effects included. *Acta Crystallographica Section A*, 31(4):507–513, 1975.

[410] W. Potrzebowski, J. Trewhella, and I. Andre. Bayesian inference of protein conformational ensembles from limited structural data. *PLoS Computational Biology*, 14(14):e1006641, 2018.

[411] M. J. D. Powell. Recent research at cambridge on radial basis functions. In Manfred W. Müller, Martin D. Buhmann, Detlef H. Mache, and Michael Felten, editors, *New Developments in Approximation Theory*, pages 215–232, Basel, 1999. Birkhäuser Basel.

[412] M. J. D. Powell. *Algorithms for Approximation*, chapter Radial Basis Functions for Multivariable Interpolation: A Review. Carendon Press, Oxford, 1987.

[413] Dhiren K. Pradhan, Shalini Kumari, Evgheni Strelcov, Dillip K. Pradhan, Ram S. Katiyar, Sergei V. Kalinin, Nouamane Laanait, and Rama K. Vasudevan. Reconstructing phase diagrams from local measurements via gaussian processes: mapping the temperature-composition space to confidence. *npj Computational Materials*, 4(1):23, 2018.

[414] Alexander J. Probst, Hoi-Ying N. Holman, Todd Z. DeSantis, Gary L. Andersen, Giovanni Birarda, Hans A. Bechtel, Yvette M. Piceno, Maria Sonnleitner, Kasthuri Venkateswaran, and Christine Moissl-Eichinger. Tackling the minority: sulfate-reducing bacteria in an archaea-dominated subsurface biofilm. *The ISME Journal*, 7(3):635–651, 2013.

[415] Apostolos F. Psaros, Xuhui Meng, Zongren Zou, Ling Guo, and George Em Karniadakis. Uncertainty quantification in scientific machine learning: Methods, metrics, and comparisons. *Journal of Computational Physics*, 477:111902, 2023.

[416] Jerome Quenum, David Perlmutter, Ying Huang, Iryna Zenyuk, and Daniela Ushizima. Lithium metal battery quality control via transformer-cnn segmentation, 2023.

[417] Paul Raccuglia, Katherine C. Elbert, Philip D. F. Adler, Casey Falk, Malia B. Wenny, Aurelio Mollo, Matthias Zeller, Sorelle A. Friedler, Joshua Schrier, and Alexander J. Norquist. Machine-learning-assisted materials discovery using failed experiments. *Nature*, 533(7601):73–76, 2016.

[418] Paul Raccuglia, Katherine C. Elbert, Philip D. F. Adler, Casey Falk, Malia B. Wenny, Aurelio Mollo, Matthias Zeller, Sorelle A. Friedler, Joshua Schrier, and Alexander J. Norquist. Machine-learning-assisted materials discovery using failed experiments. *Nature*, 533(7601):73–76, 2016.

[419] Alec Radford, Karthik Narasimhan, Tim Salimans, Ilya Sutskever, et al. Improving language understanding by generative pre-training. *Online resource* 2018.

[420] M. S. Rakitin, O. Chubar, P. Moeller, R. Nagler, and D. L. Bruhwiler. Sirepo: a web-based interface for physical optics simulations - its deployment and use at NSLS-II **(invited paper)**. In *Proc. SPIE, Advances in Computational Methods for X-Ray Optics IV (23 August 2017)*, volume 10388, page 103880R, 2017.

[421] Maksim S. Rakitin, Abigail Giles, Kaleb Swartz, Joshua Lynch, Paul Moeller, Robert Nagler, Daniel B. Allan, Thomas A. Caswell, Lutz Wiegart, Oleg Chubar, and Yonghua Du. Introduction of the Sirepo-Bluesky interface and its application to the optimization problems. In Oleg Chubar and Kawal Sawhney, editors, *Advances in Computational Methods for X-Ray Optics V*, volume

11493, pages 209–226. International Society for Optics and Photonics, SPIE, August 2020.

[422] Aditya Ramesh, Prafulla Dhariwal, Alex Nichol, Casey Chu, and Mark Chen. Hierarchical text-conditional image generation with clip latents, 2022. https://arxiv.org/abs/2204.06125

[423] Marc'Aurelio Ranzato, Sumit Chopra, Michael Auli, and Wojciech Zaremba. Sequence level training with recurrent neural networks, 2016. https://arxiv.org/abs/1511.06732

[424] H. S. Rao and A. Mukherjee. Artificial neural networks for predicting the macromechanical behaviour of ceramic-matrix composites. *Computational Materials Science*, 5(4):307–322, 1996.

[425] Ziyuan Rao, Po-Yen Tung, Ruiwen Xie, Ye Wei, Hongbin Zhang, Alberto Ferrari, T. P. C. Klaver, Fritz Körmann, Prithiv Thoudden Sukumar, Alisson Kwiatkowski da Silva, et al. Machine learning–enabled high-entropy alloy discovery. *Science*, 378(6615):78–85, 2022.

[426] Sebastian Raschka. *Python Machine Learning*. Packt Publishing - ebooks Account, 2015.

[427] Sebastian Raschka, Yuxi Hayden Liu, Vahid Mirjalili, and Dmytro Dzhulgakov. *Machine Learning with PyTorch and Scikit-Learn: Develop Machine Learning and Deep Learning Models with Python*. Packt Publishing Ltd, 2022.

[428] Mohammad Rashidi, Jeremiah Croshaw, Kieran Mastel, Marcus Tamura, Hedieh Hosseinzadeh, and Robert A. Wolkow. Deep learning-guided surface characterization for autonomous hydrogen lithography. *Machine Learning: Science and Technology*, 1(2):025001, 2020.

[429] Mohammad Rashidi and Robert A. Wolkow. Autonomous scanning probe microscopy in situ tip conditioning through machine learning. *ACS Nano*, 12(6):5185–5189, 2018.

[430] Carl Edward Rasmussen. *Gaussian Processes in Machine Learning*, volume 3176, pages 63–71. Springer Berlin Heidelberg, Berlin, Heidelberg, 2004.

[431] Carl Edward Rasmussen and Christopher K. I. Williams. *Gaussian Processes for Machine Learning*. The MIT Press, Boston, 2006.

[432] C. E. Rasmussen and C. K. I. Williams. *Gaussian Processes for Machine Learning*. The MIT Press, 2006.

[433] Scott Reed, Honglak Lee, Dragomir Anguelov, Christian Szegedy, Dumitru Erhan, and Andrew Rabinovich. Training deep neural networks on noisy labels with bootstrapping. *arXiv preprint arXiv:1412.6596*, 2014.

[434] Rommel G. Regis. Convergence guarantees for generalized adaptive stochastic search methods for continuous global optimization. *European Journal of Operational Research*, 207(3):1187–1202, 2010.

[435] Rommel G. Regis. Constrained optimization by radial basis function interpolation for high-dimensional expensive black-box problems with infeasible initial points. *Engineering Optimization*, 46(2):218–243, 2014.

[436] Rommel G. Regis. Multi-objective constrained black-box optimization using radial basis function surrogates. *Journal of Computational Science*, 16:140–155, 2016.

[437] Rommel G. Regis and Christine A. Shoemaker. Constrained global optimization of expensive black box functions using radial basis functions. *Journal of Global Optimization*, 31(1):153–171, 2005.

[438] Rommel G. Regis and Christine A. Shoemaker. A stochastic radial basis function method for the global optimization of expensive functions. *INFORMS Journal on Computing*, 19(4):497–509, 2007.

[439] Rommel G. Regis and Christine A. Shoemaker. Combining radial basis function surrogates and dynamic coordinate search in high-dimensional expensive black-box optimization. *Engineering Optimization*, 45(5):529–555, 2013.

[440] Yoram Reich and Nahum Travitzky. Machine learning of material behaviour knowledge from empirical data. *Materials & Design*, 16(5):251–259, 1995.

[441] S. Remes, M. Heinonen, and S. Kaski. Non-stationary spectral kernels. In *31st Conference on Neural Information Processing Systems (NeurIPS 2017)*, 2017.

[442] Sami Remes, Markus Heinonen, and Samuel Kaski. Non-stationary spectral kernels. *Advances in Neural Information Processing Systems*, 30, 4645–4654 2017.

[443] Fang Ren, Logan Ward, Travis Williams, Kevin J. Laws, Christopher Wolverton, Jason Hattrick-Simpers, and Apurva Mehta. Accelerated discovery of metallic glasses through iteration of machine learning and high-throughput experiments. *Science Advances*, 4(4):eaaq1566, 2018.

[444] Kristofer G. Reyes and Benji Maruyama. The machine learning revolution in materials? *MRS Bulletin*, 44(7):530–537, 2019.

[445] Graham Robinson and J. Bruce C Davies. Continuum robots-a state of the art. In *Proceedings 1999 IEEE International Conference on Robotics and Automation (Cat. No. 99CH36288C)*, volume 4, pages 2849–2854. IEEE, 1999.

[446] Loïc M. Roch, Florian Häse, Christoph Kreisbeck, Teresa Tamayo-Mendoza, Lars P. E. Yunker, Jason E. Hein, and Alán Aspuru-Guzik. Chemos: Orchestrating autonomous experimentation. *Science Robotics*, 3(19):5559, 2018.

[447] Loïc M. Roch, Florian Häse, Christoph Kreisbeck, Teresa Tamayo-Mendoza, Lars P. E. Yunker, Jason E. Hein, and Alán Aspuru-Guzik. Chemos: An orchestration software to democratize autonomous discovery. *PLoS One*, 15(4):e0229862, 2020.

[448] Robin Rombach, Andreas Blattmann, Dominik Lorenz, Patrick Esser, and Björn Ommer. High-resolution image synthesis with latent diffusion models, *Proceedings of the IEEE/CVF Conference on Computer Vision and Pattern Recognition (CVPR)*, 10684–10695, 2022.

[449] Olaf Ronneberger, Philipp Fischer, and Thomas Brox. U-net: Convolutional networks for biomedical image segmentation. In Nassir Navab, Joachim Horneger, William M. Wells, and Alejandro F. Frangi, editors, *Medical Image Computing and Computer-Assisted Intervention – MICCAI 2015*, pages 234–241, Cham, 2015. Springer International Publishing.

[450] Frank Rosenblatt. The perceptron: A probabilistic model for information storage and organization in the brain. *Psychological Review*, 65(6):386, 1958.

[451] Francesca Rossi, Peter Van Beek, and Toby Walsh. *Handbook of Constraint Programming*. Elsevier, 2006.

[452] Paul W. K. Rothemund. Folding dna to create nanoscale shapes and patterns. *Nature*, 440(7082):297–302, 2006.

[453] Frederik Ruelens, Bert J. Claessens, Stijn Vandael, Bart De Schutter, Robert Babuška, and Ronnie Belmans. Residential demand response of thermostatically controlled loads using batch reinforcement learning. *IEEE Transactions on Smart Grid*, 8(5):2149–2159, 2017.

[454] David E. Rumelhart, Geoffrey E. Hinton, and Ronald J. Williams. Learning representations by back-propagating errors. *Nature*, 323(6088):533–536, 1986.

[455] James E. Saal, Anton O. Oliynyk, and Bryce Meredig. Machine learning in materials discovery: confirmed predictions and their underlying approaches. *Annual Review of Materials Research*, 50:49–69, 2020.

[456] Logan Saar, Haotong Liang, Alex Wang, Austin McDannald, Efrain Rodriguez, Ichiro Takeuchi, and A Gilad Kusne. The legolas kit: A low-cost robot science kit for education with symbolic regression for hypothesis discovery and validation, *MRS Bulletin*, 47:881–885, 2022.

[457] J. Sacks, S. B. Schiller, and W. J. Welch. Designs for computer experiments. *Technometrics*, 31(1):41–47, 1989.

[458] R. Sadre, B. Sundaram, S. Majumdar, and D. Ushizima. Validating deep learning inference during chest x-ray classification for covid-19 screening. *Nature Scientific Reports*, 11(1):1–10, 2021.

[459] Robbie Sadre and Daniela Ushizima. Computer aided diagnostic tools for covid-19 detection via x-ray imaging. In *2020 IEEE/ACM Workshop on Urgent HPC: HPC for Urgent Decision Making at Super Computing*, page 1, Nov 2020.

[460] Farhang Sahba. Deep reinforcement learning for object segmentation in video sequences. In *2016 International Conference on Computational Science and Computational Intelligence (CSCI)*, pages 857–860. IEEE, 2016.

[461] Abdelrhman Saleh, Natasha Jaques, Asma Ghandeharioun, Judy Shen, and Rosalind Picard. Hierarchical reinforcement learning for open-domain dialog. In *Proceedings of the AAAI Conference on Artificial Intelligence*, volume 34, pages 8741–8748, 2020.

[462] Anjana M. Samarakoon, Kipton Barros, Ying Wai Li, Markus Eisenbach, Qiang Zhang, Feng Ye, V. Sharma, Z. L. Dun, Haidong Zhou, Santiago A. Grigera, Cristian D. Batista, and D. Alan Tennant. Machine-learning-assisted insight into spin ice dy2ti2o7. *Nature Communications*, 11:892, 2020.

[463] Anjana M. Samarakoon, Pontus Laurell, Christian Balz, Arnab Banerjee, Paula Lampen-Kelley, David Mandrus, Stephen E. Nagler, Satoshi Okamoto, and D. Alan Tennant. Extraction of interaction parameters for α- rucl3 from neutron data using machine learning. *Physical Review Research*, 4(2):L022061, 2022.

[464] A. L. Samuel. Some studies in machine learning using the game of checkers. *IBM Journal of Research and Development*, 3(3):210–229, 1959.

[465] Manuel Sanchez del Rio, Niccolo Canestrari, Fan Jiang, and Franco Cerrina. *SHADOW3*: a new version of the synchrotron X-ray optics modelling package. *Journal of Synchrotron Radiation*, 18(5):708–716, 2011.

[466] Benjamin Sanchez-Lengeling and Alán Aspuru-Guzik. Inverse molecular design using machine learning: Generative models for matter engineering. *Science*, 361(6400):360–365, 2018.

[467] Abhinav Sarje, Xiaoye S. Li, Slim Chourou, Elaine R. Chan, and Alexander Hexemer. Massively parallel x-ray scattering simulations. In *SC'12: Proceedings of the International Conference on High Performance Computing, Networking, Storage and Analysis*, pages 1–11. IEEE, 2012.

[468] S. Sarker, D. Chandra, M. Hirscher, M. Dolan, D. Isheim, J. Wermer, D. Viano, Marcello Baricco, Terrence J. Udovic, D. Grant, et al. Developments in the ni–nb–zr amorphous alloy membranes: A review. *Applied Physics A*, 122:1–9, 2016.

[469] Suchismita Sarker, Robert Tang-Kong, Rachel Schoeppner, Logan Ward, Naila Al Hasan, Douglas G. Van Campen, Ichiro Takeuchi, Jason Hattrick-Simpers, Andriy Zakutayev, Corinne E. Packard, et al. Discovering exceptionally hard and wear-resistant metallic glasses by combining machine-learning with high throughput experimentation. *Applied Physics Reviews*, 9(1):011403, 2022.

[470] Sreeshankar Satheeshbabu, Naveen K. Uppalapati, Tianshi Fu, and Girish Krishnan. Continuous control of a soft continuum arm using deep reinforcement learning. In *2020 3rd IEEE International Conference on Soft Robotics (RoboSoft)*, pages 497–503. IEEE, 2020.

[471] Tom Schaul, John Quan, Ioannis Antonoglou, and David Silver. Prioritized experience replay. *arXiv preprint arXiv:1511.05952*, 2015.

[472] Lucas Schneider, Philip Beck, Jannis Neuhaus-Steinmetz, Levente Rózsa, Thore Posske, Jens Wiebe, and Roland Wiesendanger. Precursors of majorana modes and their length-dependent energy oscillations probed at both ends of atomic shiba chains. *Nature Nanotechnology*, 17(4):384–389, 2022.

[473] Bernhard Schölkopf, Alexander Smola, and Klaus-Robert Müller. Nonlinear component analysis as a kernel eigenvalue problem. *Neural Computation*, 10(5):1299–1319, 1998.

[474] Bruno Schuler, Katherine A. Cochrane, Christoph Kastl, Edward S. Barnard, Edward Wong, Nicholas J. Borys, Adam M. Schwartzberg, D. Frank Ogletree, F. Javier García de Abajo, and Alexander Weber-Bargioni. Electrically driven photon emission from individual atomic defects in monolayer ws¡sub¿2¡/sub¿. *Science Advances*, 6(38):eabb5988, 2020.

[475] John Schulman, Filip Wolski, Prafulla Dhariwal, Alec Radford, and Oleg Klimov. Proximal policy optimization algorithms. *arXiv preprint arXiv:1707.06347*, 2017.

[476] Manuel Schürch, Dario Azzimonti, Alessio Benavoli, and Marco Zaffalon. Correlated product of experts for sparse Gaussian process regression. *Machine Learning*, 112:1411–1432, 2023.

[477] Nadrian C. Seeman and Hanadi F. Sleiman. Dna nanotechnology. *Nature Reviews Materials*, 3(1):1–23, 2017.

[478] Anand Seethepalli, Kundan Dhakal, Marcus Griffiths, Haichao Guo, Gregoire T. Freschet, and Larry M. York. RhizoVision Explorer: open-source software for root image analysis and measurement standardization. *AoB PLANTS*, 13(6), 09 2021. plab056.

[479] Martin Seifrid, Jason Hattrick-Simpers, Alán Aspuru-Guzik, Tom Kalil, and Steve Cranford. Reaching critical mass: Crowdsourcing designs for the next generation of materials acceleration platforms. *Matter*, 5(7):1972–1976, 2022.

[480] Claude E. Shannon. A mathematical theory of communication. *The Bell System Technical Journal*, 27(3):379–423, 1948.

[481] Gen Shirane, Stephen M. Shapiro, and John M. Tranquada. *Neutron Scattering with a Triple-Axis Spectrometer: Basic Techniques*. Cambridge University Press, 2002.

[482] Y. Sidis, S. Pailhès, B. Keimer, P. Bourges, C. Ulrich, and L. P. Regnault. Magnetic resonant excitations in high-tc superconductors. *physica status solidi (b)*, 241(6):1204–1210, 2004.

[483] David Silver, Thomas Hubert, Julian Schrittwieser, Ioannis Antonoglou, Matthew Lai, Arthur Guez, Marc Lanctot, Laurent Sifre, Dharshan Kumaran, Thore Graepel, Timothy Lillicrap, Karen Simonyan, and Demis Hassabis. A general reinforcement learning algorithm that masters chess, shogi, and go through self-play. *Science*, 362(6419):1140–1144, 2018.

[484] D. Simon. *Evolutionary Optimization Algorithms*. Wiley, 2013.

[485] Devinderjit Sivia and John Skilling. *Data Analysis: A Bayesian Tutorial*. OUP Oxford, 2006.

[486] D. S. Sivia, C. J. Carlile, W. S. Howells, and S. König. Bayesian analysis of quasielastic neutron scattering data. *Physica B: Condensed Matter*, 182(4):341–348, 1992. Quasielastic Neutron Scattering.

[487] J. C. Spall. *Introduction to Stochastic Search and Optimization: Estimation, Simulation, and Control*. John Wiley & Sons, Inc., New Jersey, 2003.

[488] G. L. Squires. *Introduction to the Theory of Thermal Neutron Scattering*. Cambridge University Press, 3 edition, 2012.

[489] Eric Stach, Brian DeCost, A. Gilad Kusne, Jason Hattrick-Simpers, Keith A. Brown, Kristofer G. Reyes, Joshua Schrier, Simon Billinge, Tonio Buonassisi, Ian Foster, et al. Autonomous experimentation systems for materials development: A community perspective. *Matter*, 4(9):2702–2726, 2021.

[490] Eric Stach, Brian DeCost, A. Gilad Kusne, Jason Hattrick-Simpers, Keith A. Brown, Kristofer G. Reyes, Joshua Schrier, Simon Billinge, Tonio Buonassisi, Ian Foster, Carla P. Gomes, John M. Gregoire, Apurva Mehta, Joseph Montoya, Elsa Olivetti, Chiwoo Park, Eli Rotenberg, Semion K. Saikin, Sylvia Smullin, Valentin Stanev, and Benji Maruyama. Autonomous experimentation systems for materials development: A community perspective. *Matter*, 4(9):2702–2726, 2021.

[491] Valentin Stanev, Corey Oses, A. Gilad Kusne, Efrain Rodriguez, Johnpierre Paglione, Stefano Curtarolo, and Ichiro Takeuchi. Machine learning modeling of superconducting critical temperature. *npj Computational Materials*, 4(1):29, 2018.

[492] A. Stein, G. Wright, K. G. Yager, G. S. Doerk, and C. T. Black. Selective directed self-assembly of coexisting morphologies using block copolymer blends. *Nature Communications*, 7(1):12366, 2016.

[493] Helge S. Stein and John M. Gregoire. Progress and prospects for accelerating materials science with automated and autonomous workflows. *Chemical Science*, 10:9640–9649, 2019.

[494] Rick Stevens, Valerie Taylor, Jeff Nichols, Arthur Barney Maccabe, Katherine Yelick, and David Brown. Ai for science: Report on the department of energy (doe) town halls on artificial intelligence (ai) for science. Technical report, Argonne National Lab.(ANL), Argonne, IL (United States), 2020.

[495] Felix M. Strnad, Wolfram Barfuss, Jonathan F. Donges, and Jobst Heitzig. Deep reinforcement learning in world-earth system models to discover sustainable management strategies. *Chaos: An Interdisciplinary Journal of Nonlinear Science*, 29(12):123122, 2019.

[496] Fazrul Razman Sulaiman and Huda Asilah Hamzah. Heavy metals accumulation in suburban roadside plants of a tropical area (jengka, malaysia). *Ecological Processes*, 7(1):28, 2018.

[497] Dajie Sun, Haruko M. Wainwright, Carlos A. Oroza, Akiyuki Seki, Satoshi Mikami, Hiroshi Takemiya, and Kimiaki Saito. Optimizing long-term monitoring of radiation air-dose rates after the fukushima daiichi nuclear power plant. *Journal of Environmental Radioactivity*, 220-221:106281, 2020.

[498] Hartmut Surmann, Christian Jestel, Robin Marchel, Franziska Musberg, Houssem Elhadj, and Mahbube Ardani. Deep reinforcement learning for real autonomous mobile robot navigation in indoor environments, 2020.

[499] Duncan R. Sutherland, Aine Boyer Connolly, Maximilian Amsler, Ming-Chiang Chang, Katie Rose Gann, Vidit Gupta, Sebastian Ament, Dan Guevarra, John M. Gregoire, Carla P. Gomes, et al. Optical identification of materials transformations in oxide thin films. *ACS Combinatorial Science*, 22(12):887–894, 2020.

[500] Ilya Sutskever, James Martens, George Dahl, and Geoffrey Hinton. On the importance of initialization and momentum in deep learning. In *International Conference on Machine Learning*, pages 1139–1147, 2013.

[501] Ilya Sutskever, Oriol Vinyals, and Quoc V Le. Sequence to sequence learning with neural networks. *Advances in Neural Information Processing Systems*, 27, (NIPS 2014).

[502] Richard Sutton. The bitter lesson. *Incomplete Ideas blog post, online resource*, 13(1), 2019.

[503] Richard S. Sutton and Andrew G. Barto. *Reinforcement Learning: An Introduction*. MIT Press, 2018.

[504] Richard S. Sutton, David McAllester, Satinder Singh, and Yishay Mansour. Policy gradient methods for reinforcement learning with function approximation. *Conference proceeding: Part of Advances in Neural Information Processing System*, 12, (NIPS 1999).

[505] Richard S. Sutton, Steven D. Whitehead, et al. Online learning with random representations. In *Part of Advances in Neural Information Processing Systems*, 12, (NIPS 1999).

[506] Johan A. K. Suykens, Carlos Alzate, and Kristiaan Pelckmans. Primal and dual model representations in kernel-based learning. *Statistics Surveys*, 4:148–183, 2010.

[507] Richard Swinburne. Bayes' theorem. *Oxford University Press, Edited by Richard Swinburne Proceedings of the British Academy*, 194(2), 2004.

[508] Jordan H. Swisher, Liban Jibril, Sarah Hurst Petrosko, and Chad A. Mirkin. Nanoreactors for particle synthesis. *Nature Reviews Materials*, 7(6):428–448, 2022.

[509] Timur Takhtaganov, Zarija Lukić, Juliane Müller, and Dmitriy Morozov. Cosmic inference: Constraining parameters with observations and a highly limited number of simulations. *The Astrophysical Journal*, 906(2):74, 2021.

[510] Matthew J. Tamasi and Adam J. Gormley. Biologic formulation in a self-driving biomaterials lab. *Cell Reports Physical Science*, 3(9):101041, 2022.

[511] Qiaoyu Tan, Ninghao Liu, and Xia Hu. Deep representation learning for social network analysis. *Frontiers in Big Data*, 2, 2019.

[512] Qiuling Tao, Pengcheng Xu, Minjie Li, and Wencong Lu. Machine learning for perovskite materials design and discovery. *npj Computational Materials*, 7(1):23, 2021.

[513] François Tardieu and Matthew P. Reynolds. Plant morphogenesis: climate and genetics interact through the "fishbone" framework. *New Phytologist*, 227(4):1147–1160, 2020.

[514] Mario Teixeira Parente, Astrid Schneidewind, Georg Brandl, Christian Franz, Marcus Noack, Martin Boehm, and Marina Ganeva. Benchmarking autonomous scattering experiments illustrated on tas. *Frontiers in Materials*, 8, 2022.

[515] J. Tersoff and D. R. Hamann. Theory of the scanning tunneling microscope. *Physical Review B*, 31:805–813, 1985.

[516] The Materials Genome Initiative. The 2021 Materials Genome Initiative Strategic Plan: https://www.mgi.gov/.

[517] John C. Thomas, Antonio Rossi, Darian Smalley, Luca Francaviglia, Zhuohang Yu, Tianyi Zhang, Shalini Kumari, Joshua A. Robinson, Mauricio Terrones, Masahiro Ishigami, Eli Rotenberg, Edward S. Barnard, Archana Raja, Ed Wong, D. Frank Ogletree, Marcus M. Noack, and Alexander Weber-Bargioni. Autonomous scanning probe microscopy investigations over ws2 and au{111}. *npj Computational Materials*, 8(1):99, 2022.

[518] John C. Thomas, Jeffrey J. Schwartz, J. Nathan Hohman, Shelley A. Claridge, Harsharn S. Auluck, Andrew C. Serino, Alexander M. Spokoyny, Giang Tran, Kevin F. Kelly, Chad A. Mirkin, Jerome Gilles, Stanley J. Osher, and Paul S. Weiss. Defect-Tolerant aligned dipoles within Two-Dimensional plastic lattices. *ACS Nano*, 9(5):4734–4742, 2015.

[519] Sebastian Thrun and Anton Schwartz. Issues in using function approximation for reinforcement learning. In *Proceedings of the Fourth Connectionist Models Summer School*, pages 255–263. Hillsdale, NJ, 1993.

[520] Thomas George Thuruthel, Egidio Falotico, Federico Renda, and Cecilia Laschi. Model-based reinforcement learning for closed-loop dynamic control of soft robotic manipulators. *IEEE Transactions on Robotics*, 35(1):124–134, 2018.

[521] Ye Tian, Yugang Zhang, Tong Wang, Huolin L. Xin, Huilin Li, and Oleg Gang. Lattice engineering through nanoparticle–dna frameworks. *Nature Materials*, 15(6):654–661, 2016.

[522] Luke Tierney. Markov chains for exploring posterior distributions. *The Annals of Statistics*, 22(4):1701–1728, 1994.

[523] Michael Tipping. Sparse kernel principal component analysis. *Conference Proceedings: Part of Advances in Neural Information Processing Systems*, 13, (NIPS 2000).

[524] Michalis Titsias and Neil D Lawrence. Bayesian gaussian process latent variable model. In *Proceedings of the Thirteenth International Conference on Artificial Intelligence and Statistics*, pages 844–851. JMLR Workshop and Conference Proceedings, 2010.

[525] Peter M. Todd and Gerd Gigerenzer. Précis of simple heuristics that make us smart. *Behavioral and Brain Sciences*, 23(5):727–741, 2000.

[526] Cormac Toher, Corey Oses, David Hicks, and Stefano Curtarolo. Unavoidable disorder and entropy in multi-component systems. *npj Computational Materials*, 5(1):69, 2019.

[527] Bryan A. Tolson and Christine A. Shoemaker. Dynamically dimensioned search algorithm for computationally efficient watershed model calibration. *Water Resources Research*, 43(1), 2007.

[528] S. Toth and B. Lake. Linear spin wave theory for single-q incommensurate magnetic structures. *Journal of Physics: Condensed Matter*, 27(16):166002, 2015.

[529] John M. Tranquada, Guangyong Xu, and Igor A. Zaliznyak. Superconductivity, antiferromagnetism, and neutron scattering. *Journal of Magnetism and Magnetic Materials*, 350:148–160, 2014.

[530] J. N. Tsitsiklis and B. Van Roy. An analysis of temporal-difference learning with function approximationtechnical. *Rep. LIDS-P-2322). Lab. Inf. Decis. Syst. Massachusetts Inst. Technol. Tech. Rep*, 1996.

[531] Vladimir A. Ukraintsev. Data evaluation technique for electron-tunneling spectroscopy. *Physical Review B*, 53:11176–11185, 1996.

[532] Utkarsh Upadhyay, Abir De, and Manuel Gomez Rodriguez. Deep reinforcement learning of marked temporal point processes. *Conference Proceedings: Part of Advances in Neural Information Processing Systems*, 31, (NeurIPS 2018).

[533] D. Ushizima, F. Araujo, and R. Silva. Searchable datasets in python: images across domains, experiments, algorithms and learning. In *Proceedings of PyData*, volume 1, pages 1–2, 2016.

[534] D. M. Ushizima and E. Chagnon. From gutenberg to bert: how transformers can change information extraction from text. In *CMStatistics*, London, UK, December 2022.

[535] Daniela Ushizima, Hanbo Chen, Manuel Alegro, Ifigeneia Ovando, Raif Eser, Jungsuk Lee, Connor Poon, Kartik Shankar, Arvind Kantamneni, Nirav Satrawada, Edson Amaro Jr., Helmut Heinsen, Duygu Tosun, and Lea T. Grinberg. Deep learning for alzheimer's disease: Mapping large-scale histological tau protein for neuroimaging biomarker validation. *NeuroImage*, 274:118876, 2022.

[536] Daniela Ushizima, Esther Singer, Elizabeth Carpenter, and Dilworth Parkinson. Multimodal Analysis of Plant Root Composition as a Function of Drought. *SPRINGER NATURE - Research Book Series: Transactions on Computational Science & Computational Intelligence*, 2021.

[537] Daniela Ushizima, Ke Xu, and Paulo Monteiro. Materials data science for microstructural characterization of archaeological concrete. *MRS Advancements - special issue: Materials Data Science*, pages 1–14, 2020.

[538] O. P. Vajk, M. Kenzelmann, J. W. Lynn, S. B. Kim, and S-W. Cheong. Magnetic order and spin dynamics in ferroelectric homno3. *Physical Review Letters*, 94(8):087601, 2005.

[539] Patricia M. Valdespino-Castillo, Ping Hu, Martín Merino-Ibarra, Luz M. López-Gómez, Daniel Cerqueda-García, González-De Zayas, Teresa Pi-Puig, Julio A. Lestayo, Hoi-Ying Holman, Luisa I. Falcón, et al. Exploring biogeochemistry and microbial diversity of extant microbialites in mexico and cuba. *Frontiers in Microbiology*, 9:510, 2018.

[540] Leslie Valiant. *Probably Approximately Correct: Natureõs Algorithms for Learning and Prospering in a Complex World*. Basic Books (AZ), 2013.

[541] Stefan Van der Walt, Johannes L. Schönberger, Juan Nunez-Iglesias, François Boulogne, Joshua D. Warner, Neil Yager, Emmanuelle Gouillart, and Tony Yu. scikit-image: image processing in python. *PeerJ*, 2:e453, 2014.

[542] Hado van Hasselt, Arthur Guez, and David Silver. Deep reinforcement learning with double q-learning. *CoRR*, abs/1509.06461, 2015.

[543] Fjodor Van Veen and Stefan Leijnen. The neural network zoo. `https://www.asimovinstitute.org/neural-network-zoo/`, 2019.

[544] Jake VanderPlas. *Python data science handbook: Essential tools for working with data.* "O'Reilly Media, Inc.", 2016.

[545] Rama K. Vasudevan, Maxim Ziatdinov, Lukas Vlcek, and Sergei V. Kalinin. Off-the-shelf deep learning is not enough, and requires parsimony, bayesianity, and causality. *npj Computational Materials*, 7(1):16, 2021.

[546] Ashish Vaswani, Noam Shazeer, Niki Parmar, Jakob Uszkoreit, Llion Jones, Aidan N. Gomez, Lukasz Kaiser, and Illia Polosukhin. Attention is all you need, 2017.

[547] Aldo V. Vecchia. Estimation and model identification for continuous spatial processes. *Journal of the Royal Statistical Society: Series B (Methodological)*, 50(2):297–312, 1988.

[548] A. Violante, M. Ricciardella, M. Pigna, and R. Capasso. Chapter 5 - effects of organic ligands on the adsorption of trace elements onto metal oxides and organo-mineral complexes. In P.M. Huang and G.R. Gobran, editors, *Biogeochemistry of Trace Elements in the Rhizosphere*, pages 157–182. Elsevier, Amsterdam, 2005.

[549] Pauli Virtanen, Ralf Gommers, Travis E. Oliphant, Matt Haberland, Tyler Reddy, David Cournapeau, Evgeni Burovski, Pearu Peterson, Warren Weckesser, Jonathan Bright, Stéfan J. van der Walt, Matthew Brett, Joshua Wilson, K. Jarrod Millman, Nikolay Mayorov, Andrew R. J. Nelson, Eric Jones, Robert Kern, Eric Larson, C. J. Carey, İlhan Polat, Yu Feng, Eric W. Moore, Jake VanderPlas, Denis Laxalde, Josef Perktold, Robert Cimrman, Ian Henriksen, E. A. Quintero, Charles R. Harris, Anne M. Archibald, Antônio H. Ribeiro, Fabian Pedregosa, Paul van Mulbregt, and SciPy 1.0 Contributors. SciPy 1.0: Fundamental Algorithms for Scientific Computing in Python. *Nature Methods*, 17:261–272, 2020.

[550] Vladimir Vovk. Kernel ridge regression. In *Empirical Inference*, pages 105–116. Springer, 2013.

[551] Haruko M. Wainwright, Akiyuki Seki, Jinsong Chen, and Kimiaki Saito. A multiscale bayesian data integration approach for mapping air dose rates around the fukushima daiichi nuclear power plant. *Journal of Environmental Radioactivity*, 167:62–69, 2017.

[552] Alex Wang, Haotong Liang, Austin McDannald, Ichiro Takeuchi, and Aaron Gilad Kusne. Benchmarking active learning strategies for materials optimization and discovery. *Oxford Open Materials Science*, 2(1):itac006, 2022.

[553] Jia Wang, Yingxue Wang, and Yanan Chen. Inverse design of materials by machine learning. *Materials*, 15(5):1811, 2022.

[554] Linda Wang, Zhong Qiu Lin, and Alexander Wong. Covid-net: a tailored deep convolutional neural network design for detection of covid-19 cases from chest x-ray images. *Scientific Reports*, 10(1):19549, 2020.

[555] Shenkai Wang, Junmian Zhu, Raymond Blackwell, and Felix R. Fischer. Automated tip conditioning for scanning tunneling spectroscopy. *The Journal of Physical Chemistry A*, 125(6):1384–1390, 2021. PMID: 33560124.

[556] Xiaomei Wang, Yingqi Li, and Ka-Wai Kwok. A survey for machine learning-based control of continuum robots. *Frontiers in Robotics and AI*, 8, 2021.

[557] Xuan Wang, Guo-Jun Xie, Ning Tian, Cheng-Cheng Dang, Chen Cai, Jie Ding, Bing-Feng Liu, De-Feng Xing, Nan-Qi Ren, and Qilin Wang. Anaerobic microbial manganese oxidation and reduction: A critical review. *Science of the Total Environment*, 822:153513, 2022.

[558] Yin-Hao Wang, Tzuu-Hseng S. Li, and Chih-Jui Lin. Backward Q-learning: the combination of sarsa algorithm and q-learning. *Engineering Applications of Artificial Intelligence*, 26(9):2184–2193, 2013.

[559] Zhi Wang, Han-Xiong Li, and Chunlin Chen. Reinforcement learning-based optimal sensor placement for spatiotemporal modeling. *IEEE Transactions on Cybernetics*, 50(6):2861–2871, 2019.

[560] Logan Ward, Alexander Dunn, Alireza Faghaninia, Nils E. R. Zimmermann, Saurabh Bajaj, Qi Wang, Joseph Montoya, Jiming Chen, Kyle Bystrom, Maxwell Dylla, et al. Matminer: An open source toolkit for materials data mining. *Computational Materials Science*, 152:60–69, 2018.

[561] Christopher J. C. H. Watkins and Peter Dayan. Q-learning. *Machine Learning*, 8:279–292, 1992.

[562] Jolan Wauters, Joris Degroote, Ivo Couckuyt, and Guillaume Crevecoeur. SAMURAI: A new asynchronous Bayesian optimization technique for optimization-under-uncertainty. *AIAA Journal*, 60(11):6133–6156, 2022.

[563] Tobias Weber. Motion planning for triple-axis spectrometers. *SoftwareX* Volume 23, 101455, DOI: 10.1016/j.softx.2023.101455, 2023.

[564] Tobias Weber. Tas-paths, *Software. Online resource*, March 2023. https://doi.org/10.5281/zenodo.4625649

[565] L. Wiegart, G. S. Doerk, M. Fukuto, S. Lee, R. Li, G. Marom, M. M. Noack, C. O. Osuji, M. H. Rafailovich, J. A. Sethian, Y. Shmueli, M. Torres Arango, K. Toth, K. G. Yager, and R. Pindak. Instrumentation for in situ/operando x-ray scattering studies of polymer additive manufacturing processes. *Synchrotron Radiation News*, 32(2):20–27, 2019.

[566] Norbert Wiener. The homogeneous chaos. *American Journal of Mathematics*, 60(4):897–936, 1938.

[567] Wikipedia. Materials science: `https://en.wikipedia.org/wiki/Materials_science`.

[568] Mark D. Wilkinson, Michel Dumontier, IJsbrand Jan Aalbersberg, Gabrielle Appleton, Myles Axton, Arie Baak, Niklas Blomberg, Jan-Willem Boiten, Luiz Bonino da Silva Santos, Philip E. Bourne, et al. The fair guiding principles for scientific data management and stewardship. *Scientific Data*, 3(1):1–9, 2016.

[569] Christopher Williams and Carl Rasmussen. *Gaussian Processes for Regression*. MIT Press, 1996.

[570] Christopher K. I. Williams and Carl Edward Rasmussen. *Gaussian Processes for Machine Learning*, volume 2. MIT Press, Cambridge, MA, 2006.

[571] Ronald J. Williams. Simple statistical gradient-following algorithms for connectionist reinforcement learning. *Reinforcement Learning*, pages 5–32, 1992.

[572] Andrew Wilson and Hannes Nickisch. Kernel interpolation for scalable structured gaussian processes (kiss-gp). In *International Conference on Machine Learning*, pages 1775–1784, 2015.

[573] Sam Wiseman and Alexander M. Rush. Sequence-to-sequence learning as beam-search optimization, 2016.

[574] Theodore Wolf. Climate change policy exploration using reinforcement learning. *arXiv preprint arXiv:2211.17013*, 2022.

[575] D. H. Wolpert and W. G. Macready. No free lunch theorems for optimization. *IEEE Transactions on Evolutionary Computation*, 1(1):67–82, 1997.

[576] Darwin L. Wood. Infrared microspectrum of living muscle cells. *Science*, 114(2950):36–38, 1951.

[577] Jing Wu and Tongwen Chen. Design of networked control systems with packet dropouts. *IEEE Transactions on Automatic Control*, 52(7):1314–1319, 2007.

[578] Qiong Wu, Xu Chen, Zhi Zhou, Liang Chen, and Junshan Zhang. Deep reinforcement learning with spatio-temporal traffic forecasting for data-driven base station sleep control. *IEEE/ACM Transactions on Networking*, 29(2):935–948, 2021.

[579] Qiuxuan Wu, Yueqin Gu, Yancheng Li, Botao Zhang, Sergey A. Chepinskiy, Jian Wang, Anton A. Zhilenkov, Aleksandr Y. Krasnov, and Sergei Chernyi. Position control of cable-driven robotic soft arm based on deep reinforcement learning. *Information*, 11(6):310, 2020.

[580] Ke Xu, Qingxu Jin, Jiaqi Li, Daniela M. Ushizima, Victor C. Li, Kimberly E. Kurtis, and Paulo J. M. Monteiro. In-situ microtomography image segmentation for characterizing strain-hardening cementitious composites under tension using machine learning. *Cement and Concrete Research*, 169:107164, 2023.

[581] Ke Xu, Anton S. Tremsin, Jiaqi Li, Daniela M. Ushizima, Catherine A. Davy, Amine Bouterf, Ying Tsun Su, Milena Marroccoli, Anna Maria Mauro, Massimo Osanna, Antonio Telesca, and Paulo JM Monteiro. Microstructure and water absorption of ancient concrete from pompeii: An integrated synchrotron microtomography and neutron radiography characterization. *Cement and Concrete Research*, 139:106–282, 2021.

[582] Dezhen Xue, Prasanna V. Balachandran, Ruihao Yuan, Tao Hu, Xiaoning Qian, Edward R. Dougherty, and Turab Lookman. Accelerated search for BaTiO3-based piezoelectrics with vertical morphotropic phase boundary using bayesian learning. *Proceedings of the National Academy of Sciences*, 113(47):13301–13306, 2016.

[583] Kevin G. Yager, Erica Lai, and Charles T. Black. Self-assembled phases of block copolymer blend thin films. *ACS Nano*, 8(10):10582–10588, 2014. PMID: 25285733.

[584] Kevin G Yager, Pawel W Majewski, Marcus M Noack, and Masafumi Fukuto. Autonomous x-ray scattering. *Nanotechnology*, 34(32):322001, 2023.

[585] Kevin G. Yager, Yugang Zhang, Fang Lu, and Oleg Gang. Periodic lattices of arbitrary nano-objects: modeling and applications for self-assembled systems. *Journal of Applied Crystallography*, 47(1):118–129, 2014.

[586] Rikiya Yamashita, Mizuho Nishio, Richard Kinh Gian Do, and Kaori Togashi. Convolutional neural networks: an overview and application in radiology. *Insights into Imaging*, 9(4):611–629, 2018.

[587] Qimin Yan, Jie Yu, Santosh K. Suram, Lan Zhou, Aniketa Shinde, Paul F. Newhouse, Wei Chen, Guo Li, Kristin A. Persson, John M. Gregoire, and Jeffrey B. Neaton. Solar fuels photoanode materials discovery by integrating high-throughput theory and experiment. *Proceedings of the National Academy of Sciences*, 114(12):3040–3043, 2017.

[588] Y. G. Yan, Joshua Martin, Winnie Wong-Ng, Martin Green, and X. F. Tang. A temperature dependent screening tool for high throughput thermoelectric characterization of combinatorial films. *Review of Scientific Instruments*, 84(11):115110, 2013.

[589] Min Yang, Qiang Qu, Kai Lei, Jia Zhu, Zhou Zhao, Xiaojun Chen, and Joshua Z. Huang. Investigating deep reinforcement learning techniques in personalized dialogue generation. In *Proceedings of the 2018 SIAM International Conference on Data Mining*, pages 630–638. SIAM, 2018.

[590] Zhenhai Yang, Kai-Tai Fang, and Samuel Kotz. On the student's t-distribution and the t-statistic. *Journal of Multivariate Analysis*, 98(6):1293–1304, 2007.

[591] Zhenze Yang, Chi-Hua Yu, and Markus J Buehler. Deep learning model to predict complex stress and strain fields in hierarchical composites. *Science Advances*, 7(15):eabd7416, 2021.

[592] Junko Yano, Kelly J. Gaffney, John Gregoire, Linda Hung, Abbas Ourmazd, Joshua Schrier, James A. Sethian, and Francesca M. Toma. The case for data science in experimental chemistry: examples and recommendations. *Nature Reviews Chemistry*, 6(5):357–370, 2022.

[593] Nathan Yee, Liane G Benning, Vernon R. Phoenix, and F. Grant Ferris. Characterization of metal- cyanobacteria sorption reactions: a combined macroscopic and infrared spectroscopic investigation. *Environmental Science & Technology*, 38(3):775–782, 2004.

[594] Hyeonseok You, EunKyung Bae, Youngjin Moon, Jihoon Kweon, and Jaesoon Choi. Automatic control of cardiac ablation catheter with deep reinforcement learning method. *Journal of Mechanical Science and Technology*, 33:5415–5423, 2019.

[595] Inmaculada Yruela. Transition metals in plant photosynthesis. *Metallomics*, 5(9):1090–1109, 2013.

[596] Xiaoze Yuan, Yuwei Zhou, Qing Peng, Yong Yang, Yongwang Li, and Xiaodong Wen. Active learning to overcome exponential-wall problem for effective structure prediction of chemical-disordered materials. *npj Computational Materials*, 9(1):12, 2023.

[597] Eliezer Yudkowsky. Pausing ai developments isn't enough. we need to shut it all down. `https://time.com/6266923/ai-eliezer-yudkowsky-open-letter-not-enough/`, 2023.

[598] Zelda B. Zabinsky and Hao Huang. *A Partition-Based Optimization Approach for Level Set Approximation: Probabilistic Branch and Bound*, pages 113–155. Springer International Publishing, Cham, 2020.

[599] Mavrik Zavarin, Elliot Chang, Haruko Wainwright, Nicholas Parham, Rahul Kaukuntla, Jadallah Zouabe, Amanda Deinhart, Victoria Genetti, Sam Shipman, Frank Bok, and Vinzenz Brendler. Community data mining approach for surface complexation database development. *Environmental Science & Technology*, 56(4):2827–2838, 2022. PMID: 35104413.

[600] Jiayong Zhang, Wah-Keat Lee, and Mingyuan Ge. Sub-10 second fly-scan nano-tomography using machine learning. *Communications Materials*, 3(1):91, 2022.

[601] Xiang Zhang, Xiaocong Chen, Lina Yao, Chang Ge, and Manqing Dong. Deep neural network hyperparameter optimization with orthogonal array tuning. In Tom Gedeon, Kok Wai Wong, and Minho Lee, editors, *Neural Information Processing*, pages 287–295, Cham, 2019. Springer International Publishing.

[602] Zhen Zhang, Dongbin Zhao, Junwei Gao, Dongqing Wang, and Yujie Dai. FMRQ-A multiagent reinforcement learning algorithm for fully cooperative tasks. *IEEE Transactions on Cybernetics*, 47(6):1367–1379, 2017.

[603] Chonghang Zhao, Cheng-Chu Chung, Siying Jiang, Marcus M. Noack, Jiun-Han Chen, Kedar Manandhar, Joshua Lynch, Hui Zhong, Wei Zhu, Phillip Maffet-tone, Daniel Olds, Masafumi Fukuto, Ichiro Takeuchi, Sanjit Ghose, Thomas Caswell, Kevin G. Yager, and Yu-chen Karen Chen-Wiegart. Machine-learning for designing nanoarchitectured materials by dealloying. *Communications Materials*, 3(1):86, 2022.

[604] Wenshuai Zhao, Jorge Peña Queralta, and Tomi Westerlund. Sim-to-real transfer in deep reinforcement learning for robotics: a survey. In *2020 IEEE Symposium Series on Computational Intelligence (SSCI)*, pages 737–744. IEEE, 2020.

[605] Zhuowen Zhao, Tanny Chavez, Elizabeth A. Holman, Guanhua Hao, Adam Green, Harinarayan Krishnan, Dylan McReynolds, Ronald J Pandolfi, Eric J Roberts, Petrus H. Zwart, et al. Mlexchange: A web-based platform enabling exchangeable machine learning workflows for scientific studies. In *2022 4th Annual Workshop on Extreme-scale Experiment-in-the-Loop Computing (XLOOP)*, pages 10–15. IEEE, 2022.

[606] Kaiyang Zhou, Yu Qiao, and Tao Xiang. Deep reinforcement learning for un-supervised video summarization with diversity-representativeness reward. In *Proceedings of the AAAI Conference on Artificial Intelligence*, volume 32, 2018.

[607] Zhenpeng Zhou, Steven Kearnes, Li Li, Richard N. Zare, and Patrick Riley. Optimization of molecules via deep reinforcement learning. *Scientific Reports*, 9(1):1–10, 2019.

[608] Zhenpeng Zhou, Xiaocheng Li, and Richard N. Zare. Optimizing chemical reactions with deep reinforcement learning. *ACS Central Science*, 3(12):1337–1344, 2017.

[609] Yuke Zhu, Roozbeh Mottaghi, Eric Kolve, Joseph J. Lim, Abhinav Gupta, Li Fei-Fei, and Ali Farhadi. Target-driven visual navigation in indoor scenes using deep reinforcement learning, *Conference proceedings: Published in: 2017 IEEE International Conference on Robotics and Automation (ICRA)* 2016.

[610] Maxim Ziatdinov, Ondrej Dyck, Artem Maksov, Xufan Li, Xiahan Sang, Kai Xiao, Raymond R. Unocic, Rama Vasudevan, Stephen Jesse, and Sergei V.

Kalinin. Deep learning of atomically resolved scanning transmission electron microscopy images: Chemical identification and tracking local transformations. *ACS Nano*, 11(12):12742–12752, 2017. PMID: 29215876.

[611] Maxim Ziatdinov, Dohyung Kim, Sabine Neumayer, Rama K. Vasudevan, Liam Collins, Stephen Jesse, Mahshid Ahmadi, and Sergei V. Kalinin. Imaging mechanism for hyperspectral scanning probe microscopy via gaussian process modelling. *npj Computational Materials*, 6(1):21, 2020.

[612] Maxim Ziatdinov, Yongtao Liu, Kyle Kelley, Rama Vasudevan, and Sergei V. Kalinin. Bayesian active learning for scanning probe microscopy: From gaussian processes to hypothesis learning. *ACS Nano*, 16(9):13492–13512, 2022. PMID: 36066996.

[613] Maxim A. Ziatdinov, Ayana Ghosh, and Sergei V. Kalinin. Physics makes the difference: Bayesian optimization and active learning via augmented gaussian process. *Machine Learning: Science and Technology*, 3(1):015003, 2022.

[614] Pawel Ziolkowski, Matthias Wambach, Alfred Ludwig, and Eckhard Mueller. Application of high-throughput seebeck microprobe measurements on thermoelectric half-heusler thin film combinatorial material libraries. *ACS Combinatorial Science*, 20(1):1–18, 2018.

Index